Visual Psychophysics

Visual Psychophysics

From Laboratory to Theory

Zhong-Lin Lu and Barbara Dosher

The MIT Press
Cambridge, Massachusetts
London, England

MIT Press books may be purchased at special quantity discounts for business or sales promotional use. For information, please email special_sales@mitpress.mit.edu or write to Special Sales Department, The MIT Press, 55 Hayward Street, Cambridge, MA 02142.

This book was set in Syntax and Times New Roman by Toppan Best-set Premedia Limited, Hong Kong. Printed and bound in the United States of America.

Library of Congress Cataloging-in-Publication Data

Lu, Zhong-Lin.
Visual psychophysics : from laboratory to theory / Zhong-Lin Lu and Barbara Dosher.
pages cm
Includes bibliographical references and index.
ISBN 978-0-262-01945-3 (hardcover : alk. paper)
1. Vision. 2. Psychophysics. 3. Visual perception. 4. Image processing. I. Dosher, Barbara, 1951– II. Title.
QP475.L776 2013
612.8'4—dc23
2013002296

10 9 8 7 6 5 4 3 2 1

As a matter of course, we cannot in general deny that the mind is subject to quantitative principles. This is because apart from distinguishing stronger and weaker sensations, we can also distinguish stronger and weaker intensities of drive, higher and lower degrees of attention or vividness of recollections and fantasies, as well as different stages of consciousness and different intensities of individual thoughts. . . . Consequently, the higher mental processes can—in much the same way as sensory processes—be measured quantitatively, and the activity of the mind can be measured quantitatively in its entirety as well as in its components.
—Gustav Theodor Fechner, 1860

Contents

Preface

We wrote this book for anyone who wants to set up or work in a visual psychophysics laboratory and to develop theories and models of visual functions. Our aim was to write a book useful for beginning researchers and students and for experienced researchers or clinicians in allied areas.

Vision science is one of the most important areas of biomedical research. Practitioners in this area include those in basic science disciplines such as psychology, neuroscience, and computer sciences and those in clinical and applied areas of optometry, ophthalmology, and biomedical engineering. Visual psychophysical techniques are one of the foundational methodologies for this research enterprise.

Our own experience is that every time a new person joins our laboratories, whether he or she is an undergraduate, graduate student, or postdoc from another training area, we need to teach this new colleague a set of skills and techniques for visual psychophysics. We have done this again and again on an individual basis. Although we pointed to books and articles for our new laboratory members to consult, it was difficult to find integrated coverage of all the information and all the skills that were needed to carry out psychophysical research independently.

Our intention in writing this book was to provide the coverage for teaching a new vision scientist the necessary techniques in visual psychophysics and to orient him or her to the basic tools and the capabilities and limitations of these methodologies. Visual psychophysics is a classical field but has widespread applications in modern vision science. Our hope is that a new vision scientist, after studying this book, will have the skills to apply visual psychophysics to cutting edge vision research.

In this book, we focus on bridging the gap between theory and practice with sufficient detail and coverage to take the reader from the initial design and practicalities of understanding display devices and how to program an experiment all the way to analysis of psychophysical data in the context of example models and theories. We cover the basic software and hardware specifications for visual psychophysical experiments, including an extensive treatment of calibration procedures. Issues of precise timing and integration of displays with measurements of brain activities (electroencephalography, functional magnetic resonance imaging) and other relevant techniques are considered. A section on psychophysical methodologies includes many classic and modern experimental designs and data analysis techniques. The treatment of experimental design incorporates the most commonly used psychophysical paradigms and the extension of these basic paradigms in sophisticated, theory-driven psychophysical experiments. All these procedures are considered within a signal-detection theory framework. We also cover representative data analysis, model fitting, and model comparison, as well as sampling methods and computations for estimating the variability of parameters and the statistical power in complex psychophysical designs. We analyze many representative examples of new adaptive testing methods and provide a general framework for the development of adaptive testing designs. We end the book with sample applications and potential future directions in combining psychophysics with neuroscience techniques, or neuropsychophysics.

In writing the book, we aimed to integrate the practical with the theoretical. We use contemporary examples and provide about 100 sample programs written in MATLAB. The experimental programs also use the subroutines provided by Psychtoolbox Version 3 (Psychtoolbox-3), which is a public-domain toolbox for real-time experimental control. Psychtoolbox is used in the generation and presentation of stimuli, in timing and synchronization, and in the collection and recording of data.

We designed the book to be used interactively by alternately reading, working through, and executing sample programs. The reader will need access to a computer with MATLAB and Psychtoolbox-3. Psychtoolbox-3 can be downloaded and installed, and procedures to do this are described on the Psychtoolbox website (http://psychtoolbox.org/HomePage). The sample programs provided in the book are available for downloading from the website mitpress.mit.edu/visualpsychophysics. This should allow the reader to see some experiments in action. It also provides a programming core that can be modified for the reader's own use. As each new topic is covered, we have tried to incorporate citations to serve as pointers to the relevant literature and as the starting point for more extensive investigations.

Several items, described and provided on the book's web page, need to be downloaded to execute the programs provided in the book (i.e., routines `ReadKey.m`, `WaitTill.m`, the image `Church.jpg`, and the video file `introhigh.mpg`). We encourage the reader to make extensive use of the online help functions in MATLAB. Many useful discussions and information related to the real-time functions of Psychtoolbox are available on the Psychtoolbox website.

This new book project has taken us almost two years to complete. Writing the book has left us more excited about the future of visual psychophysics and the related theoretical and computational approaches than when we began. Psychophysics had its origins more than a century ago, and, like the integration of mathematics into many areas of science, the methods and theories of visual psychophysics have been integrated into many areas of biomedical research related to human behavior. The role of visual psychophysics has not diminished but has become stronger through the integration with other methods and modern measurement tools. From the use of visual psychophysical paradigms in the measurement of brain activity to applications in clinical testing and system design, we are seeing new applications of visual psychophysics in areas not even imagined by early practitioners. We hope that by assisting in the training of new researchers, this book will make a small contribution to the exciting new frontiers in biomedical research.

Acknowledgments

We salute the contributions of the many scientists who developed the field of visual psychophysics and the applications of signal-detection theory. These include the seminal contributions of historical figures but also the inspiring work of many contemporary scientists. In a book like this, we cannot possibly have cited all of the relevant sources. We request the tolerance of those whose citations we may have overlooked.

We especially acknowledge the contributions of the developers of Psychtoolbox, Dennis Pelli, David Brainard, and the large and active community of current Psychtoolbox programmers.

Zhong-Lin would like to thank Sam Williamson and Lloyd Kaufman for their introduction to scientific research. Barbara would like to thank Wayne Wickelgren for insights and training in the process of developing a theory and the sophisticated application of signal-detection theory in human memory and cognition. We both wish to take this opportunity to especially thank George Sperling for his introduction to the world of visual psychophysics and his many inspiring contributions to the field.

Many individuals assisted in the completion of the book project itself. Special thanks are due to Xiangrui Li for his help with experimental programs and with several figures. Fang Hou also tested some of the experimental programs. We thank Jongsoo Baek who re-drew several figures and assisted in the reformatting of the text for submission. Several individuals assisted in the creation of single figures, including Greg Appelbaum, Yukai Zhou, and Mae Lu. Rachel Miller provided useful comments on several chapters of an earlier draft. Lina Hernandez and Stephanie Fowler coordinated printouts and practical details.

We wish to thank the highly professional staff of the MIT Press, including Bob Prior, executive editor, and Susan Buckley, acquisitions editor. We also wish to thank those who assisted with the copyediting and production, including Christopher Curioli, Katherine A. Almeida, Sharon Deacon Warne, and Kate Elwell.

We would each also like to thank the current and former members of our laboratories for their many contributions to our research and for the opportunity to work with them. Our interactions with them provided much of the inspiration for this book.

Zhong-Lin Lu and Barbara Dosher

Thanks to my parents, Daode Zhong and Hongchun Lu, for their unconditional love. Thanks to my wife, Wei Sun, son, James, and daughter, Mae, for their understanding, tolerance, and accommodation of my busy work schedule. I would also like to acknowledge all my collaborators and my colleagues of the Departments of Psychology of The Ohio State University and of the University of Southern California for numerous scientific discussions and for support.

Zhong-Lin Lu

I would like to recognize my colleagues in the Department of Cognitive Sciences at the University of California, Irvine, for their high intellectual standards and for creating an extraordinarily congenial research home. Finding the focus and time for research writing while being a dean would not have been possible without the moral and practical support of the people in the dean's office, especially Assistant Dean David Leinen, Associate Deans Bill Maurer and Mark Petracca, and the many others who make everything work.

Thanks to Eileen Kowler and Norma Graham for their friendship and for the examples of excellence they provide in the pursuit of vision science. I would also like to thank Kristen Monroe, other close personal friends, and my parents Anne and Guy Dosher and sister Catherine Tonkins for many years of fun and support.

Last, but certainly not least, I thank my wonderful son, Joshua Sperling, for his intellectual engagement, encouragement, and support.

Barbara Dosher

I INTRODUCTION TO VISUAL PSYCHOPHYSICS

1 Understanding Human Visual Function with Psychophysics

Humans are visual creatures. The amazing abilities of human vision far exceed the capabilities of the most sophisticated machines and may inspire new machine design. Human visual capabilities and processes can only be known through an interdisciplinary approach that combines visual psychophysics, understanding of brain responses, and computational models of visual function. Visual psychophysics studies the relationship between the physical stimulus in the outside world and how that is connected to human performance. It has played a central role in the understanding of human visual capabilities and the brain. In addition, the assessment of human vision through applications of human visual psychophysics has been the core discipline behind the design of standards for visual devices and has many other practical applications.

1.1 Humans as Visual Animals

Survival and reproduction of living organisms requires interactions with other organisms and adaptation to environments. Millions of years of evolution have developed five senses in humans and other higher organisms that provide the doorways to interaction with the environment. Of the five major senses—taste, smell, touch, hearing, and sight—sight is perhaps the most highly developed in humans and the most important.[1–4] More than 50% of the cerebral cortex of the human brain is involved with processing of visual inputs.[5–7]

Vision begins with light that impinges on the eyes. In humans, the components of the visual system are the eye, including the retina, the optic nerves, the lateral geniculate nucleus (LGN), primary visual cortex, and other visual association cortices (figure 1.1, plate 1).[8,9]

The cornea and lens of the eye act as a compound lens that projects an inverted image of the visual input onto photoreceptors on the retina at the back of the eye.[10] Light-sensitive photoreceptors are more densely packed close to the fovea or center of gaze where vision is best.[11] The visible light spectrum for humans corresponds to approximately 390- to 750-nm wavelength of electromagnetic spectrum. Rods respond to most

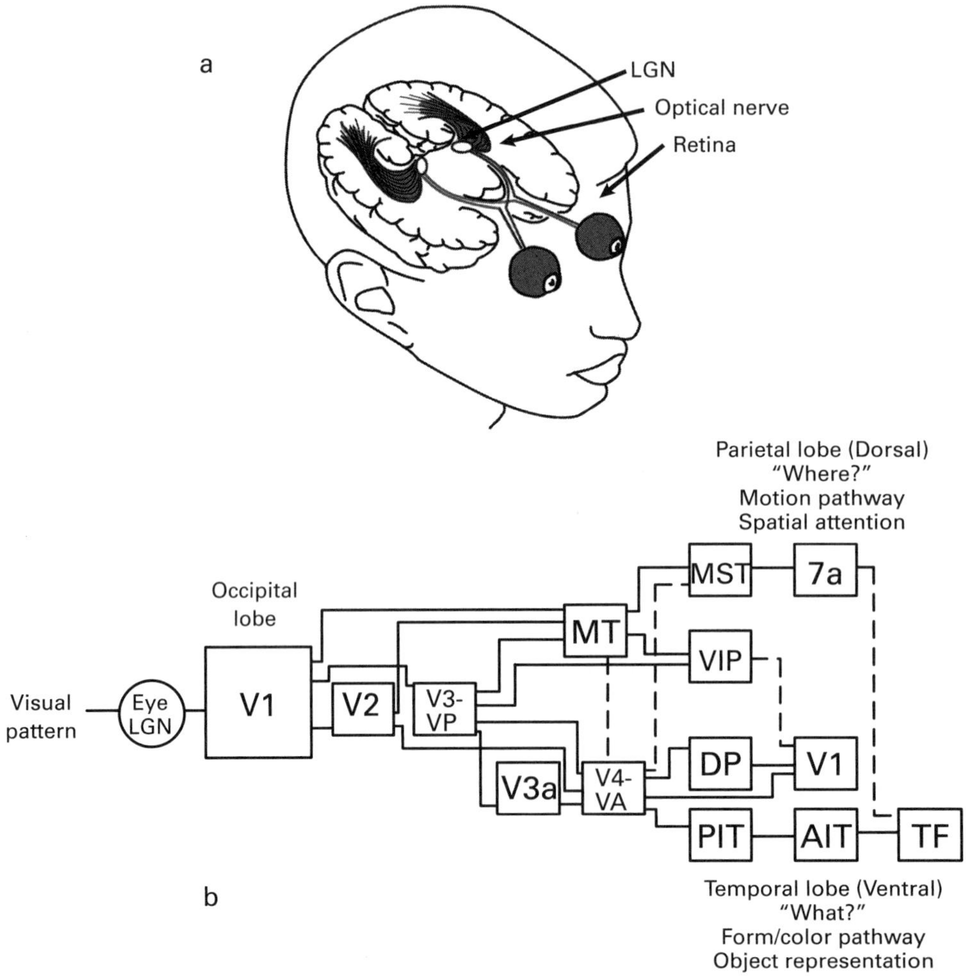

Figure 1.1 (plate 1)
(a) A picture of the visual system starting from the eyes and going through to various regions in visual cortex. (b) A two-pathway architecture for visual processing (adapted from Movshon[21]). Each box in the diagram refers to a functional region of the visual processing pathway. V1, V2, V3-VP, V3a, V4-VA, and 7a all refer to areas of visual cortex, each with distinct sensitivities. Higher processing areas involved in further processing, decision, and memory include the medial superior temporal cortex (MST), the ventral intraparietal cortex (VIP), the posterior inferior temporal cortex (PIT), the dorsal prelunate gyrus (DP), the anterior inferior temporal cortex (AIT), and area TF of parahippocampal cortex. The dashed lines refer to connections across the dorsal and ventral processing pathways thought to subserve analyses of "where" and "what," respectively.

visible wavelengths of light and serve vision in low illumination, while several families of cone receptors with differential wavelength or color sensitivity support vision in daylight conditions.[12–14]

The photoreceptors convert light energy into electrical signals that are further processed in the retina and then transmitted through the axons of approximately one million neurons to the cell bodies of neurons of the LGN in thalamus.[15] Axons from the LGN distribute information to the primary visual cortex in the striate cortex of the occipital lobe and then onward to a series of interconnected stages of visual processing.[16] Much of the primary visual cortex has a retinotopic organization that preserves the spatial organization or topology of the visual image.[17–20] In creating a cortical representation of the world, the visual system represents the images that stimulate adjacent parts of the retina by adjacent neurons. Figure 1.1b illustrates a diagram of our current understanding of the stages of visual processing.[21] Each stage of visual processing represents different computations and is sensitive to different aspects of the stimulus. Surprisingly, the visual system first breaks an image into features or cues and then binds the encoded aspects of objects—such as color, depth, or motion—back together to form our percepts of the world.[22]

Vision supports many critical functions. Animals who can see well are better adapted for finding food and mates and for avoiding dangers.[23] Visual cues are used in signaling for reproduction and attractiveness.[24,25] Vision is important in navigation through the local terrain and integrates with motor systems in the efficient guidance of reaching, grasping, or pushing in physical interaction with the world.[26,27]

The importance of clear vision can be even better appreciated from the challenges for individuals with major visual impairments. Imagine navigating even your own home in complete darkness. Try making a cup of tea with your eyes closed. Visual diseases challenge individuals and create medical and economic challenges for society. An estimated 2.4 million Americans over the age of 40 years have low vision, and an estimated 1 million Americans over the age of 40 are legally blind. The total annual economic impact of adult vision problems in the United States exceeds $50 billion.[28,29]

Vision also supports many higher-level mental processes. Often, visual images or visual details are integral to our memory of places and people.[30–32] Humans use visual imagery to construct mental models of the environment.[33,34] Mental imagery may also be used to help manipulate abstract relations and concepts.[35–38]

For humans, vision is increasingly important for communication through reading, visual media, and device interaction.[39,40] For most of us, the modern world involves interaction with visual devices such as iPhones or computers. Vision drives much of the human appreciation of art and the enjoyment of the environment.

For all these and other reasons, vision is the subject of extensive basic, applied, and clinical research. How the brain processes visual stimuli and integrates the visual world with the auditory and tactile inputs, what kinds of displays lead to what kinds of perception, the nature of the impairments in low vision, and many other issues are actively studied today.

1.2 Understanding Vision

Vision research is one of the important areas of biomedical research. Practitioners in this area include those in basic science disciplines such as psychology, neuroscience, and computer science and those in clinical and applied areas of optometry, ophthalmology, and biomedical engineering. Visual psychophysical techniques are one of the foundational methodologies in this area.

1.2.1 The Stimulus Environment

The interaction of light and objects in the environment provides rich cues for visual perception.[3,41] What you can see from the environment depends upon the spatial distribution, composition, and quantity of light. The human visual system perceives over an enormous range of illumination. We can see visual features of a starlit sky with luminance as low as 10^{-3} candelas per square meter (cd/m^2) up to luminance values as high as 10^5 cd/m^2 in brightest sunlight.[42,43] The composition of light varies depending on the source. Laser lights have a single wavelength, whereas bright sunlight includes all visible wavelengths. The light source determines what you can see, as you know if you have ever developed film in a darkroom with a red bulb or experienced a strobe light.

When light hits an opaque surface, it is absorbed and reflected. The amount of absorption versus reflection depends upon the wavelength of light and is a property of the material.[44] Objects with dark surfaces mostly absorb light and reflect relatively little, whereas bright surfaces reflect more light and absorb relatively little. The set of wavelengths that are reflected from surfaces determines the perceived color.[12–14] The materials of surfaces determine the patterns and distributions of reflectance in space. The distributions of random variations in reflected light are perceived as material properties, such as metallic or ceramic or furry.[45,46] Spatial patterns of reflected light at a larger scale are perceived as texture patterns.[47]

Objects have surfaces with different geometric relationships to the light source and to the eye. These relationships result in light being reflected in different directions over the visible portion of the object, and the array of light over space is decoded to create a sense of three-dimensional object shape.[48,49] The array of light also conveys the spatial relationships between objects, which supports the perception of depth.[50,51] When objects move, the changes in the light array over time provide important cues for the perception of motion.[52,53] When people move relative to objects in the environment, the systematic changes in the light array give rise to optic flow that is key to stable and efficient locomotion through the environment.[54,55]

Figure 1.2 (plate 2)
An image of the coffee shop where we are writing, with several processed versions emphasizing luminance, edges, and color blocks. The upper left panel shows the original digital photograph; the upper right is the same image rendered in grayscale (black and white); the lower left is a version of the grayscale image that shows edges (thresholded); the lower right shows the large regions of color in the image without detailed texture (filtered).

All of these cues to the visual world are encountered in every instant of our waking lives. As we write this chapter in our local coffee shop (figure 1.2, plate 2), all of these cues, and many more, create our perception of the environment.

The light in the shop combines full-spectrum daylight from a cloudless California summer sky passing through partly shaded windows behind the image and light from a collection of large and small standard indoor floodlights. The illumination is reflected from mustard yellow walls, matte-black surfaces, and metal fixtures. There are specular reflections from glass and matte reflections from the clothing, hair, and skin of the customers. Each object has spatial textures characteristic of those materials. Larger texture patterns, such as the woven chair back, are seen in larger-scale spatial patterns of the light array. Clusters in the light array are recognizable as objects or people. We have purposely not included faces—to guarantee anonymity for reasons of privacy. The spatial arrangements of objects are conveyed through the spatial array of light interpreted through perspective and the complex decoding of light reflections between objects. As we enter or leave the coffee shop, our locomotion splashes our retinas with systematic flows of the spatial array, or optic flow.

Plate 2 shows the original color photograph of the coffee shop (upper left). A black-and-white version preserves the luminance distribution without color (upper right). Edges extracted from the spatial array of light approximate the outlines of objects (lower left). The average color in each region emphasizes color without detailed texture (lower right). The visual system processes the visual image in these ways and many more. Then, the visual system pieces the puzzle back together.

1.2.2 Interdisciplinary Approaches in Visual Sciences: A Historical Example

Our contemporary understanding of color vision provides a good illustration of the interdisciplinary nature of investigations in visual science. The most important breakthrough in color vision begins with Isaac Newton's observation that a beam of white daylight can be split through a prism into different colors, or wavelengths, and that recombining several hues recovers white light.[56]

The next step was the recognition in printing and in various visual matching experiments that human color perception could be based on the sensations created by a mixture of as few as three color bases. The so-called trichromate theory of color provided an explanation for the observations that correctly selected combinations of three primary light colors could perfectly match the perceived color of any "pure" (restricted wavelength) light. This was formalized as the now-famous Young–Helmholtz theory of trichromacy.[57,58]

Focusing on color appearance—largely ignored by the proponents of color trichromacy, who were mostly physicists—led Ewald Hering to a competing view of color vision.[59] He observed that certain colors do not co-occur in appearance; you never see reddish greens or yellowish blues. This led him to propose that an opponent system combines these inputs in opposition to each other. The opponent system was also seen in the yellow afterimages from staring at a blue pattern or red afterimages from staring at a green pattern.

These analyses based on the early psychophysics of color provided a motivating structure for biological investigations. Trichromate theory led scientists to seek and find evidence for three kinds of cones in the human retina by measuring the light reflected from different retinal pigments (retinal densitometry) to infer what lights are absorbed by photosensitive cells.[60] Scientists have gone on to investigate the chemical and molecular mechanisms underlying these different cone classes.[61]

The pioneering work of De Valois and his colleagues used single-cell recordings in monkey and found evidence of opponent color processing in physiology.[62] Certain cells in LGN, for example, responded positively to red inputs but negatively to green inputs and were labeled +R –G cells, on so on. So both models identified through visual psychophysics were implemented in biology, trichromacy in photopigments of the retina and opponent processing combining these inputs in the LGN. In both cases, the functional understanding from psychophysics set the stage for the biological investigation.

Computational models of color vision incorporate our understanding of biological aspects of color and aim to predict major phenomena of color appearance, including color matching, context-dependence of color perception, and color constancy, which is the constancy of color perception in the face of changing lighting.[63–65]

As you can see, the case of color vision starts with the physics of the stimulus and ends with functional human behavior. Although our understanding of color vision is not perfect and color vision is still under active investigation, discoveries from visual psychophysics and physiology have been tremendously useful in application to technology. These have led to international standards of color, namely the 1931 Commission Internationale de l'Éclairage (CIE) standards[66] based on the trichromatic theory of color vision. The CIE standards motivate the use of three phosphors (red, green, blue) in color computer monitors, three-color bases of some printing processes, and so forth.

Precise characterization of the physical stimulus, psychophysical experimentation, neurophysiology, and computational modeling are all important tools used toward understanding of human visual function. A good understanding of human visual function can lead to powerful applications.

1.2.3 Understanding Visual Functions

Visual function begins with the visual stimulus in all its complexity. Light images impinge on the visual senses, are recoded and processed in various stimulus representations, and some of these are the basis for generating goal-directed behavior. Understanding of vision involves understanding all of the components and their interactions in this process of creating and interacting with the visual world. This is a complex and challenging investigation that will only be solved by the converging efforts of several scientific disciplines each approaching the problem at a different level.[49]

Visual psychophysics starts with the precise characterization of the relevant physical properties of the stimulus environment using a variety of tools for presentation and measurement. It goes on to measure the quantitative relationship between physical stimulus properties and the functional behavior of the organism. Measurement of these relationships relies on tasks that serve as tools for identifying lawful relationships in some perceptual domain and of understanding the decisions for particular actions or task goals. The outcome is an understanding of the capabilities and limitations in functional behavior.

Biological investigations of visual function study how visual stimuli are processed and represented at different stages of the visual pathways. Other investigations study how vision interacts with other sensory modalities and other brain areas to achieve goal-directed behavior. A multiplicity of methods and tools are used in biological investigation of the visual brain. These range from molecular biology of subcellular reactions to brain-imaging techniques that measure the responses of populations of neurons in different spatial regions of the brain. Each one of these techniques contributes to understanding the puzzle of visual perception leading to detailed descriptions of anatomy, behavior, and function of the "biological hardware" that implements visual functions.

Computational models are also critical in elaborating our understanding of visual function. At the psychophysical level, they provide quantitative descriptions of the relationship between the physical environment and human behavior. At the neuronal level, they provide candidate representations of stimuli and relate these representations to behavior. One major role played by computational models is to bridge the gaps between different levels of description and to provide an understanding of how different systems, or populations of units, work together to bring about behavior. These models make explicit the interactions between representation and activity at different levels of the system and how that information is used to determine behavior. Computational models must be constrained by all the known facts of both biological and psychophysical experimentation. They should provide testable predictions.

To summarize, the goal of scientific investigation of visual function is to understand the response of a functioning goal-directed organism operating in the visual environment. In many ways, these investigations begin and end with visual psychophysics. It is visual psychophysics that presents the first functional understanding of behavior. Biological measurements are increasingly made in the context of awake and behaving organisms to guarantee that the measured properties of the hardware are relevant to the behavior. Finally, the goal of computational models is to predict the functional architecture of the visual system and the laws of response that are inferred from the visual psychophysics.

1.3 Approaches to Visual Psychophysics

The publication of *Elemente der Psychophysik* by Gustav Theodor Fechner in 1860 marked the official beginning of a new scientific study of psychology that Fechner called "psychophysics."[67] The new scientific discipline incorporated contributions of many other scientists, most famously including the physiologist Ernst Heinrich Weber,[68] the physiologist and surgeon Hermann Ludwig Ferdinand von Helmholtz,[69] and Fechner's psychologist colleague Wilhelm Maximillian Wundt.[70] In counterpoint to philosophers such as Kant[71] who questioned whether psychology was accessible to scientific study, from the very beginning the aim of psychophysics was to formulate a quantitative relationship between the physical world and human perception. It is the quantification of perceptual experience, or the relationship between perception and the physical world, that is the foundational goal of the psychophysics enterprise.

Scaling methods were introduced to measure the perceived intensity or perceived magnitude of physical variation. At the same time, methods were developed to infer the minimal threshold levels to perceive a stimulus, to perceive the identity of the stimulus, or to perceive a difference between two stimuli.[67] Over the subsequent 150 years, these methods and new measurement techniques were developed.[72–76] The methods, analyses of data, theories, and models of psychophysics have also become core to the scientific study of many areas of psychology. They have influenced the study of learning and memory, the measurement of attitudes and beliefs, and social and affective psychology. Psychophysical methods have also been extended and applied in areas of engineering, quality control, and risk management—areas far from the origins in human perception. Visual psychophysics has had direct applications in areas of human factors, individual differences, and clinical vision (see chapter 13).

From its inception, pioneers in psychophysics recognized the strong relationship between physiology and behavior. Fechner recognized the importance not just of the neurophysiology (the relationship between the stimulus and neural activity) but also of what he termed the "inner psychophysics" (the relationship between neural activity and perception).[67] With the modern advancement of techniques in neurophysiology and brain imaging, the study of the relation extending from physical stimulus to neural activity to response, which we term *neuropsychophysics,* is entering a new phase of scientific inquiry into brain and mind.

1.3.1 Scaling

Scaling is a collection of psychophysical methods that quantify the relationship between perceptual intensity or perceptual qualities and physical manipulations of the stimulus.[77] For example, people might be asked to assign a number from 1 to 100 to reflect the perceived intensity of stimuli varying in lightness.[78] Psychophysical approaches to scaling have been applied not just in vision but also in many other modalities to measure perceptions of sound, weight of objects, taste, or even emotion.

In the simplest cases, such as the perception of light intensity, the stimulus and the perception both vary along a single dimension, or are unidimensional. Almost all such cases show lawful input–output relationships, such as power functions.[79] In other cases, physical variations along a single physical dimension create perceptions that are more complicated and must be characterized in multidimensional space. One example is in color perception (see earlier), where physical variations along a single dimension of the wavelength of light are perceived by three classes of receptors and so perception of color is encoded in a three-dimensional space.[80] In other cases, physical manipulations along multiple dimensions result in a perception that is characterized in a single dimension. An example of this is the manipulation of visual intensity and duration that combine together, for short durations, to a unidimensional percept of intensity.[81] In the most general case, stimulus variations along multiple dimensions are mapped into perceptions that are characterized in multidimensional spaces. For example, faces can differ in age, gender, and eye width, and judgments of similarity between faces reflect distances in a multidimensional similarity space.[82] In cross-modal scaling, observers are asked to relate perception in one modality to that in another modality. Observers may judge whether a light intensity is more or less than the intensity of a sound.[83]

Psychophysical scaling investigations have led to discoveries that have been widely incorporated into everyday life. For example, understanding of the color space of perception is directly reflected in the paint "color wheels" that relate color chips by hue and by saturation, where a particular strip shows colors of increasing saturation or intensity of color, and adjacencies in the wheel of strips express color similarities.

The visual psychophysics of scaling provides a quantitative specification of the lawful relationships between physical variations of the stimulus and of the percept.[67,79] It also helps to identify the relevant dimensions in both physical stimulus and in the perceptual system.[14]

1.3.2 Sensitivity

Another major focus in psychophysics is measurement of the sensitivity of the perceptual system. Sensitivity refers to the minimum physical stimulus that is just detectable or the minimum physical difference between stimuli that is just detectable—the absolute or difference thresholds.[67,68] Sensitivity measurements and theories about sensitivity and signal detection have been a foundational part of psychophysics that estimates and describes the limitations of the sensory system.[84] Most people are familiar with visual acuity tasks such as eye charts that are used by optometrists or ophthalmologists. These charts measure the finest detail a person can see. Visual acuity tests are still the most widely used measures of functional vision in practical applications.

Often, sensitivities or thresholds are measured as a function of systematic manipulations of the physical characteristics of the stimulus, and it is this more detailed description of the operation under variation that has become fundamental to understanding of sensory systems. In the laboratory, perhaps the best-known example of a sensitivity *function* is the contrast sensitivity function (CSF).[85,86] Contrast is the ratio between the luminance deviation from the background and mean luminance; it reflects the size of the differences between the dark and light parts of a stimulus. The CSF measures contrast threshold as a function of the spatial frequency of sinusoidal gratings. This is the minimum visible contrast as a function of the coarseness or fineness of the pattern. The CSF provides a good test of the limitations of the early visual system. It is used in modeling behavior in complex tasks and can summarize the properties and limitations of the first stages of visual neurons.[87,88] Measurements of the CSF also provide one of the best ways to characterize loss of visual capability or deficits in functional vision due to visual diseases.[89]

Sensitivity or thresholds may also be measured in multiple physical dimensions to evaluate a *sensitivity surface* as a function of two or more variables. For example, we may measure sensitivity for combinations of spatial frequency and temporal frequency (time variation) of the stimulus.[90]

The form or shape of a sensitivity function or of a sensitivity surface provides a systematic and quantitative description of how the observer's system functions under significant variation in the stimulus. It provides the basis for predicting visual performance under different circumstances. Empirical results of this kind can place very strong constraints on candidate neural representations or activities and on computational models of the sensory system.

Sensitivity functions or surfaces depend on the state of the observer. An individual may be less able to detect stimuli under conditions of low attention, high fatigue, or when walking into a dark environment from full sunlight.[91] People may become more sensitive after practice or training.[92] Measurement of sensitivity functions under these different states allows you to make predictions about how people function in different states for a range of stimuli and conditions. These comparisons also inform both neural and computational models.

1.3.3 Neuropsychophysics

With major advancements in neuroscience and brain imaging, it is increasingly possible to identify candidate neural systems that mediate perception of the stimulus, selection and generation of response, and implementation of a goal structure for behavior. This leads to the possibility of controlling the physical variations of the stimulus, measuring the neural responses to those variations, and also measuring the overt behavior of the animal or the person. This provides a platform to conduct *neuropsychophysics*, which seeks to understand the complete relationship between the physical stimulus, the neural responses, and the behavior. Moreover, in some cases it may be possible to directly manipulate aspects of the neural responses, thereby creating new tests of causal relationships in this chain.[93,94]

Single-unit recording measures the neural responses of a few selected neurons. Multielectrode arrays measure the simultaneous neural responses of sets of neurons. Brain-imaging techniques such as electroencephalography (EEG),[95] magnetoencephalography (MEG),[96] and functional magnetic resonance imaging (fMRI)[97] can measure responses of populations of neurons in psychophysical experiments. Current injection techniques in animals or transcranial magnetic stimulation (TMS) or transcranial direct current stimulation (tDCS) may alter the response properties of either neurons or brain regions providing direct causal tests of cognitive or neuropsychophysical models of the brain.[98,99]

By combining psychophysical methods, physiological investigation, and computational modeling in neuropsychophysics, we can understand the relationship between the physical stimulus and the internal representation and the relationship between the internal representation and the behavior—closing the full loop from stimulus to response.

1.4 Overview of the Book

This book, *Visual Psychophysics: From Laboratory to Theory*, aims to provide a comprehensive treatment of visual psychophysics. Bridging the gap between theory and practice, the book begins with practical information about how to set up a vision lab and integrates this with the creation, manipulation, and display of visual images, experimental designs, and estimation of behavioral functions. The treatment of experimental design incorporates the most commonly used psychophysical paradigms, applications of these basic paradigms in sophisticated, theory-driven psychophysical experiments, and the analysis of these procedures and data in a signal-detection theory framework.

The book provides examples of commonly used psychophysical paradigms, extensions to modern adaptive methods, and procedures to measure some of the most important visual sensitivity functions and surfaces.

The book treats the theoretical underpinnings of data analysis and scientific interpretation. Data analysis approaches include model fitting, model comparison, and examples of optimized adaptive testing methods. The treatment also includes methods to synchronize visual displays with measurements of brain activities such as EEG or fMRI. The book includes many sample programs in MATLAB[100] with functions from Psychtoolbox,[101,102] which is a public-domain toolbox for real-time experimental control.

The book ends with a discussion of the important questions that can be addressed using the laboratory methods and theoretical testing principles covered in the book, how to extend and integrate the methodology discussed in this book with other tools to study important questions in human factors, and biomedical and clinical research.

References

1. Gregory RL. *Eye and brain: The psychology of seeing*. Princeton, NJ: Princeton University Press; 1997.

2. Wandell BA. *Foundations of vision*. Sunderland, MA: Sinauer; 1995.

3. Bruce V, Green PR, Georgeson MA. *Visual perception: Physiology, psychology, & ecology*. New York: Psychology Press; 2003.

4. Palmer SE. *Vision science: Photons to phenomenology*. Cambridge, MA: MIT Press; 1999.

5. Zeki SM. 1978. Functional specialisation in the visual cortex of the rhesus monkey. *Nature* 274(5670): 423–428.

6. Allman JM, Kaas JH. 1974. The organization of the second visual area (V II) in the owl monkey: A second order transformation of the visual hemifield. *Brain Res* 76(2): 247–265.

7. Felleman DJ, Van Essen DC. 1991. Distributed hierarchical processing in the primate cerebral cortex. *Cereb Cortex* 1(1): 1–47.

8. Polyak S, Kluver H. *The vertebrate visual system*. Chicago: University of Chicago Press; 1968.

9. Hubel DH, Wensveen J, Wick B. *Eye, brain, and vision*. New York: Scientific American Library; 1988.

10. Atchison DA, Smith G. *Optics of the human eye*. Boston: Butterworth-Heinemann Medical; 2000.

11. Curcio CA, Sloan KR, Kalina RE, Hendrickson AE. 1990. Human photoreceptor topography. *J Comp Neurol* 292(4): 497–523.

12. Cornsweet TN. *Visual perception*. New York: Academic Press; 1974.

13. Kaiser PK, Boynton RM, Swanson WH. *Human color vision*, 2nd ed. Washington, DC: Optical Society of America, 1996.

14. Wyszecki G, Stiles WS. *Color science: Concepts and methods, quantitative data and formulas*. New York: Wiley; 1967.

15. Brodal P. *The central nervous system: Structure and function*. Oxford: Oxford University Press; 2003.

16. Schiller PH. 1986. The central visual system. *Vision Res* 26(9): 1351–1386.

17. Holmes G. 1918. Disturbances of visual orientation. *Br J Ophthalmol* 2(9): 449–468.

18. Holmes G. 1945. Ferrier Lecture: The organization of the visual cortex in man. *Proc R Soc Lond B Biol Sci* 132: 348–361.

19. Horton JC, Hoyt WF. 1991. The representation of the visual field in human striate cortex: A revision of the classic Holmes map. *Arch Ophthalmol* 109(6): 816–824.

20. Engel SA, Glover GH, Wandell BA. 1997. Retinotopic organization in human visual cortex and the spatial precision of functional MRI. *Cereb Cortex* 7(2): 181–192.

21. Movshon A. Visual processing of moving images. In: Barlow H, Blakemore C, Weston-Smith M, eds. *Images and understanding: Thoughts about images, ideas about understanding*. New York: Cambridge University Press; 1990: pp. 122–137.

22. Treisman A. 1996. The binding problem. *Curr Opin Neurobiol* 6(2): 171–178.

23. Goldsmith TH. 2006. What birds see. *Sci Am* 295(1): 68–75.

24. Walster E, Aronson V, Abrahams D, Rottman L. 1966. Importance of physical attractiveness in dating behavior. *J Pers Soc Psychol* 4(5): 508–516.

25. Thornhill R, Gangestad SW. 1999. Facial attractiveness. *Trends Cogn Sci* 3(12): 452–460.

26. Goodale MA, Milner AD. 1992. Separate visual pathways for perception and action. *Trends Neurosci* 15(1): 20–25.

27. Loomis JM, Da Silva JA, Fujita N, Fukusima SS. 1992. Visual space perception and visually directed action. *J Exp Psychol Hum Percept Perform* 18(4): 906–921.

28. Rein DB, Zhang P, Wirth KE, et al. 2006. The economic burden of major adult visual disorders in the United States. *Arch Ophthalmol* 124(12): 1754–1760.

29. Frick KD, Gower EW, Kempen JH, Wolff JL. 2007. Economic impact of visual impairment and blindness in the United States. *Arch Ophthalmol* 125(4): 544–550.

30. Perky CW. 1910. An experimental study of imagination. *Am J Psychol* 21(3): 422–452.

31. Kosslyn SM. *Image and mind*. Cambridge, MA: Harvard University Press; 1980.

32. Pylyshyn ZW. 1973. What the mind's eye tells the mind's brain: A critique of mental imagery. *Psychol Bull* 80(1): 1–24.

33. McNamara TP. 1986. Mental representations of spatial relations. *Cognit Psychol* 18(1): 87–121.

34. Maguire EA, Gadian DG, Johnsrude IS, et al. 2000. Navigation-related structural change in the hippocampi of taxi drivers. *Proc Natl Acad Sci USA* 97(8): 4398–4403.

35. Piaget J. 1972. Intellectual evolution from adolescence to adulthood. *Hum Dev* 15(1): 1–12.

36. Miller AI. *Imagery in scientific thought: Creating 20th-century physics*. Cambridge, MA: MIT Press; 1986.

37. Lakoff G, Johnson M. *Philosophy in the flesh: The embodied mind and its challenge to western thought*. New York: Basic Books; 1999.

38. Davidson D, Block N. Mental events. In: Block NJ, ed. *Mental events. Readings in philosophy of psychology*, Vol. 1. Cambridge, MA: Harvard University Press; 1980;1:107–119.

39. Boyer MC. *Cybercities: Visual perception in the age of electronic communication*. New York: Princeton Architectural Press; 1996.

40. Raupp G, Cranton W. *Handbook of visual display technology*. New York: Springer; 2009.

41. Gibson JJ. *The ecological approach to visual perception*. Hillsdale, NJ: Lawrence Erlbaum; 1986.

42. Barlow HB. 1981. The Ferrier Lecture, 1980: Critical limiting factors in the design of the eye and visual cortex. *Proc R Soc Lond B Biol Sci* 212(1186): 1–34.

43. Barlow HB. Dark and light adaptation: Psychophysics. In: L. Hurvich and D. Jameson, eds. *Handbook of sensory physiology*, Vol. VII/4. New York: Springer-Verlag; 1972; pp. 1–28.

44. Born M, Wolf E, Bhatia AB. *Principles of optics*. Oxford: Pergamon Press; 1970.

45. Motoyoshi I, Nishida S, Sharan L, Adelson EH. 2007. Image statistics and the perception of surface qualities. *Nature* 447(7141): 206–209.

46. Adelson EH. On seeing stuff: The perception of materials by humans and machines. In: Society of Photo-Optical Instrumentation Engineers SPIE Conference series, Vol. 4299. 2001: pp. 1–12.

47. Julesz B. 1981. Textons, the elements of texture perception, and their interactions. *Nature*. 290(5802): 91–97.

48. Ullman S. *High-level vision: Object recognition and visual cognition*. Cambridge, MA: MIT Press; 2000.

49. Marr D. *Vision*. New York: WH Freeman; 1982.

50. Julesz B. *Foundations of cyclopean perception*. Chicago: University of Chicago Press; 1971.

51. Kaufman L. *Sight and mind: An introduction to visual perceptio*n. Oxford: Oxford University Press; 1974.

52. Kolers PA. *Aspects of motion perception*. Oxford: Pergamon Press; 1972.

53. Wuerger S, Shapley R, Rubin N. 1996. On the visually perceived direction of motion by Hans Wallach: 60 years later. *Perception* 25: 1317–1368.

54. Gibson JJ. 1958. Visually controlled locomotion and visual orientation in animals*. *Br J Psychol* 49(3): 182–194.

55. Warren WH, Jr. 1998. Visually controlled locomotion: 40 years later. *Ecol Psychol* 10(3–4): 177–219.

56. Newton I. *Opticks or, a treatise of the reflexions, refractions, inflexions and colours of light. Also two treatises of the species and magnitude of curvilinear figures*. London: Royal Society; 1704.

57. Helmholtz H. *Treatise on physiological optics. III. The perceptions of vision*. Southall JPC, ed. New York: Optical Society of America; 1925.

58. Young T. 1802. On the theory of light and colour. *Phil Trans R Soc Lond* 92: 12–48.

59. Hering E. *Outlines of a theory of the light sense*. Cambridge, MA: Harvard University Press; 1964.

60. Marks WB, Dobelle WH, MacNichol EF. 1964. Visual pigments of single primate cones. *Science* 143(3611): 1181–1183.

61. Pugh EN. 1999. Molecular mechanisms of vertebrate photoreceptor light adaptation. *Curr Opin Neurobiol* 9(4): 410–418.

62. De Valois RL, Abramov I, Jacobs GH. 1966. Analysis of response patterns of LGN cells. *JOSA* 56(7): 966–977.

63. D'Zmura M, Lennie P. 1986. Mechanisms of color constancy. *JOSA A* 3(10): 1662–1672.

64. Maloney LT, Wandell BA. 1986. Color constancy: A method for recovering surface spectral reflectance. *JOSA A* 3(1): 29–33.

65. Brainard DH, Freeman WT. 1997. Bayesian color constancy. *JOSA A* 14(7): 1393–1411.

66. Commission Internationale de l'Eclairage. *Proceedings of the Eighth Session*. Cambridge, UK: Cambridge University Press; 1931.

67. Fechner G. *Elemente der psychophysik*. Leipzig: Breitkopf & Härtel; 1860.

68. Weber EH. *De pulsu, resorptione, auditu et tactu*: Leipzig: Koehler; 1834.

69. Von Helmholtz H. *Handbuch der physiologischen Optik: Mit 213 in den Text eingedruckten Holzschnitten und 11 Tafeln*. Leipzig: Voss; 1866.

70. Wundt WM. *Lectures on human and animal psychology*. Creighton JG, Titchener EB, trans. London: Allen. Translation of Wundt, 1863.

71. Kant I. *Critique of pure reason*. Riga: Johann Friedrich Hartknoch; 1781.

72. Gescheider GA. *Psychophysics: Method and theory*. Hillsdale, NJ: Lawrence Erlbaum; 1976.

73. Falmagne JC. *Elements of psychophysical theory*. Oxford: Oxford University Press; 2002.

74. Kingdom FAA, Prins N. *Psychophysics: A practical introduction*. New York: Academic Press; 2009.

75. Green DM, Swets JA. *Signal detection theory and psychophysics*. Melbourne, FL: Robert E. Krieger; 1974.

76. Stevens SS. *Psychophysics: Introduction to its perceptual, neural, and social prospects*. New Brunswick, NJ: Transaction Publishers; 1975.

77. Krantz DH, Luce RD, Suppes P, Tversky A. *Foundations of measurement: Additive and polynomial representations*, Vol. 1. New York: Academic Press; 1971.

78. Stevens SS. 1946. On the theory of scales of measurement. *Science* 103(2684): 677–680.

79. Stevens SS. 1957. On the psychophysical law. *Psychol Rev* 64(3): 153–181.

80. Derrington AM, Krauskopf J, Lennie P. 1984. Chromatic mechanisms in lateral geniculate nucleus of macaque. *J Physiol* 357(1): 241–265.

81. Kahneman D, Norman J. 1964. The time-intensity relation in visual perception as a function of observer's task. *J Exp Psychol* 68(3): 215–220.

82. Valentine T. 1991. A unified account of the effects of distinctiveness, inversion, and race in face recognition. *Q J Exp Psychol* 43(2): 161–204.

83. Krantz DH. 1972. A theory of magnitude estimation and cross-modality matching. *J Math Psychol* 9(2): 168–199.

84. Graham NVS. *Visual pattern analyzers*. Oxford: Oxford University Press; 2001.

85. Campbell FW, Robson JG. 1968. Application of Fourier analysis to the visibility of gratings. *J Physiol* 197(3): 551–566.

86. Enroth-Cugell C, Robson JG. 1966. The contrast sensitivity of retinal ganglion cells of the cat. *J Physiol* 187(3): 517–552.

87. Movshon JA, Thompson ID, Tolhurst DJ. 1978. Spatial and temporal contrast sensitivity of neurones in areas 17 and 18 of the cat's visual cortex. *J Physiol* 283(1): 101–120.

88. Watson AB, Ahumada AJ. 2005. A standard model for foveal detection of spatial contrast. *J Vis* 5(9): 717–740.

89. Schwartz SH. *Visual perception: A clinical orientation*. New York: McGraw-Hill Medical; 2009.

90. Kelly DH. 1979. Motion and vision. II. Stabilized spatio-temporal threshold surface. *JOSA* 69(10): 1340–1349.

91. Lu ZL, Dosher BA. 2008. Characterizing observers using external noise and observer models: assessing internal representations with external noise. *Psychol Rev* 115(1): 44–82.

92. Fahle M, Poggio TA. *Perceptual learning*. Cambridge, MA: The MIT Press; 2002.

93. Salzman CD, Britten KH, Newsome WT. 1990. Cortical microstimulation influences perceptual judgements of motion direction. *Nature* 346(6280): 174–177.

94. Shadlen MN, Newsome WT. 2001. Neural basis of a perceptual decision in the parietal cortex (area LIP) of the rhesus monkey. *J Neurophysiol* 86(4): 1916–1936.

95. Nunez PL, Srinivasan R. *Electric fields of the brain: The neurophysics of EEG*. Oxford: Oxford University Press; 2006.

96. Lu ZL, Kaufman L. *Magnetic source imaging of the human brain*. Hillsdale, NJ: Erlbaum; 2003.

97. Huettel SA, Song AW, McCarthy G. *Functional magnetic resonance imaging*. Sunderland, MA: Sinauer; 2004.

98. Nitsche MA, Cohen LG, Wassermann EM, et al. 2008. Transcranial direct current stimulation: State of the art 2008. *Brain Stimulat* 1(3): 206–223.

99. Wassermann E, Epstein CM, Ziemann U. *The Oxford handbook of transcranial stimulation*. Oxford: Oxford University Press; 2008.

100. The MathWorks Inc. MATLAB [computer program]. Natick, MA: The MathWorks Inc.; 1998.

101. Pelli DG. 1997. The VideoToolbox software for visual psychophysics: Transforming numbers into movies. *Spat Vis* 10(4): 437–442.

102. Brainard DH. 1997. The psychophysics toolbox. *Spat Vis* 10(4): 433–436.

2 Three Examples from Visual Psychophysics

We introduce the fundamental components of visual psychophysics by dissecting three example experiments from scaling and sensitivity estimation. The basic elements of visual psychophysics include the selection of what to measure, the method of measurement, visual displays, response collection, and data analysis. This is designed to create a concrete overview that will be the basis of more detailed treatment of these issues later in the book.

2.1 Example 1: Estimate Perceived Intensity of a Light Patch

One of the classical questions in visual psychophysics is quantification of the relationship between physical intensity and perceived intensity.[1,2] To measure perceived intensity of a particular visual stimulus, you must generate and display the stimulus, show it to the observer, and collect a response.[3] In this first example experiment in visual psychophysics, we demonstrate one way of measuring the perceived intensity of a light patch—direct magnitude estimation. In this experiment, the observer is shown a series of patches with different physical light intensities and asked to assign a numerical value to his or her perceived intensity. Before the experiment, the observer is provided with a "standard" light intensity along with a numerical value, 10, as a guide to the rating scale. Typically, the same set of light intensities are tested several times in random order. The average rating is the estimate of the perceived intensity of that stimulus. These values can be used to create a graph of perceived intensity as a function of physical intensity, or the *psychophysical function* (figure 2.1). Psychophysical functions can often be described as power functions of physical intensity.[2,4–6] Power functions quantify the extent to which the response is a nonlinear function of the input quantity. They take the form $f(x) = kx^a$.

2.2 Example 2: Measure Contrast Threshold

Contrast threshold represents the minimum stimulus energy that can be detected by the observer.[2,7] To measure threshold for a particular visual stimulus, you need to try many different levels of stimulus energy and determine the particular level at which the stimulus

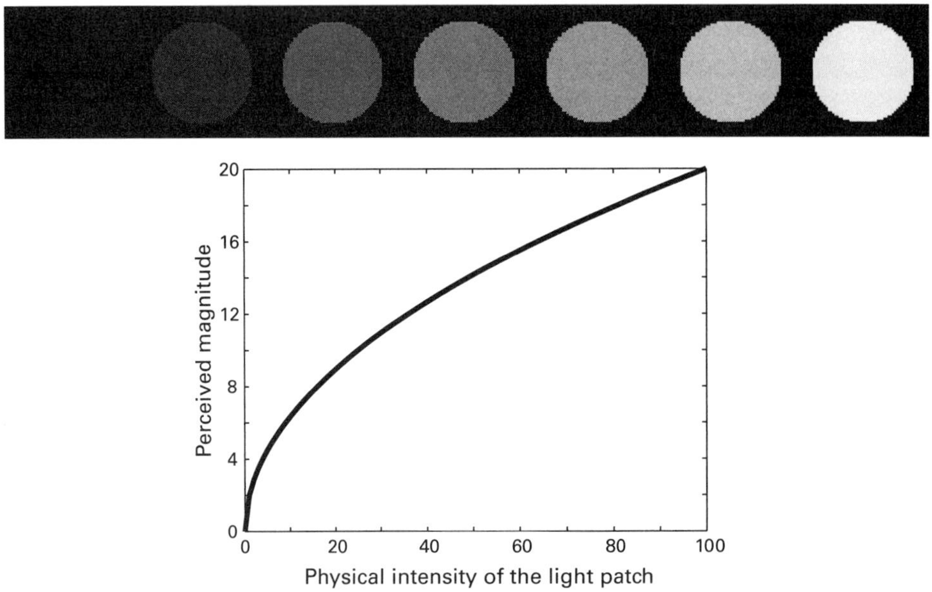

Figure 2.1
Magnitude estimation for patches of different light intensity.

is barely visible. In a formal experiment, we show observers a set of stimuli varying in energy, each many times, and use the accuracy of their judgments to estimate the lowest stimulus energy in which the stimulus can be detected.[8]

An example experiment measures the contrast threshold for a sine wave grating. A sine wave grating is generated by a computer and displayed on a computer monitor. We test several different levels of energy in the sine wave grating, so that the lowest is invisible and the highest is clearly visible. On every trial of testing, the computer selects one of the energy levels. Each trial consists of two time intervals during which the stimulus may be presented. The computer chooses randomly in which of the two intervals to display the sine pattern. The observer uses the computer keyboard or response keys to indicate in which interval they think they saw the sine pattern, and the computer records the stimulus condition and the observer's response on each trial. After the experiment measures a number of trials at each of the energy levels, the computer tabulates the accuracy of selecting the correct interval and graphs this accuracy as a function of the stimulus energy (figure 2.2).

The function that describes the relationship between the accuracy of performance and stimulus energy is the *psychometric function*. We can fit a smooth curve to this function, and from this curve we can determine exactly how much stimulus energy is required for

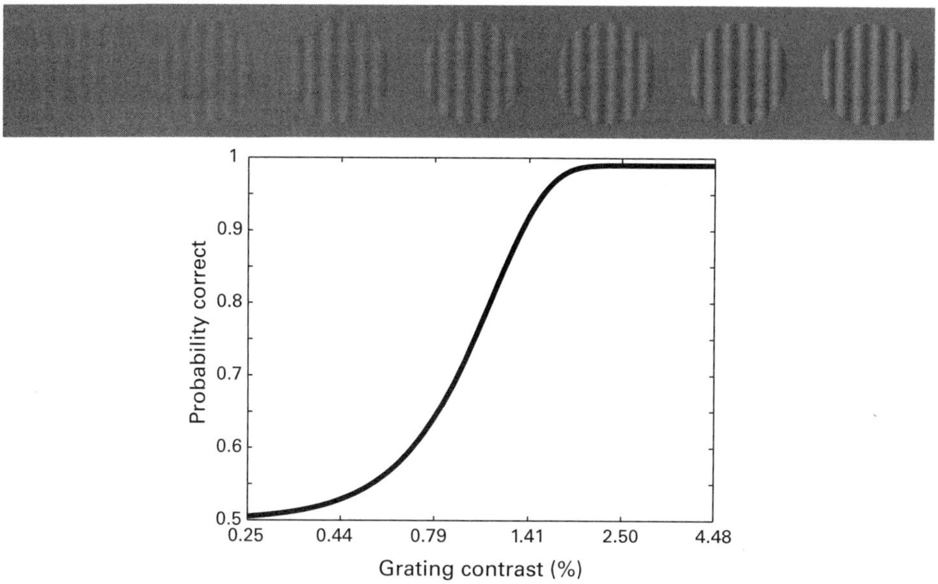

Figure 2.2
Psychometric function for detecting sine-wave gratings of different contrasts.

the observer to achieve 75% correct, which is the usual criterion for threshold when guessing is 50% correct.

2.3 Example 3: Measure the Contrast Sensitivity Function

In the next experiment, we use measurements of contrast threshold as basic building blocks to construct a more sophisticated measure of visual sensitivity: the contrast sensitivity function.[9,10] A contrast sensitivity function is a description of the sensitivity of the visual system to sine-wave patterns of different spatial frequencies, from coarse to fine. In example 2, we used a sine-wave pattern of a single spatial frequency and only varied the contrast energy of the sine pattern. Here in example 3, both the contrast energy of the pattern and its spatial frequency are varied (figure 2.3).

We choose a number of spatial frequencies from the coarsest to the finest. For each spatial frequency, we choose contrast levels to measure the psychometric function at that spatial frequency. On each trial, the computer selects a particular sine pattern and a particular contrast level to show to the observer. For each combination of frequency and contrast, we test a reasonable number of trials—so this experiment has as many times the number of trials as example 2 as there are tested spatial frequencies.

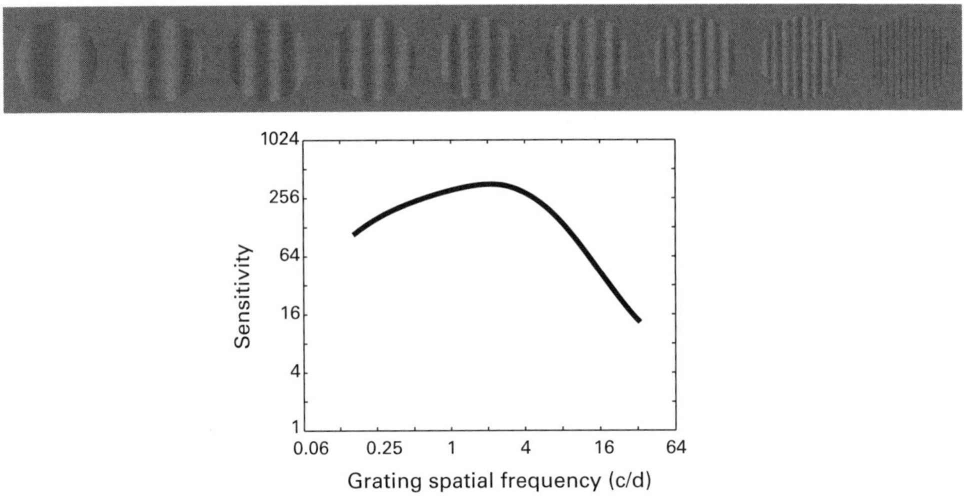

Figure 2.3
Contrast sensitivity function.

In the end, the experiment produces a psychometric function showing accuracy as a function of contrast for each sine-wave frequency. Each of the psychometric functions is fit with a descriptive function, and the 75% correct threshold is estimated. We graph the inverse of each threshold—which is a measure of sensitivity—as a function of the spatial frequency of the tested patterns. These data, too, can be fit with another smooth function describing the contrast sensitivity function,[11] from which we can determine the maximum sensitivity and the maximum resolution of the visual system (figure 2.3). A demonstration of the contrast sensitivity function is shown in figure 2.4. Vision scientists have used these measurements to model basic visual functions.[11,12]

2.4 Discussion

These three example experiments illustrate the fundamental elements of visual psychophysics. They represent some of the simplest paradigms and procedures that make up the toolkit of the visual psychophysicist. Psychophysicists have developed many different standardized methods of quantifying the relationship between perception and the physical environment.[8] Example 1 shows one of the simplest experiments in psychophysical scaling, the method of magnitude estimation. Examples 2 and 3 illustrate one classical method of threshold measurement, the method of constant stimuli. This method uses prior information to select a set of relevant stimulus conditions that are fixed throughout the experiment. All stimulus conditions are tested a given number of times, which provides one or a set of psychometric functions under a range of conditions. This in turn allows us to derive

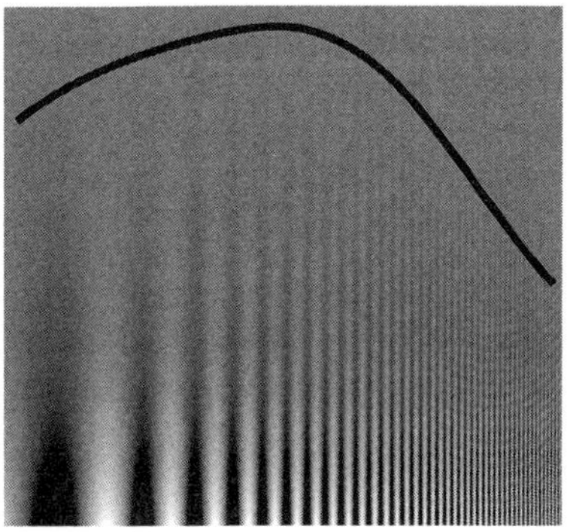

Figure 2.4
Demonstration of a contrast sensitivity function with increasing spatial frequency from left to right and decreasing contrast from bottom to top.

measures of visibility (or discriminability) such as thresholds that can be used to characterize human sensitivity over different kinds of stimulus variations. There are, however, other methods to accomplish scaling such as magnitude production or similarity ratings.[3] There are also other methods to estimate thresholds, such as method of limits, method of adjustment, and various adaptive methods.[2]

The quality of the visual psychophysics is only as good as the physical control and precision of the displays. The experimenter must make sure that the physical properties of the stimuli are precisely known and quantified. Often in visual psychophysics, the stimuli are displayed on a computer monitor. For example, in our second and third experiments, the physical variations involved contrast for stimuli of given spatial frequency and timing. Contrast thresholds in many of these conditions are remarkably low contrasts, perhaps at 0.1% of the full contrast range of the displays, and computers often generate only 256 programmable levels of intensity, so specialized methods are required to expand the number of programmable levels in certain restricted ranges. The displays also require precise control over the time of onset and duration of the displays. All of these properties of a proper display can be challenging for the beginning experimenter.

Measurement of human performance in visual psychophysics involves the collection of responses or judgments. A range of standardized methods may be used to collect these responses on standard computer equipment, such as the keyboard, computer mouse, joysticks, or touch-screens. If very precise and accurate measurements are required, such as for response times, specialized equipment might be involved. In addition, investigators might choose to measure other behavior such as eye movements, EMG, EEG, MEG, or fMRI. In all of these cases of specialized equipment, a critical consideration is synchronization with the stimulus displays.

Once the psychophysical experiment is complete, the data that have been collected must be tabulated, usually through simple statistical analysis. Then, these data are used to estimate some critical measure, such as the psychological scaling function, or the psychometric function and threshold, or the contrast sensitivity function. In example 2, we collected responses in the method of constant stimuli for a set of stimulus conditions, from which we construct a table of response accuracy as a function of the signal contrast in different stimulus conditions. We then quantify the empirical psychometric functions using descriptive functions from which we then estimate the contrast thresholds by interpolation. The set of contrast threshold measurements allow us to construct the visibility window of vision.

In the remainder of the book, we take you to the laboratory, where you will see how to construct an experiment. Then, we provide a tour of methods and models in psychophysics that allow theoretical interpretation of empirical findings. And finally, we integrate psychophysics—the laboratory methods and theoretical testing principles covered in the book—with applications to important questions in visual function in applied and clinical settings.

References

1. Stevens SS. 1946. On the theory of scales of measurement. *Science* 103(2684): 677–680.

2. Fechner G. *Elemente der psychophysik*. Leipzig: Breitkopf & Härtel; 1860.

3. Gescheider GA. 1988. Psychophysical scaling. *Annu Rev Psychol* 39(1): 169–200.

4. Krantz DH, Luce RD, Suppes P, Tversky A. *Foundations of measurement: Additive and polynomial representations*, Vol. 1. New York: Academic Press; 1971.

5. Falmagne JC. *Elements of psychophysical theory*. Oxford: Oxford University Press; 2002.

6. Stevens SS. 1957. On the psychophysical law. *Psychol Rev* 64(3): 153–181.

7. Weber EH. *De pulsu, resorptione, auditu et tactu*. Leipzig: Koehler; 1834.

8. Gescheider GA. *Psychophysics: Method and theory*. Hillsdale, NJ: Lawrence Erlbaum; 1976.

9. Campbell FW, Robson JG. 1968. Application of Fourier analysis to the visibility of gratings. *J Physiol* 197(3): 551–566.

10. Enroth-Cugell C, Robson JG. 1966. The contrast sensitivity of retinal ganglion cells of the cat. *J Physiol* 187(3): 517–552.

11. Watson AB, Ahumada AJ. 2005. A standard model for foveal detection of spatial contrast. *J Vis* 5(9): 717–740.

12. Movshon JA, Thompson ID, Tolhurst DJ. 1978. Spatial and temporal contrast sensitivity of neurones in areas 17 and 18 of the cat's visual cortex. *J Physiol* 283(1): 101–120.

II IN THE VISUAL PSYCHOPHYSICS LAB

3 Generating Stimuli

The implementation of experiments generally starts with the definition of what is displayed to the observer. This chapter is designed to take the reader through a simple introduction to digital images and to show how to generate typical psychophysical stimuli. The methods include creation of simple geometric patterns and more complex images. The examples also provide a tour of common image manipulations. Taken together, these methods will allow the creation of the most commonly used classes of psychophysical stimuli.

3.1 Simple Computer Graphics

3.1.1 Digital Images

Computer graphics involves the creation and manipulation of images in computers.[1] In computer graphics, images are commonly represented digitally in files or in computer memory as bitmaps or pixmaps—spatial maps of bits or pixels (picture elements) representing the color of the image at a each point in a grid of *x*, *y* locations. In some contexts, the term *bitmap* means 1 bit per pixel, whereas *pixmap* is used for images with multiple bits per pixel. A *bit* is the unit in a binary representation. Each bit takes on two values, a 0 or 1, and larger numbers require more bits to represent. Eight bits can represent numbers from 0 to 255. We use the term *bitmap* in the broader sense of pixmap in this book.

Commonly, the bitmap of an image is a two-dimensional array of *pixels* with an associated number of rows and columns. Each pixel element gives an intensity value for the pixel in the corresponding row and column location in the image. In typical bitmaps, image pixels are generally stored with 1, 4, 8, 16, 24, 32, 48, or 64 bits per pixel. The number of bits/pixel is called *color depth.* Binary images are 1 bit/pixel with each pixel having one of two values, either 0 or 1, for example, black and white. Grayscale images typically have 8 bits per pixel and so can take on values between 0 and 255. Pixels of 8 bits are typically used to represent either grayscale, an image containing only graded intensity information, or indexed color, in which the numbers code color selection from a color palette. Color images typically have 24 bits per pixel, 8 bits for each of red, green, and blue channels. All colors can be created by combining red, green, and blue values defined

Display 3.1

```
%%% Program Checkboard1.m
M = [1 2 1 2 1 2 1 2
     2 1 2 1 2 1 2 1
     1 2 1 2 1 2 1 2
     2 1 2 1 2 1 2 1
     1 2 1 2 1 2 1 2
     2 1 2 1 2 1 2 1
     1 2 1 2 1 2 1 2
     2 1 2 1 2 1 2 1];
```

Figure 3.1 (plate 3)
A checkerboard pattern shown with three different color lookup tables.

by an RGB triplet. A fourth channel, the *alpha channel*, stores a separate bitmap that when multiplied with the three primary channels determines the intensity of the RGB image by a factor from 0 to 1 at each location.[2] So, the alpha channel can be used to emphasize some regions of the image while hiding other regions. Addition of an alpha channel expands 24-bit images to 32 bits per pixel.

The bitmap of an 8×8 checkerboard (figure 3.1, plate 3) is given in display 3.1. (Notice that although 1-bit bitmaps normally take on values of 0 or 1 and 8-bit bitmaps take on values from 0 to 255, in MATLAB[3] a 1-bit image is represented by 1 and 2 and an 8-bit image is represented by numbers from 1 to 256 because these values are actually used to index locations in a colormap with index beginning at 1.[4,5])

3.1.2 Image Coordinates

A bitmap of an image M of $I \times J$ pixels is represented as a matrix of I rows and J columns. The matrix establishes a coordinate system where each entry in the matrix $M(i, j)$ is indexed by its location in the matrix, (i, j). In matrix format, the coordinate associated with the upper left corner of an image is (1, 1), the upper right corner of an image is $(1, J)$, the

Display 3.2

```
%%% Program Checkboard2.m
for i = 1:8
    for j = 1:8
        M(i, j) = mod(i + j, 2) + 1; % mod(v, 2) returns the
                                     % residual of v divided by 2
    end
end
```

Display 3.3

```
%%% Program Checkboard3.m
[x, y] = meshgrid(-4:3, 4:-1:-3);
M = mod(x + y, 2) + 1;
```

bottom left corner is (I, 1), and the bottom right corner is (I, J). Once a coordinate system is established, assigning an intensity value to each matrix location defines the image.

A different way of computing the bitmap of the checkerboard image in figure 3.1 (plate 3) is shown in display 3.2.

It is also possible to transform the image coordinate system, for example to place (0, 0) in the center of an image. MATLAB provides a function called `meshgrid` that sets up coordinate systems for a matrix of size [I, J]. A call to `meshgrid` that places (0, 0) at the center of the image is:

```
[x, y]=meshgrid((-J/2):(J/2-1), (I/2):-1:(-I/2+1))          (3.1)
```

where x is a matrix that contains the x coordinates of all the pixels that goes from ($-J/2$) to ($J/2 - 1$) from left to right, and y contains the y coordinates that goes from ($-I/2+1$) to ($I/2$) from bottom to top.

The bitmap of the same checkerboard from display 3.1 can be also constructed using the `meshgrid` function as shown in display 3.3.

3.1.3 Color Lookup Table

To display an image that is in a file or in computer memory, the bitmap that represents the image—a collection of numbers—must be translated into the codes for physical intensities on display devices, such as the computer screen. For this reason, digital images are often stored together with color lookup tables (CLUTs; sometimes also referred to as *colormaps*), which provide the translation from the intensities defined in bitmaps into physical levels on a display device. The CLUT is like a switchboard that connects a phone number to a physical address in a phone network. The digital image intensity for a pixel together with the lookup table determines the color or luminance displayed on the screen at the location of that pixel. In some cases, the experimenter may change what is displayed on the screen by leaving the image unchanged but altering the lookup table, or CLUT. Changing the CLUT rather than the contents of the displayed image can sometimes be a very useful tool in controlling experimental displays.

The checkerboard image defined earlier, with digital values of 1 or 2, could be shown as black and white, or red and green, or different levels of gray. Specifically, the CLUT may translate 1 to black and 2 to white; or the CLUT may translate 1 to red and 2 to green; or it may translate 1 to one gray level and 2 to another. One could reverse the checkerboard that is actually displayed not by rewriting the entire $I \times J$ pattern of 1s and 2s in the matrix, but rather by changing the CLUT to translate 1 to green and 2 to red, which requires only changes in two values of the CLUT.

In MATLAB, a matrix that codes a colormap or CLUT may have any number of rows, but it must have exactly three columns. Each row is interpreted as a color, with the first element specifying the intensity of red light, the second green, and the third blue. The *i*th row of the colormap specifies the color translation of intensity value *i* in the image bitmap. For example, the fifth row of the colormap specifies the color translation of intensity value 5 in the bitmap. Color intensity is specified on the interval 0.0 to 1.0. For example, [0 0 0] is black, [1 1 1] is white, [1 0 0] is pure red, [0.5 0.5 0.5] is gray, and [127/255 1 212/255] is aquamarine.

In display 3.4, we show how to use different color maps to display the same checkerboard image *M* computed in displays 3.1–3.3. The results are shown in figure 3.1 (plate 3).

Display 3.4

```
%%% Program ThreeCLUT.m
figure;                 % set up a new figure window
image(M);               % display an image described in matrix M
axis('square');         % make the aspect ratio of the image
                        % square
axis('off');            % turn off all axis lines and markings
lut1 = [0 0 0;  1 1 1]; % define a two row lookup table for
                          % the color translation of 1,2
colormap(lut1);         % apply the lut to the displayed image
keyboard;               % stops execution of the program and
                        % gives control to the user's keyboard
                        % type "return" at the "K>>" prompt,
                        % the program will continue
lut2 = [1 0 0;  0 1 0]; % a new lookup table
colormap(lut2);
keyboard;
lut3 = [0.6 0.6 0.6;  0.4 0.4 0.4]; %another lookup table
colormap(lut3);
```

3.2 Generating Images

This section illustrates a range of images and how to create them. Examples include simple images, such as dots, lines, and letters, but also other images that are commonly used in experiments, such as sine-wave gratings, external noise images, and textures.

Before we start, we define a new MATLAB function, `showImage`, to display images. The function, shown in display 3.5, should be saved with the file name "`showImage.m`" and included in the PATH of MATLAB. The `showImage` code is a new function that displays images without altering the aspect ratio using either a gray lookup table or MATLAB's default colormap or lookup table. A call to the MATLAB function `image` displays an image *M* in an unnecessary axis frame and the default colormap. Calls to the `showImage` function eliminate repetitive parts in subsequent programs in this chapter.

3.2.1 A Dot on a Gray Background

We begin by drawing a black dot with a radius of 15 pixels, centered in a gray image of size 256 × 256 pixels, with the center of the image indexed as (0, 0). Display 3.6 shows the program that creates the image seen in figure 3.2.

Display 3.5

```
%%% Program showImage.m
function gca = showImage(M, lut_type)
gca = figure;     % Opens a graphics window
image(M);         % Display image M in the graphics window
axis('image');    % Sets axis properties of the image in the
                  % graphics window
axis('off');      % Turn off the axes

if strcmp(lut_type,  'grayscale') % Check if lut_type is
                                  % grayscale
   lut = [(0:255)' (0:255)' (0:255)']/255;
   colormap(lut);
else
   colormap('default'); % Use Matlab's default colormap
end
```

Display 3.6

```
%%% Program FixationDot.m
[x,  y] = meshgrid(-128:127,  128:-1:-127);
M = (x.^2 + y.^2  >= 15^2)*127 + 1;
showImage(M,  'grayscale');
```

Figure 3.2
A fixation dot in the center of the display.

3.2.2 A Fixation Cross

The next example draws a fixation cross of two lines of 1 pixel width, also centered in the image. Display 3.7 shows the program that creates the image seen in figure 3.3.

3.2.3 Sine Wave and Gabor Patterns

The sine wave is a very important basic pattern displayed in visual psychophysics[6] (see figure 3.4a). Here we show an example sine wave programmed to appear in a 256 × 256 pixel image. This sine wave is tilted 35° from horizontal and has a frequency of 1/32 pixels. The equation to create the tilted sine wave is

$$l(x,y) = l_0\left(1.0 \pm c\sin\{2\pi f[y\sin(\theta) + x\cos(\theta)]\}\right), \tag{3.2}$$

where $l(x, y)$ is the gray level for a pixel at location (x, y) in the image, l_0 is the mean gray level, f is the frequency of the sine wave in (1/pixels), and θ is the angular tilt of the sine wave. In this example, f is 1/32 and θ is 35°.

We also show how to create a Gabor, which is a sine-wave patch windowed by a 2D Gaussian (normal or bell-shaped) function. The Gabor has a well-defined spatial frequency range and is contained in a windowed region of space (figure 3.4b). It is another of the most frequently used stimuli in vision research.[7]

In this example, the Gabor is also tilted 35° from horizontal. An equation that creates this Gabor is

$$l(x,y) = l_0\left(1.0 \pm c\sin\{2\pi f[y\sin(\theta) + x\cos(\theta)]\} \times \exp\left[\frac{x^2 + y^2}{2\sigma^2}\right]\right), \tag{3.3}$$

Display 3.7

```
%%% Program FixationCross.m
[x,   y] = meshgrid(-128:127,   128:-1:-127);
M = 127*(1- ((y == 0 & x > -8 & x < 8) ...
     |(x == 0 & y > -8 & y < 8))) + 1;
showImage(M,    'grayscale');
```

Figure 3.3
A fixation cross in the center of the display.

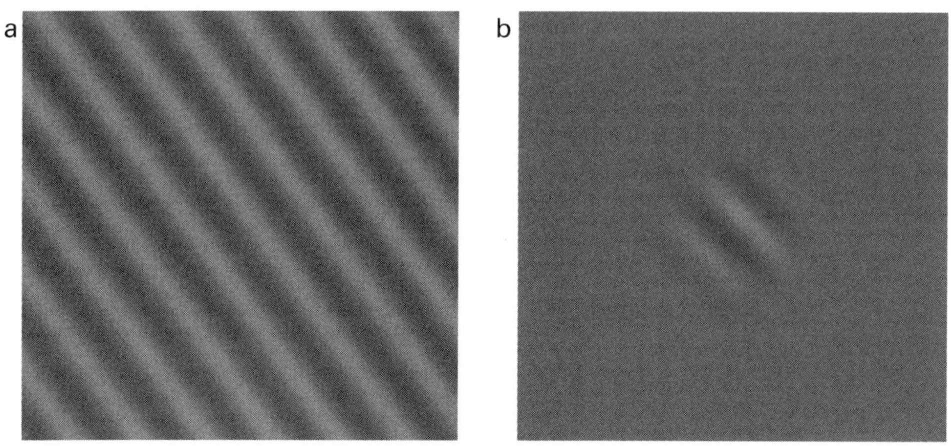

Figure 3.4
A sine-wave pattern (a) and a Gabor pattern (windowed sine-wave grating) (b).

Display 3.8

```
%%% Program SinewaveGratingGabor.m
c = 0.25;           % contrast of the Gabor
f = 1/32;           % spatial frequency in 1/pixels
t = 35*pi/180;      % tilt of 35 degrees into radians
s = 24;             % standard deviation of the spatial
                    % window of the Gabor
[x, y] = meshgrid(-128:127, 128:-1:-127);
M1 = uint8(127*(1 + c*sin(2.0*pi*f*(y*sin(t) + x*cos(t)))));
        % uint8 converts the elements of the array into unsigned
        % 8-bit integers. Values outside this range are mapped
        % to 0 or 255.
showImage(M1, 'grayscale');
M2 = uint8(127*(1 + c*sin(2.0*pi*f*(y*sin(t) + x*cos(t))) ...
    .*exp(-(x.^2 + y.^2)/2/s^2)));
showImage(M2, 'grayscale');
```

where $\sigma = 24$ pixels is the standard deviation of the spatial window that defines the Gabor patch. Display 3.8 shows the program to create both these stimuli.

3.2.4 A White Noise Image

Another common image used in visual psychophysics experiments is the noise image. Noise images are often used to change image quality or stop visual processing.[8–10] Noise images can be binary; that is, setting each pixel randomly to full black or full white. Alternatively, Gaussian *white noise* is created by setting each pixel to a value drawn randomly from a normal distribution with mean equal to the mid-value, or neutral gray, and with standard deviation set to a fraction of the range of achievable intensity values. A sample Gaussian white noise image created with the program in display 3.9 is seen in figure 3.5.

3.2.5 A Contrast-Modulated Sine Wave on a Noise Texture Carrier

Another class of stimuli in visual psychophysics reflects second-order processing. In the image seen in figure 3.6, a fine spatial pattern is defined by random noise in which pixels are either white or black, creating a noise texture. Another, *second-order* pattern emerges when the contrasts of the noise pixels are altered or "modulated" by the contrast of a larger-scale sine-wave pattern. This second-order pattern is determined by changes in texture contrast over space, whereas first-order patterns are determined by changes in luminance over space. Any local patch of this image that encompasses a collection of pixels has the same expected luminance as the background. Second-order processing has been widely studied in visual perception over the past several decades.[11,12] The image shown in figure 3.6 is generated using the program in display 3.10.

Display 3.9

```
%%% Program WhiteNoiseImage.m
M = 127 + 42*randn(64);
M = uint8(M)  +  1;
showImage(M,   'grayscale');
```

Figure 3.5
A Gaussian white noise image.

3.2.6 A Random Dot Stereogram

One important cue to depth, or the distance of an object from the viewer, arises because the retinas of the two eyes receive projections of the world from slightly different angles. We can mimic this in a laboratory situation by arranging—through goggles, special glasses, or mirror systems—to present different images to the left and right eyes that mimic the different viewing angles for objects at different distances (see section 5.2.4 in chapter 5). These differences, or horizontal disparities, between the view of identical features in the left and right eyes provide *stereopsis* cues to depth perception.

A special form of stimulus called the *random dot stereogram* tests stereopsis in the absence of visible objects in either eye alone. This pure test of stereo, extensively used by Julesz,[13] eliminates other cues to depth (such as perspective) as no objects are visible in a single eye's view, which consists of only a random pattern of black and white pixels. Observers see either single eye view alone as a flat field of noise elements. If the left and right eye images are identical, there is no disparity, and no depth pattern is visible. Horizontally shifting the position of a patch of dots in opposite directions in the two images generates horizontal disparity cues as the texture patterns of the patch are matched in the process of stereo perception. Larger perceived depth offsets of the patch correspond to

Display 3.10

```
%%% Program ContrastModulatedGrating.m
c = 0.50;
f = 1/32;
[x,  y] = meshgrid(-64:63,  64:-1:-63);
N = 2*(rand(128,128) > 0.5) - 1;
M = uint8(127*(1 + N.*(0.5 + c*sin(2.0*pi*f*x))));
showImage(M, 'grayscale');
```

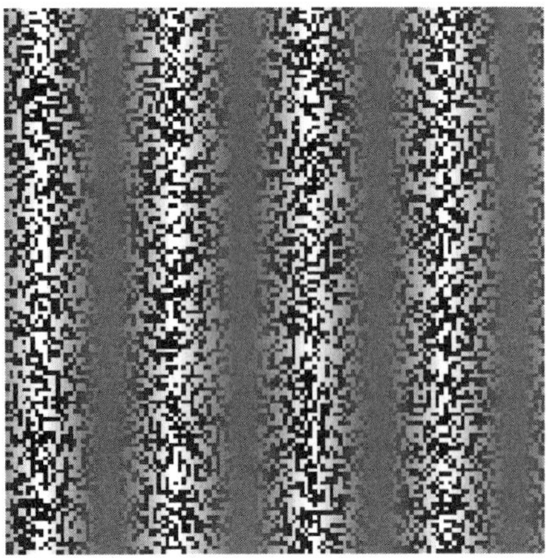

Figure 3.6
A contrast modulated sine-wave grating.

larger disparities or shifts in the image in the two eyes. Some people, "free fusers," are able to process and combine two separated images, one for the left eye and one for the right eye, without a stereo viewer or other specialized laboratory equipment.

Display 3.11 shows the generation of a stereo pair of 256 × 256 binary noise images in which a common 64 × 64 binary noise image is embedded at slightly different locations near the middle. The two are offset by 2 pixels of disparity in this example. The pair of images for the random dot stereogram is shown in figure 3.7. An individual who can fuse the two images without devices by relaxing the convergence of the eye—or someone who is viewing the two images through a stereo device—will see a square floating above the background.

3.2.7 A True Color Image

A *true color image* specifies the color values for each pixel directly as RGB triplets. In MATLAB, an RGB color image is represented by an $I \times J \times 4$ matrix in which the first

Display 3.11

```
%%% Program RandomDotStereogram.m
N1 = 128 + 127*(2*(rand(256, 256) > 0.5) - 1);
N2 = 128 + 127*(2*(rand(64,  64) > 0.5) - 1);
M1 = N1;
M2 = N1;
M1(97:160,  96:159) = N2;    % replace a region of N1 with N2
                             % in the left eye image
M2(97:160,  98:161) = N2;    % replace a region of N1 with N2
                             % in the right eye image
showImage(M1,   'grayscale');
showImage(M2,   'grayscale');
```

Figure 3.7
A random dot stereogram made of a left-eye and a right-eye image.

three $I \times J$ submatrices represent intensity values in the R, G, and B channels. We show an example in display 3.12 and figure 3.8 (plate 4). The program creates a 256 × 256 zero-centered image with a colorful circular bull's-eye target by defining different RGB combinations within certain radii, r, from the center point at (0, 0).

3.2.8 Creating an Image from a MATLAB Graph

Unfortunately, MATLAB does not have a predefined function that creates an image out of text directly. However, MATLAB has predefined functions to add text to a graph. So one way to generate an image with text is first to make a MATLAB figure that is empty, and use `image` to display a uniform gray background. Then add text to the figure using

Display 3.12

```
%%% Program TrueColorImage.m
[x,  y] = meshgrid(-128:127,  128:-1:-127);
r = sqrt(x.^2 + y.^2);
RI = 128 + (r <32 )*127 + (r >= 128)*127;
GI = 128 + (r >= 28 & r < 80)*127 + (r >= 128)*127;
BI = 128 + (r >= 72)*127;
M = cat(3,  RI,  GI,  BI)/255; % cat takes 3 2-dimensional
                               % RGB matrices to create
                               % an image with 3 color planes
showImage(M,   ' ');
```

Figure 3.8 (plate 4)
A true color image defined with RGB values.

the function `text`. Finally, `getframe` is used to create an image of the content of the figure, and the relevant subset is extracted. This same approach can be used to draw lines, or other patterns, using predefined plot or graphics functions in MATLAB and then extracting them as images for your experiments. Display 3.13 shows an example program for creating an image of a letter, shown in figure 3.9.

3.2.9 Wedge and Ring

Other standard but somewhat more complicated images, often used in functional magnetic resonance imaging (fMRI), are a wedge and a ring of checkerboards in which the check sizes increase as a function of eccentricity, or distance from the fovea. These images are used to stimulate different regions of the visual cortex that have retinotopic maps of space[14,15] and so can be used to localize different retinotopic brain areas. The program that

Display 3.13

```
%%% Program CaptureGraph.m
I = 65; J = 65;
M = 128*ones(I,J);
hf  =  figure;     % define a new figure
image(M);          % create an image of M in the figure
clut = [(0:255)' (0:255)' (0:255)']/255;
colormap(clut);    % apply a grayscale clut to the figure

%define the coordinates in the image of M
set(gca, 'units', 'pixels', 'position', ...
    [1 1 J-1 I-1], 'visible', 'off');
text('units', 'pixels', 'position', [round(J/2 - 15) ...
    round(I/2)], 'color', [0.1 0.1 0.1], 'fontsize', ...
    30, 'string', 'A');
drawnow;              % Draw letter "A"
tim = getframe(gca); % Capture the image from the graph window
close(hf);
tim2  = tim.cdata(1:(I-1), 1:(J-1), :); % Extract the image
                                          % of "A" from cdata.
showImage(tim2,  'grayscale');
```

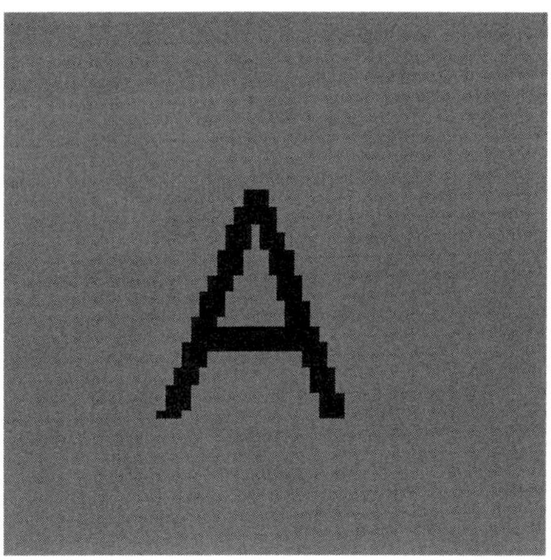

Figure 3.9
Image of a character extracted from a MATLAB figure.

defines these wedge and ring images in figure 3.10 is shown in display 3.14. To represent fine geometric shape well, we used 2048 × 2048 images. Regions of alternating intensity were defined by different radii and angled lines in the pattern. To do this, we first translated the pixel locations in [*x*, *y*] into a polar coordinate system [*r*, theta] of the radius and angle.

3.2.10 Making a Texture Pattern

In all of the previous cases, we have shown you simple patterns. In many experiments, the stimulus is composed of a collection of elements made of simple patterns. We can create textures from simple patterns by tiling smaller images onto a larger image. Figure 3.11 shows an example in which the texture consists of semiregularly spaced small images, with random jitter in the exact positions. Each little subimage shows a randomly oriented Gabor. The program to create this example is shown in display 3.15.

3.3 Manipulating Images

One of the powerful uses of computer graphics is the transformation or alteration of images. If you have been interested in digital photography, for example, you are probably familiar with transformations of your images in Adobe Photoshop that alter contrast or

Display 3.14

```
%%% Program WedgeRing.m
[x, y] = meshgrid(-1024:1023, 1024:-1:-1023);
theta = atan2(y, x); % translates x, y to polar angle
r = sqrt(x.^2 + y.^2); % and radius
sWidth = pi/8;          % width of each wedge in polar angle
mask1 = 2*round((sin(theta*2*(2*pi/sWidth)) + 1)/2) - 1;
                        % make wedge pattern
r0=[ 64 96 144 208 288 400 528 672 832 1024];
                        % radii of different rings
mask2 = (r < r0(1));
for i = 2:10
  mask2 = mask2 + (2*mod(i,2) - 1)*(r >= r0(i-1) & r <
r0(i));
end
mask3 = mask1.*mask2;
Wmask = (theta  > -pi/6 & theta < pi/6);
Rmask = (r > 64 & r <= 144) + (r > 672 & r < 1024);
Wedge = Wmask.*mask3*127 + 128;
Ring = Rmask.*mask3*127 + 128;
showImage(Wedge,   'grayscale');
showImage(Ring,   'grayscale');
```

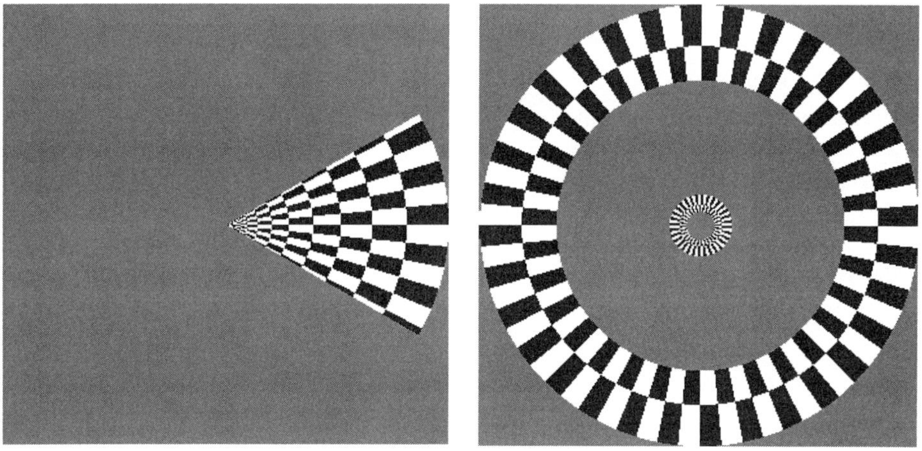

Figure 3.10
A wedge (left) and a ring pattern (right)are frequently used in fMRI visual experiments to measure retinotopic areas in visual cortex.

color profiles, blur edges, cartoon, and many other effects to achieve a creative goal. In visual psychophysics, we are interested in quantifying the relationship between image properties and human performance. This requires us to manipulate images to control for certain properties, such as size or intensity, in order to test ideas about human visual processing. For these purposes, it is important that we control image manipulations exactly and can characterize the resulting images that are displayed. Starting with images stored in a particular format, we use MATLAB routines to create new images with exact properties.

3.3.1 Image File Format

Every digital image is stored in a particular file format. *Image file formats* are standardized means of organizing and storing digital images. Image files are composed of either pixels, vectors, or a combination of the two. Regardless of the format, the files are converted (rasterized) into pixels when the image is displayed.

Each image file format follows certain conventions that determine how the information is stored and includes general information about the image in a *header* followed by the data of the image in either pixel or vector format. The header is the decoder key for the image. It defines the image size, pixel depth, CLUT, and other information needed to interpret and display the image. The pixel/vector map is a collection of numbers that are related to image intensity in space.

There are many image file formats, some of which you will already recognize by their computer extensions, such as *bmp* (windows bitmap), *hdf* (hierarchical data format), *gif*

Display 3.15

```
%%% Program TexturePattern.m
c = 0.50;           % contrast of Gabor
f = 1/8;            % spatial frequency of Gabor in 1/pixels
t = [22.5 67.5 112.5 157.5]*pi/180; % 4 different Gabor tilts
                                    % in polar angles
s = 4;              % standard deviation of Gabor window
[x, y] = meshgrid(-16:15, 16:-1:-15);
for i = 1:4
    M(:, :, i) = 127*(1 + c*sin(2.0*pi*f*(y*sin(t(i)) + …
                 x*cos(t(i)))).*exp(-(x.^2 + y.^2)/2/s^2));
end

[Tx,  Ty] = meshgrid(1:40:320, 1:40:320);
T = 127*ones(320,  320);
for i = 1:8
    for j = 1:8
        M_index = round(3*rand(1)) + 1;
        Tx(i,j) = Tx(i,j) + round(rand(1)*6);
        Ty(i,j) = Ty(i,j) + round(rand(1)*6);
        T(Ty(i,j):(Ty(i,j)+31), Tx(i,j):(Tx(i,j) + 31)) = …
            M(:, :,  M_index);
    end
end

showImage(T,   'grayscale');
```

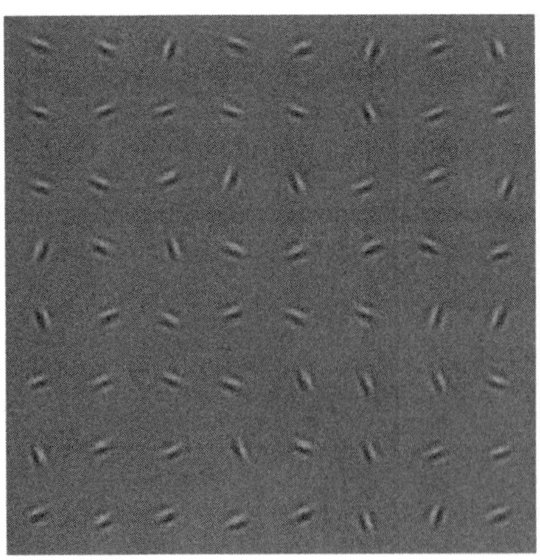

Figure 3.11
A texture made of many Gabors.

(graphics interchange format), *jpeg* (joint photographic experts group), *pbm* (portable bitmap), *pgm* (portable graymap), *png* (portable network graphics), *ppm* (portable pixmap), or *tiff* (tagged image file format). The different formats were developed by different user groups or companies for somewhat different purposes. The various formats follow different conventions in storing images. Some of them are used to represent true color images; some of them are used to represent indexed color images. Some represent color in RGB space (e.g., JPEG); some represent color in CMYK color space (e.g., TIFF), which is a special four-color model used in color printing where four inks (cyan, magenta, yellow, and black) are used.

3.3.2 Reading in an Image

MATLAB provides routines to read in preexisting images. In this example, we show how preexisting images are read into a matrix and a corresponding colormap using `imread` with a specified input format. In display 3.16, a digital photograph Church.jpg, which can be downloaded from the book's web page, is read in as a true color image in JPEG format. The corresponding image of the church is shown in plate 5 (figure 3.12).

Display 3.16

```
%%% Program ReadInImage.m
[M, map] = imread('Church.jpg', 'jpeg');
      % M is a true color image in a 1944x2896x3 matrix
showImage(M, ' ');
```

Figure 3.12 (plate 5)
A digital color photograph of a church.

3.3.3 Image Type Conversions

Image type conversion is a frequent operation in computer graphics. Often, applications make it necessary to convert an image stored in one image type to a different image type. For example, some applications may expect black and white images, or certain programs for controlled display of stimuli expect images to be in an 8-bit (256 level) format. There are a number of available conversion functions within MATLAB. Or, other programs such as Adobe Photoshop also allow you to alter images. MATLAB allows you to convert from any color space to another using the function `applycform`. However, special routines have been developed for commonly used conversions. A few examples of special routines are illustrated next.

A true RGB image may be converted to a grayscale image through the use of the MATLAB function `rgb2gray`, which eliminates information about hue and saturation of the colors while retaining the intensity information. In display 3.17, we show how to use `rgb2gray` to convert the color photograph of the church in plate 5 into a grayscale image shown in figure 3.13.

The function `im2bw` reads in and transforms an image to binary black and white (without grayscale). This function converts an indexed, intensity, or RGB image to a binary image by first converting an RGB image to a grayscale image and then converting the grayscale

Display 3.17

```
%%% Program Color2Grayscale.m
[M,  map] = imread('Church.jpg',   'jpeg');
M2 = rgb2gray(M);
showImage(M2,   'grayscale');
```

Figure 3.13
A grayscale image converted from the color photograph of the church.

Display 3.18

```
%%% Program Color2BW.m
[M,  map] = imread('Church.jpg',   'jpeg');
M3 = im2bw(M)*255 + 1; % The output of im2bw consists of 0's
                       % and 1's. The values are converted
                       % into 1 and 256 for display.
showImage(M3,   'grayscale');
```

Figure 3.14
A black and white image converted from the color photograph of the church by thresholding.

image to binary by applying a threshold. The binary image contains values of 1 for white, for all pixels with a value greater than the threshold, and values of 0 for black for all other pixels. The conversion `im2bw` returns values of 0 or 1, which are converted to 1 and 256 for display corresponding to the lowest and highest value in the lookup table. Display 3.18 shows a program using `im2bw` to convert the RGB color image of the church into a binary black and white image seen in figure 3.14.

In addition to transforming RGB images to grayscale images, it may be useful to convert between the two forms of grayscale image—between index and intensity images. The MATLAB function `ind2gray` converts an indexed image and the associated colormap into an intensity image. The MATLAB function `gray2ind` takes a grayscale image and creates an indexed image and a colormap with a specified number of gray levels. In display 3.19, we show an example program that combines a grayscale image M1 with a new lookup table to generate a new grayscale image M2 with an inverted color lookup table (figure 3.15). We then covert M2 into an indexed image M3 that includes a lookup table.

Display 3.19

```
%%% Program IndexGrayscale.m
[M,   map] = imread('Church.jpg',   'jpeg');
M1 = rgb2gray(M);

clut1 = [(255:-1:0)'   (255:-1:0)'   (255:-1:0)']/255;
M2 = ind2gray(M1,   clut1);
showImage(M2,    'grayscale');

[M3,   clut2] = gray2ind(M2,   256);
        % gray2ind returns a grayscale image M3 with a
        % color lookup table clut2
```

Figure 3.15
A new grayscale image created by combining an indexed image with a color lookup table.

3.3.4 Geometric Transformations

Geometric transformations change the spatial layout of the image. This includes some simple cases such as resizing, rotation, and cropping. The program showing these three manipulations of the church image is in display 3.20, with the corresponding images shown in figure 3.16.

3.3.5 Intensity Transformations

One of the key manipulations in visual psychophysics is the variation of contrast, as seen in chapter 2 for sine-wave gratings. Here we show an example of reducing the contrast of a natural image by half, with the program in display 3.21 and the resulting low-contrast grayscale image of the church in figure 3.17.

Display 3.20

```
%%% Program GeometricTransformations.m
[M,   map] = imread('Church.jpg',   'jpeg');
M1 = rgb2gray(M);

M2 = imresize(M1,  0.25,  'bicubic');
        % scale the image by 0.25 using cubic interpolation.
showImage(M2,  'grayscale');

M3 = imrotate(M2,  45,  'bicubic');
        % rotate M2 counter-clock wise by 45 degrees.
showImage(M3,   'grayscale');

M4 = imcrop(M3,  [185 335 340 327]);
showImage(M4,   'grayscale');
```

Figure 3.16
Resized, rotated, and cropped images of the church photograph.

3.3.6 Edge Extraction

People are very good at recognizing objects from line drawings, which have far less information than photographs.[16] Here, we show one standard method of extracting edges from a photograph that uses a simple MATLAB subroutine, `edge`. The internal processing in the `edge` routine uses mathematically defined "filters" developed in computer graphics that mimic some processing in early visual system. Other routines use more sophisticated edge extraction algorithms. Display 3.22 shows the program, and figure 3.18 shows the edges extracted from the church image.

Display 3.21

```
%%% IntensityTransformation.m
[M,  map] = imread('Church.jpg',  'jpeg');
M1 = rgb2gray(M);
bkgrd = mean2(M1);      % mean2  computes the matrix mean.
M2 = uint8(0.50*(M1 - bkgrd) + bkgrd);
                        % Reduce image contrast by 50%
showImage(M2,   'grayscale');
```

Figure 3.17
A digital photograph of the church in lower contrast.

Display 3.22

```
%%% Program EdgeExtraction.m
[M,  map] = imread('Church.jpg',  'jpeg');
M1 = rgb2gray(M);
M3 = 255*edge(M1,  'log',  0.005)+1;
    % compute edges of M1 using Laplacian of Gaussian method
showImage(M3,   'grayscale');
```

Figure 3.18
Extracted edges of the church photograph.

3.3.7 Selecting a Region of Interest

Human processing of visual input usually focuses on specific areas of the visual scene by fixating on a location of interest. The region around the fixation is clear, and less information is available away from fixation where there are fewer receptors—although we may not always be aware of it.[17] People with *macular degeneration*, a degenerative eye disease observed in the elderly, suffer from loss of visibility near fixation that is especially challenging for visual function.[18] Here we show programs for grayscale images that mimic selection of a region of interest in a large display and reductions in contrast away from the center that illustrate loss of information away from the fovea. In another case, we illustrate selective loss of information at the center of the region of interest. Programs are shown in display 3.23, and the resulting images are shown in figure 3.19. Figure 3.19a shows the image through a circular window, figure 3.19b uses a Gaussian spatial window, and figure 3.19c mimics macular degeneration by filtering out a central region around a hypothetical fixation point.

3.3.8 Adding Two Images

Some key applications in visual psychophysics—such as visual masking—use images that are composed of two superimposed images that are combined through addition. In the case of noise masking, we combine a target image, such as the church, with a noise image, as seen in the program in display 3.24 and the resulting masked image of the church in figure 3.20. This noise image draws the intensity value of each pixel from a normal distribution centered at 0 with some standard deviation s. Noise standard deviations near 0 barely disrupt the underlying image, whereas high standard deviations show more visible masking.

Display 3.23

```
%%% Program RegionOfInterest.m
[M,  map] = imread('Church.jpg',  'jpeg');
M1 = rgb2gray(M);
mI = mean2(M1);
Sz = size(M);
[x,  y] = meshgrid(-Sz(2)/2:(Sz(2)/2-1),  ...
          Sz(1)/2:-1:-(Sz(1)/2-1));

% A circular window mask with radius s
s = 400;
Gmask1 = ((x + 600).^2 + (y - 100).^2  <  s^2);
          % center mask at (-600,+100)
M2 = uint8((double(M1) - mI).*Gmask1 + mI);
showImage(M2,    'grayscale');

% A Gaussian mask with standard deviation s
Gmask1 = exp(-((x + 600).^2 + (y - 100).^2)/2/s^2);
M3 = uint8((double(M1) - mI).*Gmask1 + mI);
showImage(M3,    'grayscale');

% AMD (age related macular degeneration example)
Gmask1 = 1 - exp(-((x + 600).^2 + (y - 100).^2)/2/s^2);
M4 = uint8((double(M1) - mI).*Gmask1 + mI);
showImage(M4,    'grayscale');
```

Figure 3.19
(a) A windowed region, (b) a Gaussian windowed region, and (c) a notch windowed region of the church photograph.

Display 3.24

```
%%% Program NoiseMasking.m
[M,  map] = imread('Church.jpg',  'jpeg');
M1 = rgb2gray(M);
Mn = mean2(M1);
Sz = size(M);
% Generating a Gaussian white noise image with standard
% deviation s
s = 16;
noiseI = s*randn(Sz(1),  Sz(2));
M2 = uint8(double(M1) + noiseI);  % noise is double, so make M1
                                   % double, too
showImage(M2,   'grayscale');

s = 96;
noiseI = s*randn(Sz(1),  Sz(2));
M3 = uint8(double(M1) + noiseI);
showImage(M3,   'grayscale');
```

Figure 3.20
Gaussian white noise masked church photograph.

Noise masking manipulates the signal-to-noise ratio in images and thereby controls the quality of the images. Noise masking has been a major instrument in understanding visual processing.[8,9] A similar process can be used to combine any two images.

3.3.9 Filtering

Filtering is used to remove or reduce certain features or components of an image.[19] A filter is a device or a computational procedure that removes or attenuates some components or features from an image. Filtering is used in some practical situations to reduce noise in a stimulus. In the vision laboratory, spatial frequency filtering is used to create specialized test stimuli.

In many applications, this is accomplished through the *fast Fourier transform* (`fft2`), which codes the quantity of each spatial frequency in the image. This mimics the operation of the early visual system that analyses images into spatial frequency components.[5,20] The Fourier analysis describes an image in terms of its spatial frequency content. Low spatial frequencies correspond with slow undulations in intensity over space, whereas high spatial frequencies correspond with rapid changes in intensity over space. Visibility of different spatial frequencies was shown in the earlier discussion of the contrast sensitivity function (CSF) in section 1.3.2 in chapter 1 and section 2.3 in chapter 2.

A Fourier transform takes an input image and redescribes the image in terms of coefficients representing the amount, or magnitude, of each spatial frequency and orientation and its phase. The Fourier representation of the input image makes it easy to alter the spatial frequency and orientation content of images. For example, in filtering, some spatial frequencies are reduced or filtered out while others are left intact. Finally, the filtered image is reconstructed through the inverse fast Fourier transform (`ifft2`).

Display 3.25 shows a program that carries out low-pass, high-pass, and band-pass filtering of the church image. Filtering removes certain frequencies from the grayscale church image by multiplying the coefficients of the original image "magnitude spectrum" and the filter in the domain of the Fourier representation. Figure 3.21 shows the original church image and images resulting from low-pass, high-pass, and band-pass transformations from top to bottom. A low-pass filter attenuates high spatial frequencies while "passing" (keeping) low spatial frequencies, which blurs the original image. A high-pass filter does the opposite; it attenuates low spatial frequencies while keeping high spatial frequencies, retaining the sharper edges in the image. Because low spatial frequencies are less important in object recognition, the high-pass image appears similar to the original to us. A band-pass filter attenuates both very low and very high spatial frequencies. Band-pass filters can be either broad or narrow. Figure 3.21 shows a narrow band-pass filter that is the part in common between the low- and high-pass examples—which has extracted the more obvious edges in the image.

The middle column of figure 3.21 shows the log of the Fourier spectra of the images in the left column. (It is necessary to take the log to make small coefficient values for many spatial frequencies more visible in displaying Fourier spectra.) In these spectra, orientation is depicted by the radial angle, and the spatial frequency is low in the center and increasing with distance from the center. The spectra of the filtered images were created by multiplying the Fourier spectrum of the original church images and the spatial frequency filters shown in the right column of figure 3.21. The filters are white (multipliers of 1) where frequencies are being "passed" or retained and black (multipliers of 0) where frequencies are being filtered out.

The mathematical treatment of the discrete fast Fourier transform[21] is not provided here, but these transformed images, composition of the `fft` spectra, and filters should provide

Display 3.25

```
%%% FourierFiltering.m
[M,  map] = imread('Church.jpg',  'jpeg');
M1 = rgb2gray(M);
fM = fftshift(fft2(M1 - mean2(M1))); %Fourier transformation
maxfM = max(max(log(abs(fM))));
showImage(uint8(256*log(abs(fM))/maxfM),  'grayscale');
Sz = size(M1);
[x,  y] = meshgrid(-Sz(2)/2:(Sz(2)/2-1),  ...
          Sz(1)/2:-1:-(Sz(1)/2-1));
r = sqrt(x.^2 + y.^2);
r0 = 64;
s1 = 10;

% Lowpass filtering
lowpass = (r <= r0) + (r > r0).*exp(-(r - r0).^2/2/s1^2);
        % Construct a lowpass filter
showImage(uint8(256*lowpass),  'grayscale');
fMl = fM.*lowpass;  % Apply the lowpass filter to the
                    % Fourier spectrum
showImage(uint8(256*log(abs(fMl) + 1)/maxfM),  'grayscale');
M2 = uint8(ifft2(fftshift(fMl)) + mean2(M1) );
        % Inverse Fourier transformation
showImage(M2,  'grayscale');

% Highpass filtering
highpass = (r >= r0) + (r < r0).*exp(-(r - r0).^2/2/s1^2);
         % Construct a highpass filter
showImage(uint8(256*highpass),  'grayscale');
fMh = fM.*highpass; % Apply the highpass filter to the
                    % Fourier spectrum
showImage(uint8(256*log(abs(fMh) + 1)/maxfM),  'grayscale');
M3=uint8(ifft2(fftshift(fMh)) + mean2(M1) );
        % Inverse Fourier transformation
showImage(M3,  'grayscale');

% Bandpass filtering
bandpass = exp(-(r - r0).^2/2/s1^2);
        % Construct a bandpass filter
showImage(uint8(256*bandpass),  'grayscale');
fMb = fM.*bandpass; % Apply the bandpass filter to the
                    % Fourier spectrum
showImage(uint8(256*log(abs(fMb) + 1)/ maxfM),  'grayscale');
M4 = uint8(ifft2(fftshift(fMb)) + mean2(M1) );
        % Inverse Fourier transformation
showImage(M4,  'grayscale');
```

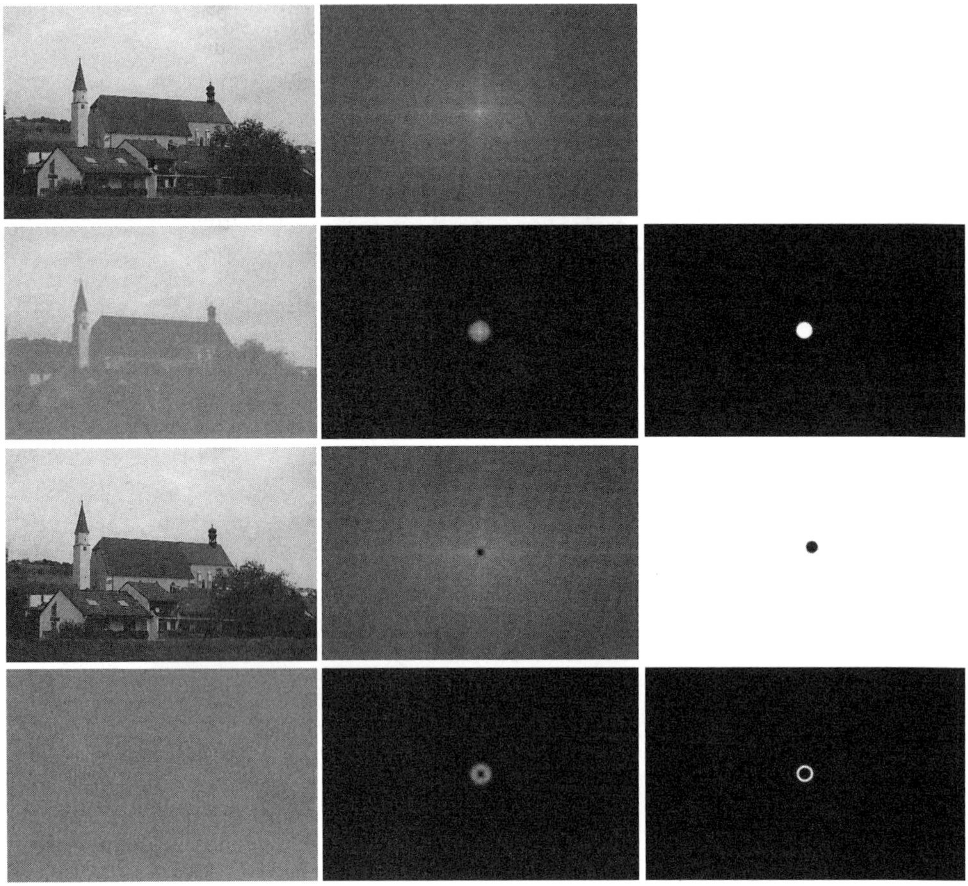

Figure 3.21
The original church image and its Fourier spectrum (top row). A low-pass filtered church image, its Fourier spectrum, and the low-pass filter (second row). A high-pass filtered church image, its Fourier spectrum, and the high-pass filter (third row). A band-pass filtered church image, its Fourier spectrum, and the band-pass filter (bottom row).

the reader with a strong intuition about the importance of different spatial frequencies in the perception of objects and how images can be transformed for visual experiments.

The program in display 3.25 reads in the church picture, which is in color. It then transforms it to grayscale using `rgb2gray`. It then computes the `fft2` of the image after removing the "DC" component, which is the mean value. The Fourier spectrum is rearranged with `fftshift` so that the low spatial frequencies are in the center and orientations are radial. Multiplying the coefficients of the shifted spectrum with different masks carries out different filtering operations. Taking the inverse Fourier transformation with `ifft2` and adding back the "DC" reconstructs the new filtered image.

3.3.10 Phase Scrambled Image

The Fourier description of an image includes both a magnitude spectrum and a phase spectrum. The magnitude spectrum represents the magnitude (amount) of each spatial frequency component, and the phase spectrum describes the positioning of the sine-wave frequency components. The combination of the magnitude spectrum and the phase spectrum provides a complete description of an image. An image can be re-created by combining its magnitude and phase spectra through the inverse Fourier transformation. Phase scrambled images preserve all the magnitude information in an image, but randomize the location of the Fourier components within the image. Phase scrambled images are no longer recognizable. So, for perception of objects within images, phase information is absolutely critical.[22] However, phase scrambled images elicit essentially the same distribution of brain activity in the early visual system while changing the recognizability of the image[23] (but see Ref. 24). For this reason, phase scrambled images are used as a control stimulus for activation in early visual areas in fMRI applications.

The program in display 3.26 computes the Fourier representation of the grayscale church image and extracts its magnitude spectrum. It then generates a random phase spectrum (from $-\pi$ to π) and then creates a new image. The Fourier spectrum of the new image combines the original magnitude spectrum of the church with the random phase spectrum and uses the inverse transformation to produce the resulting image shown in figure 3.22.

3.3.11 Convolution

Signal processing of images is often used to find features in the image—and some of these image transformations may mimic the processing of the visual system. Many important forms of image transformation are carried out by a mathematical operation called *convolution*. Convolution computes the outputs of a spatial array of detectors looking for a particular spatial profile or pattern. The convolution operation computes the dot product or match between the detector and the image in every spatial location.

Display 3.27 shows a program that computes a convolution of the image with a so-called difference of Gaussians (DOG) detector at each spatial position. The DOG detector looks

Display 3.26

```
%%% Program PhaseScrambledImage.m
[M, map] = imread('Church.jpg', 'jpeg');
M1 = rgb2gray(M);
fM = fftshift(fft2(M1 - mean2(M1))); %Fourier transformation
Sz = size(M1);
m_fM = abs(fM);       % extract the magnitudes spectrum
R = 2*rand(Sz)-0.5; % generate a random image of the same size
fR = fftshift(fft2(R - mean2(R)));
        % perform Fourier transformation on the random image
p_fR = angle(fR);     % extract the phase spectrum of
                      % the random image
fM2 = complex(m_fM.*cos(p_fR), m_fM.*sin(p_fR));
        % combine the original magnitude spectrum with the
        % random phase spectrum
M2 = real(ifft2(fftshift(fM2))) + mean2(M1);
showImage(M2, 'grayscale');
```

Figure 3.22
A phase scrambled image of the grayscale church photograph.

Display 3.27

```
%%% Program DOGConvolution.m
[M,   map] = imread('Church.jpg',   'jpeg');
M1  = rgb2gray(M);
Sz  = size(M1);

% Construct a DOG image
[x,   y] = meshgrid(-32:31,   32:-1:-31);
s1  = 4;     % standard deviation of the Gaussian center
s2  = 16;    % standard deviation of the Gaussian surround
DOGmask  = exp(-(x.^2 + y.^2)/2/s1^2) - ...
           (s1/s2)^2*exp(-(x.^2 + y.^2)/2/s2^2);

%Convolve M1 with the DOG image
M2  = conv2(double(M1),   DOGmask);
showImage(uint8(127 + 126*DOGmask),   'grayscale');

% Crop the image to eliminate edge effects
M2  = uint8(M2(17:(Sz(1)  + 15),   17:(Sz(2)  + 15)));
showImage(M2,    'grayscale');
```

Figure 3.23
DOG convolved church photograph.

for transitions in intensity. If the DOG is located over a patch where all intensities in the original image are the same, the correlation is 0. The DOG convolution provides a new image that highlights the edges in the pattern of the original image, as seen in figure 3.23.

3.4 Summary

The basic concepts of computer graphics and digital manipulation of images described in this chapter provide a basis for generating some of the most commonly used stimuli in psychophysics. The examples were selected to illustrate important principles and tools that can be used to generate many other kinds of visual images for display. Although all of the current examples create individual images, compositions of multiple images can be used to create more complex displays. And a series of images can be used to generate dynamic sequences in time. The display of multiple images in space or in time is considered in chapter 4.

References

1. Foley JD. *Computer graphics: Principles and practice*. New York: Addison-Wesley Professional; 1996.

2. Porter T, Duff T. 1984. Compositing digital images. *ACM SIGGRAPH Computer Graphics* 18(3): 253–259.

3. The MathWorks Inc. MATLAB [computer program]. Natick, MA: MathWorks; 1998.

4. Gonzalez RC, Woods RE, Eddins SL. *Digital image processing using MATLAB*, 2nd ed. Knoxville, TN: Gatesmark Publishing; 2009.

5. Marchand P, Holland OT. *Graphics and GUIs with MATLAB*. Boca Raton, FL: CRC Press; 2003.

6. Campbell F, Robson J. 1968. Application of Fourier analysis to the visibility of gratings. *J Physiol* 197(3): 551–566.

7. Porat M, Zeevi YY. 1988. The generalized Gabor scheme of image representation in biological and machine vision. Pattern Analysis and Machine Intelligence. *IEEE Trans Pattern Anal Mach Intell* 10(4): 452–468.

8. Pelli DG, Farell B. 1999. Why use noise? *J Opt Soc Am A Opt Image Sci Vis* 16(3): 647–653.

9. Lu ZL, Dosher BA. 2008. Characterizing observers using external noise and observer models: Assessing internal representations with external noise. *Psychol Rev* 115(1): 44–82.

10. Sperling G. 1965. Temporal and spatial visual masking. I. Masking by impulse flashes. *JOSA* 55(5): 541–559.

11. Chubb C, Sperling G. 1989. Two motion perception mechanisms revealed through distance-driven reversal of apparent motion. *Proc Natl Acad Sci USA* 86(8): 2985–2989.

12. Cavanagh P, Mather G. 1989. Motion: The long and short of it. *Spatial Vision* 4 (2/3): 103–129.

13. Julesz B. *Foundations of cyclopean perception*. Chicago: University of Chicago Press; 1971.

14. Engel SA, Glover GH, Wandell BA. 1997. Retinotopic organization in human visual cortex and the spatial precision of functional MRI. *Cereb Cortex* 7(2): 181–192.

15. Sereno M, Dale A, Reppas J, Kwong K, Belliveau J, Brady T, Rosen B, Tootell R. 1995. Borders of multiple visual areas in humans revealed by functional magnetic resonance imaging. *Science* 268(5212): 889–893.

16. Biederman I, Ju G. 1988. Surface versus edge-based determinants of visual recognition. *Cognit Psychol* 20(1): 38–64.

17. Geisler WS, Perry JS. 1998. A real-time foveated multi-resolution system for low-bandwidth video communication. SPIE Proceedings 3299: 294–305.

18. de Jong PTVM. 2006. Age-related macular degeneration. *N Engl J Med* 355(14): 1474–1485.

19. Jain AK. *Fundamentals of digital image processing*. Englewood Cliffs, NJ: Prentice-Hall; 1989.

20. Enroth-Cugell C, Robson JG. 1966. The contrast sensitivity of retinal ganglion cells of the cat. *J Physiol* 187(3): 517–552.

21. Nussbaumer HJ. *Fast Fourier transform and convolution algorithms*. Berlin: Springer-Verlag; 1982.

22. Piotrowski LN, Campbell FW. 1982. A demonstration of the visual importance and flexibility of spatial-frequency amplitude and phase. *Perception* 11(3): 337–346.

23. Olman CA, Ugurbil K, Schrater P, Kersten D. 2004. BOLD fMRI and psychophysical measurements of contrast response to broadband images. *Vision Res* 44(7): 669–683.

24. Rainer G, Augath M, Trinath T, Logothetis NK. 2001. Nonmonotonic noise tuning of BOLD fMRI signal to natural images in the visual cortex of the anesthetized monkey. *Curr Biol* 11(11): 846–854.

4 Stimulus Presentation

Typical experiments in visual psychophysics present a sequence of one or more images called a "movie" to the observer and collect responses. In this chapter, we introduce the basic terminology and issues in stimulus presentation. We dissect a simple experimental program written in Psychtoolbox[1,2] and then provide experimental implementations of several representative stimuli and tasks.

4.1 From Digital Images to Movies

4.1.1 Movies

A movie is a sequence of images shown in succession over time. The movie definition for each image M in the sequence consists of the display time t, a display duration, a particular screen position x, y, with magnification factors *zoomx*, *zoomy*, and a color lookup table (*CLUT*). This definition is repeated for each image in the movie sequence.

Whether computed in the user program or loaded from an archive on a computer disk, M is an image matrix that is stored in computer memory. To display M, it must be copied from computer memory to a special memory on a graphics card along with the instructions about when, where, and how to display the image. The display device, usually a computer monitor, is a passive device that displays the contents of a currently active *frame buffer*. A frame buffer is a section of memory on a graphics device that drives the contents of one image on the display device.

Modern graphics cards usually have large memory capacity that can store multiple image frames of a movie. During a given display sequence, the user program specifies which section of graphics memory is the active frame buffer being displayed. In this multi-buffering mode, one can display an active frame while simultaneously loading other image frames into graphics memory.

4.1.2 Display Timing

Display devices, such as computer monitors, re-display the contents of the active frame buffer many times per second, usually 60 to 200 times per second. The frequency of re-display is called the *refresh rate*, which is either a preset property or a programable property of the display device. The duration of each movie frame is therefore defined in terms of the number of refreshes during which an image frame is displayed. The beginning of each image frame coincides with the refresh of the display device. The contents of a frame buffer must be completely specified before it becomes active. If a frame buffer is still in the process of being constructed while it is active, mergers of several images, visible discontinuities, or "tears" in those images may occur. We describe tools to test for and avoid these failures of display synchronization in chapter 5.

4.2 Programming Experiments in Psychtoolbox

This book includes many example programs for psychophysical experiments. All the sample experimental programs are written in MATLAB with Psychtoolbox Version 3 extensions (subroutines). Psychtoolbox Version 3 is a collection of MATLAB and GNU/Octave functions for generating and presenting accurately controlled visual and auditory stimuli and collecting responses from the observer.[3] GNU/Octave is a high-level interpreted language for numerical computations. The Psychtoolbox package is freeware and has more than 15,000 active users. It continues to be developed on a voluntary basis by a large community of users. The Psychtoolbox wiki http://psychtoolbox.org/wiki contains a forum and information for downloading and installing the package and about system requirements.

Our intention is not to provide full coverage and instruction for all of the many Psychtoolbox functions. Instead, we focus on those functions that are essential to the psychophysical laboratory and provide a general orientation to basic functions and what they are designed to accomplish. The Psychtoolbox website includes a general overview and introduction as well as various tutorials (http://psychtoolbox.org/PsychtoolboxTutorial). Readers may also use the help function in MATLAB to access the document pages concerning toolbox functions.

Unlike many software packages developed to run experiments, the philosophy of Psychtoolbox is not to write and run a small number of highly structured experiments through a descriptive language. Rather, it provides the interface between a MATLAB interpretive programming environment and devices such as graphics devices, sound devices, clocks, keyboards, joysticks, and touchpads by providing function calls to control the devices. The use of general programming in a MATLAB environment allows the user to define any number of different experiments in a very flexible manner.

To speed the execution of time-critical operations during an experiment, key Psychtoolbox routines are written in C code that are called as functions from MATLAB. The functions present a simple software interface that provides full control of the various devices to the user. This allows experiments programmed in MATLAB with Psychtoolbox functions to access all of the display, timing, and response collection hardware with millisecond timing accuracy. In the comments of the programs in this book, Psychtoolbox is sometimes referred to as PTB.

Our first sample program has some calls to Psychtoolbox functions that are used in all experiments but that may not be fully explained until later chapters in this book. In these cases, we aim to provide a brief functional description of what a set of calls accomplishes and refer the reader to the relevant subsequent section that treats this topic. In most cases, the specific details that cannot be understood in full now will be explained or become clear as you proceed through the book.

Each experimental program in this book consists of three modules: the Display Setup, Experimental, and System Reinstatement modules. The first module of the program performs general actions that set up and specify the experiment. The second module specifies and runs the experiment. The last module resets the computer and screens to the state in which they were found before the experiment.

The Display Setup module of each sample experimental program includes the definition of basic aspects of the experimental setup such as properties of the screen (e.g., size and resolution), the distance of the screen from the viewer, and a function to carry out something called *Gamma correction* (see chapter 5) in order to guarantee that what is shown on the screen corresponds precisely to what is programmed for display in terms of luminance or relative contrast.

The Experimental module of each sample experimental program will define and compute the stimuli to be shown to the observer, perform randomization or ordering of testing trials, generate visual and/or auditory displays, supervise the delivery and timing of the displays, and collect responses. In some cases, the Experimental module includes a simple data summary table at the end. In all cases, the relevant information about the experimental trials and collected data are saved in a raw data trace that could be used in subsequent programs to provide more detailed analyses.

The final module of the sample programs carries out all operations to return the computer and display to their original state so that others can use it.

Our philosophy in developing these sample experimental programs has been to provide as much portability to new computer and display setups as feasible. For this reason, we generally set up and define visual stimuli in terms of their size on the retina, or degrees of visual angle. Sine-wave gratings are defined by their spatial frequencies in cycles per degree (c/d) of visual angle, or the number of periods of the sine wave in 1° of visual angle. Timing is defined in terms of seconds. A few basic facts about a particular experimental setup, such as the physical size of a pixel (or of a 100-pixel line or square), the viewing distance of the observer, and the refresh rate of the graphics display card, specify the experimental setup. These key parameters can be used to express pixel sizes of images from sizes in degrees of visual angle and duration of displays that correspond to a number of screen refreshes. Once some of these basic parameters are set for your display and system, the program should run in a portable way.

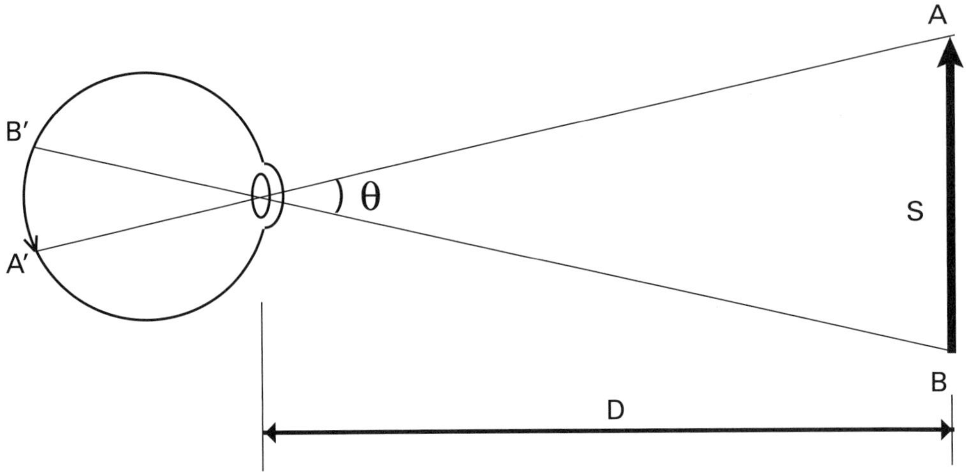

Figure 4.1
Illustration of the concept of visual angle.

Figure 4.1 shows some of the key parameters of the display setup that determine the translation between pixel size, viewing distance, and degrees of visual angle. In viewing angle, the size of an object is expressed as the degrees of arc that it subtends on the retina. The visual angle at the retina is a function of the size of the image, S, and the distance from the eye, D:

$$\begin{aligned}\text{Visual angle} &= 2\arctan\left[S/(2D)\right] \\ &\approx 57.3\frac{S}{D}\ \text{(degrees), if } S < D.\end{aligned} \tag{4.1}$$

The larger the object and the smaller the distance, the larger the visual angle subtended by the object on the retina.

Finally, many key parameters of the experiment in the sample programs are defined and saved in a MATLAB structure. Structures are a way to group variables together. For example, instead of defining individual variables `ScreenDistance`, `ScreenHeight`, and so forth, they may be defined as fields in a structure: `p.ScreenDistance`, `p.ScreenHeight`, and so on. The advantage of storing variables of different kinds in a structure is that they can be grouped together even if they are of different lengths, sizes, or types. This allows the experimenter to save all of the relevant variables together in a single command. For example, after storing the list of parameter variables in the structure `p`, you may save them all together as: `save('p.mat','p')`. The structure of variables can be reloaded together as: `load p.mat`.

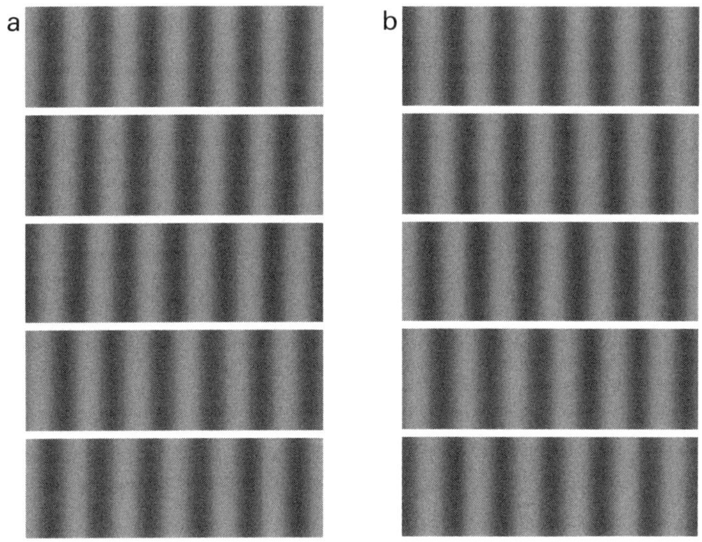

Figure 4.2
A motion stimulus defined by successive image frames. Five frames of a leftward (a) and a rightward (b) moving sine-wave grating.

4.3 A Detailed Sample Program

In this section, we explain a very simple program that runs an experiment to measure the observer's ability to see motion direction of a drifting sine-wave grating (figure 4.2).

The example program is divided into several modules. A Display Setup module defines the various screen parameters, initializes the display screen for experimental displays, loads the color lookup table, and specifies the screen font. The Experimental module creates images for the movies, executes the movies, collects responses, and saves data. The System Reinstatement module restores the settings of the screen and the computer to the original state at the beginning of the program, which leaves it in a standard state for other users or applications. Each module is considered in turn.

The experiment tests motion perception by showing a small number of frames of a moving stimulus and asking the observer to identify the direction of motion as left or right. Figure 4.2 shows five image frames of a moving sine-wave grating with the first frame at the top and frames that would be shown in successive time intervals going down. Leftward motion is shown on the left and rightward motion on the right.

The sample program is a very simple example of an experiment with 100 trials. Fifty of the 100 trials, randomly chosen, present a sine-wave grating moving to the left. The remaining 50 trials present a rightward moving sine-wave grating. The experiment measures the observer's ability to see motion direction of a drifting 8×6 degree sine-wave grating with 20% contrast, a spatial frequency of 1 c/d, and a temporal frequency of 4 Hz. Each trial starts with a 250-ms fixation, followed by the presentation of the moving grating (250 ms or 1 temporal cycle). The observer presses the left or right arrow on the keyboard to indicate the perceived direction of motion after each stimulus presentation. The program saves the observer's response along with the stimulus condition for each trial.

The general approach in this program is to specify the stimulus in terms of its physical parameters and then compute the corresponding images that are required with respect to the screen size, pixel resolution, and screen refresh rate of your display. For example, the number of pixels of a sine-wave grating of a given size and spatial frequency are computed from the viewing distance and the known physical parameters of the display screen. The temporal frequency of 4 Hz corresponds with a period—or full repeat—of the position of the sine wave four times per second, or cycle per 250 ms. The required phase shift of the sine wave in each new frame of the movie is computed based on the known refresh rate of the display screen.

4.3.1 Display Setup Module

In the Display Setup module (display 4.1), we define which screen is to be used for the experimental presentation. Often, a computer setup will have multiple monitors or screens, each referenced by a number. The program uses the command `Screen('screens')` to find the screens available on the system and chooses the one with the highest reference number to be used. Alternatively, a specific one may be preset in the code. The initialization module then defines a number of physical parameters about the display setup, including the viewing distance of the observer from the screen, `p.ScreenDistance`, and the height of the screen, `p.ScreenHeight`. These are measured properties of a specific experimental setup.

Some basic characteristics of the visual display device are also specified in the program. The parameter specifying the characteristic function that relates the luminance on the monitor to the programmed intensity values, or the `p.ScreenGamma`, is determined by a separate calibration procedure (see chapter 5). The maximum luminance, or `p.maxluminance`, is specified based on calibration measurements. The standard background value of the display, `p.ScreenBackground`, is selected to be the midpoint of the luminance range, and so is set to 0.5. The background value determines the default value or luminance of the areas of the screen apart from the currently displayed image(s). In different experiments, the background value could, for example, be the minimum (0) or the maximum (1) or the midpoint (0.5), selected here, or any other value.

First, we clean up the system by closing any preexisting Psychtoolbox windows. Next, a series of actions are taken to set up a new window to be used to display stimuli in the experiment. Details of the new display window are set by a series of calls to the function `PsychImaging`. This function has many options and can carry out many actions. You

Display 4.1

```
%%% Program: DriftingSinewaveGrating.m
function DriftingSinewaveGrating

%% Display Setup Module
% Define display parameters
whichScreen = max(Screen('screens'));
p.ScreenDistance = 30;   % in inches
p.ScreenHeight = 15;     % in inches
p.ScreenGamma = 2;       % from monitor calibration
p.maxLuminance = 100;    % from monitor calibration
p.ScreenBackground = 0.5;

% Open the display window, set up lookup table, and hide the
% mouse cursor
if exist('onCleanup', 'class'), oC_Obj = onCleanup(@()sca); end
        % close any pre-existing PTB Screen window
% Prepare setup of imaging pipeline for onscreen window.
PsychImaging('PrepareConfiguration');    % First step in starting
                                         % pipeline
PsychImaging('AddTask', 'General', …
       'FloatingPoint32BitIfPossible');
        % set up a 32-bit floatingpoint framebuffer
PsychImaging('AddTask', 'General', …
       'NormalizedHighresColorRange');
        % normalize the color range ([0, 1] corresponds
        % to [min, max])
PsychImaging('AddTask', 'General', 'EnablePseudoGrayOutput');
        % enable high gray level resolution output with
        % bitstealing
PsychImaging('AddTask','FinalFormatting', …
       'DisplayColorCorrection','SimpleGamma');
        % setup Gamma correction method using simple power
        % function for all color channels
[windowPtr p.ScreenRect] = PsychImaging('OpenWindow', …
        whichScreen, p.ScreenBackground);
        % Finishes the setup phase for imaging pipeline
        % creates an onscreen window, performs all remaining
        % configuration steps
PsychColorCorrection('SetEncodingGamma', windowPtr, …
        1/ p.ScreenGamma);
% set Gamma for all color channels
HideCursor;   % Hide the mouse cursor

% Get frame rate and set screen font
p.ScreenFrameRate = FrameRate(windowPtr);
        % get current frame rate
Screen('TextFont', windowPtr, 'Times');
        % set the font for the screen to Times
Screen('TextSize', windowPtr, 24);  % set the font size
                                    % for the screen to 24
```

could learn more about the PsychImaging function by using the MATLAB help command. In this particular program, we set up a 32-bit floating-point frame buffer and set the range of color values from 0 to 1. We also enable high gray-level resolution with bit stealing and set up the method of Gamma correction as a simple power function for all color channels. Bit stealing and Gamma correction are discussed in detail in chapter 5. Each one of these choices is seen in a line of the code in display 4.2 along with a comment. The purpose of this series of PsychImaging commands is to create displays with calibrated high gray-level resolution.

The last call to PsychImaging opens the display window:

```
PsychImaging('OpenWindow',whichScreen,p.ScreenBackground).
```

This call returns a pointer to the display area in the variable windowPtr. It returns the coordinates of the upper left and lower right of the screen in the parameter p.ScreenRect. The values in the p.ScreenRect correspond to selected resolution of the display (i.e., [0 0 1024 768]). The following few lines of code set up the Gamma correction, compute and load the color lookup table, and make the cursor invisible.

A call to FrameRate returns the current frame rate setting of the experimental window identified by windowPtr and stores the result in p.ScreenFrameRate.

Finally, a font for alphanumeric character displays, chosen by the experimenter, is specified in the command Screen('TextFont', windowPtr, 'Times'). The font size is loaded by a call to Screen('TextSize', windowPtr, 24).

4.3.2 Experimental Module

The Experimental Module begins by specifying general experiment parameters, in this case the number of trials as 100. A random seed number is selected based on the clock time. This provides a unique random start number that will be used in trial randomization.

Next, the program specifies some aspects of the drifting sine-wave stimulus, including the size (p.stimSize), the duration (p.stimDuration), the intertrial interval (p.ITI), the contrast (p.contrast), and temporal and spatial frequency (p.tf and p.sf). All these are provided in physical units of degrees of visual angle, duration and ITI in seconds, temporal frequency in hertz and spatial frequency in cycles per degree.

The next section of the program takes these physical definitions and computes the relevant properties in units of the graphical display device. For example, the p.ScreenDistance and p.ScreenHeight and the height in pixels in the screen resolution, or p.ScreenRect[4], are used to compute the number of pixels per degree of visual angle. The stimulus duration and frame rate (p.stimDuration and p.ScreenFrameRate) are used to compute the duration of the stimulus in number of frames (nFrames). The size of the visible sine-wave region is translated to pixels. The temporal frequency is translated into a phase shift of the sine wave for each frame. And so forth.

Display 4.2

```
%% Experimental Module

% Specify general experiment parameters
nTrials = 100;              % number of trials
p.randSeed = ClockRandSeed;   % use clock to set random
                              % number generator
% Specify the stimulus
p.stimSize = [8 6];         % horizontal and vertical stimulus size
                            % in visual angle
p.stimDuration = 0.250;     % stimulus duration in seconds
p.ISI = 0.5;   % duration between response and next trial onset
p.contrast = 0.2;           % grating contrast
p.tf = 4;                   % drifting temporal frequency in Hz
p.sf = 1;                   % spatial frequency in cycles/degree

% Compute stimulus parameters
ppd = pi/180 * p.ScreenDistance / p.ScreenHeight * ...
      p.ScreenRect(4);    % pixels per degree
nFrames = round(p.stimDuration * p.ScreenFrameRate);
                            % # stimulus frames
m = 2 * round(p.stimSize * ppd / 2);% horizontal and vertical
                                    % stimulus size in pixels
sf = p.sf / ppd;            % cycles per pixel
phasePerFrame = 360 * p.tf / p.ScreenFrameRate;
                            % phase drift per frame
fixRect = CenterRect([0 0 1 1] * 8, p.ScreenRect);
                            % 8 x 8 fixation
params = [0 sf p.contrast 0];
        % parameters for DrawTexture: initial phase, spatial
        % frequency, contrast, 0

% procedural sinewavegrating on GPU sets up a texture for
% the sine wave
tex = CreateProceduralSineGrating(windowPtr, m(1), m(2), ...
      [1 1 1 0] * 0.5, [], 0.5);
        % CreateProceduralSineGrating is a psychtoolbox
        % function that creates a procedural texture that
        % allows you to
        % draw sine grating stimulus patches in a very fast
        % and efficient manner on modern graphics hardware.

% Initialize a table to set up experimental conditions
p.recLabel = {'trialIndex' 'motionDirection' 'respCorrect' ...
             'respTime'};
```

Display 4.2 (continued)

```
rec = nan(nTrials, length(p.recLabel));
        % matrix rec is nTrials x 4 of NaN
rec(:, 1) = 1 : nTrials;
        % label the trial type numbers from 1 to nTrials
rec(:, 2) = -1;              %  -1 for left motion direction
rec(1 : nTrials/2, 2) = 1 ; % half of the trials set to +1 for
                             % right motion Direction
rec(:, 2) = Shuffle(rec(:, 2));
        % randomize motion direction over trials

% Prioritize display to optimize display timing
Priority(MaxPriority(windowPtr));

% Start experiment with instructions
str = sprintf('Left/Right arrow keys for direction.\n\n …
      Press SPACE to start.');
DrawFormattedText(windowPtr, str, 'center', 'center', 1);
        % Draw instruction text string centered in window
Screen('Flip', windowPtr);
        % flip the text image into active buffer
WaitTill('space');            % wait till space bar is pressed
Screen('FillOval', windowPtr, 0, fixRect);
        % create fixation box as black (0)
Secs = Screen('Flip', windowPtr);
        % flip the fixation image into active buffer
p.start = datestr(now); % record start time

% Run nTrials trials
for i = 1 : nTrials
    params(1) = 360 * rand; % set initial phase randomly
    Screen('DrawTexture', windowPtr, tex, [], [], 0, …
          [], [], [], [], [], params);
        % call to draw or compute the texture pointed to by tex
        % with texture parameters of the initial phase, the
        % spatial frequency, the contrast, and fillers required
        % for 4 required auxiliary parameters
    t0 = Screen('Flip', windowPtr, Secs + p.ISI);
        % initiate first frame after p.ISI secs
    for j = 2 : nFrames % For each of the next frames one by one
        params(1) = params(1) - phasePerFrame * rec(i, 2);
              % change phase
        Screen('DrawTexture', windowPtr, tex, [], [], 0, …
              [], [], [], [], [], params);
        % call to draw/compute the next frame
```

Display 4.2 (continued)

```
        Screen('Flip', windowPtr); % show frame
                 % each new computation occurs fast enough to show
                 % all nFrames at the framerate
    end

    Screen('FillOval', windowPtr, 0, fixRect);
        % black fixation for response interval
    Screen('Flip', windowPtr);

    [key Secs] = WaitTill({'left' 'right' 'esc'});
        % wait till response
    if iscellstr(key), key = key{1}; end
        % take the 1st key in case of multiple key presses
    if strcmp(key, 'esc'), break; end
        % stop the trial sequence if keypress = <esc>
    respCorrect = strcmp(key, 'right') == (rec(i, 2) == 1);
        % compute if correct or incorrect
    rec(i, 3 : 4) = [respCorrect Secs-t0];
        % record correctness and RT in rec
    if rec(i, 3), Beeper; end % beep if correct
end

p.finish = datestr(now); % record finish time

% Save Results
save DriftingSinewave_rst.mat rec p;     % save the results
```

Next, the program creates sine-wave grating images using a function called `CreateProceduralSineGrating`. This function creates a texture on the graphics processing unit (GPU) for sine-wave images and returns the pointer to `tex`. Programming the sine wave on a GPU using this function has the advantage of rapid computation. Your system must include a graphics card that supports it—a recent Direct 3D-10/11 capable or OpenGL-3/4 capable system (see Psychtoolbox help files for a list of capable graphics cards).

If your system is older and does not have a compatible graphics card, it is straightforward to use standard MATLAB programming to create the images (using `meshgrid`) and compute the luminance of each image pixel for the appropriate sine waves (see chapter 3 for examples).

The experiment itself is set up in a 100 × 4 matrix `rec` representing the conditions to be tested and the responses to be stored for the 100 trials. The labels of each column are saved in `p.recLabel`. Each of the 100 rows of the matrix `rec` contains four columns.

The first column of the matrix holds the trial number, and the second column assigns exactly 50 trials for left-drifting sine waves and 50 trials for right-drifting sine waves (–1, 1). The third and fourth columns are set up in advance to store the response accuracy and the response time for the trial once it has been run. The routine Shuffle is used to randomize the testing order of the different trials, and so randomizes the direction of drift over trials.

Just before the experiment is actually run, the priority of the display window is set to highest priority using the Psychtoolbox function call `Priority(MaxPriority(windowPtr))`. It is quite important to give maximum priority to the experimental display window in order to improve the accuracy of the timing of the display. If priority is not given to the experiment, random actions by the operating system may delay or distort the timing of displays.

First, at the beginning of the experimental session, instructions are shown on the screen indicating the arrow keys to be used for responses, and then providing a fixation point as a simple centered rectangle.

Next, the trials are presented and data are collected. Each trial has two parts. The first computes and displays the movie of the moving sine wave. The second collects the response and reads the response time for that response.

Each frame of the movie corresponds with a call to the sine-wave procedure on the GPU. These were set up earlier in the program using `CreateProceduralSineGrating`. The function call to `Screen('DrawTexture', windowPtr, tex, [], [], 0, [], [], [], [], [], params)` computes and stores each new frame with the parameters in `params` using the sine-wave procedure pointed to by `tex`. The call is: `Screen('DrawTexture', windowPointer, texturePointer [,sourceRect] [,destinationRect] [,rotationAngle])`, where [ ] is optional. Use the `help` function for descriptions and to see more options of the function `Screen`.

Each frame of the movie is shown one by one at the frame rate by a call to `Screen('Flip')`. This "flips" what is shown on the display from one image to the next with a timing that is synchronized with the frame rate of the display device. Finally, at the end of the `nFrames` of the drifting sine wave, we show the fixation display. The response of the observer is collected and recorded along with the response time from the beginning of the trial and stored in the third and fourth columns of the matrix `rec`.

After all the trials are completed, a date string is recorded in `p.finish`. At this point, the record of all trials in the matrix `rec` contains—in the order of running—a trial number (1 to 100), the trial type (left or right motion), whether that trial was correct or incorrect, and the associated response time. The trial records are saved along with all of the parameters of the experiment in `DriftingSinewave_rst.mat`.

Display 4.3

```
%% System Reinstatement Module
Priority(0);   % restore priority
sca; % close display window and textures,
     % and restore the original color lookup table
```

4.3.3 Reinstatement Module

The Reinstatement module is shown in display 4.3. After completing all 100 trials of the experiment, it is often useful to clean up the system by returning it to its state before the experiment. This includes resetting the priority of the display to the default (`Priority(0)`). In addition, `sca` closes the display window and textures and restores the original color lookup table.

4.4 Other Sample Programs

In section 4.3, we reviewed in detail a sample program that carried out a simple experiment in motion perception. In this section, we provide a number of sample programs to carry out other experiments, including measurement of magnitude estimation, contrast psychometric function, contrast sensitivity function, rapid serial visual presentation for a memory task, playing a movie read in from an outside source, and measurement of retinotopy of visual cortex for functional magnetic resonance imaging (fMRI). These examples are selected to reflect a representative sample of important paradigms and to include often used programming requirements and functions. Taken together, this set of examples should provide the basis for devising and running a very wide range of psychophysical and cognitive experiments in the MATLAB plus Psychtoolbox environment.

Although the example experiments differ in many details, there are many similarities between the different programs. The similarities mean that understanding the first example with detailed annotation in section 4.3 should make it easy to understand the new experiments. For the programs and experiments here, we show only the Experimental module—the Display Setup module and System Reinstatement module would be the same as the first example. Each program here includes inline comments. The help command for Psychtoolbox functions could be used to understand more details about individual function calls as the reader works through each experimental program.

4.4.1 Magnitude Estimation

The program in display 4.4 implements the magnitude estimation procedure[4,5] described in section 2.1 of chapter 2. Magnitude estimation asks the observer to provide a number corresponding with the subjective magnitude of stimuli with different intensities in order

Display 4.4

```
%%% Program MagnitudeEstimation.m
function MagnitudeEstimation

%% Experimental Module

%specify general experiment parameters
p.nTrialsPerBlock = 14; % 14 trials/block; we show the
                        % reference luminance in the
                        % beginning of each block
nBlocks = 10;           % number of blocks
nTrials = p.nTrialsPerBlock * nBlocks;
                        % Total number of trials
p.randSeed = ClockRandSeed;
        % use clock to set random number generator

% Specify the stimulus
p.stimSize = 6;         % diameter of the test stimulus
                        % disk size in visual angle
p.stimDuration = 0.2;   % stimulus display duration in seconds
p.luminance = [1 11 21 31 41 51 61];
                        % 7 luminance levels in cd/m^2
p.refLuminance = 25;    % reference luminance; shown once
                        % every 14 trials
p.ISI = 0.5;            % interval (secs) between response
                        % and next trial

% Compute stimulus parameters
grayLevels = p.luminance / p.maxLuminance;
refGrayLevel = p.refLuminance / p.maxLuminance;
ppd = pi/180 * p.ScreenDistance / p.ScreenHeight * …
      p.ScreenRect(4);    % compute pixels per degree
m = round(p.stimSize * ppd);    % stimulus size in pixels
stimRect = CenterRect([0 0 1 1] * m, p.ScreenRect);
                          % center within ScreenRect
nLevels = numel(p.luminance); % number of conditions from
                              % number of p.luminance
KbName('UnifyKeyNames');  % set up keyboard functions to use
                          % the same labels on different
                          % computer platforms
% Initialize a table to set up experimental conditions
p.recLabel = {'trialIndex' 'luminanceIndex' …
              'reportedLuminance'};
rec = nan(nTrials, length(p.recLabel));
                          % matrix rec is nTrials x 3 of NaN
rec(:, 1) = 1 : nTrials; % set trial numbers from 1 to nTrials
```

Display 4.4 (continued)

```
luminanceIndex = repmat((1 : nLevels)', …
                  p.nTrialsPerBlock/nLevels, nBlocks);
        % set the luminance index 1 to 7 repeatedly into
        % matrix using repmat (replicate array)
luminanceIndex = Shuffle(luminanceIndex);
        % shuffle each block (column) of matrix
rec(:, 2) = luminanceIndex(:);
        % randomized (shuffled) luminance indexes
        % into rec(:,2)

% Prioritize display to optimize display timing
Priority(MaxPriority(windowPtr));

% Start experiment with instructions
str = sprintf(['Input the perceived intensity of the disk ' …
               'for each trial.\n\n' 'Press Backspace or '…
               'Delete to remove the last input.\n\n' …
               'Press Enter to finish your input.\n\n\n' …
               'Press SPACE to start the experiment.']);
DrawFormattedText(windowPtr, str, 'center', 'center', 1);
        % Draw Instruction text string centered in window
        % onto frame buffer
Screen('Flip', windowPtr);           % flip the text image into
                                     % active buffer
WaitTill('space');        % wait till space bar is pressed
Secs = Screen('Flip', windowPtr); % flip the background image
                                  % into active buffer
p.start = datestr(now); % record start time

% Instruction for trial and reference intensity
trialInstruction = 'The perceived intensity of the disk is';
refInstruction = 'This is the reference with an intensity of …
               10.\n\nPress SPACE to proceed.';

% Run nTrials trials
for i = 1 : nTrials
    % Show the reference luminance once every 14 trials
    if mod(i, p.nTrialsPerBlock) == 1
        Screen('FillOval', windowPtr, refGrayLevel, stimRect);
              % make stimulus disk
        t0 = Screen('Flip', windowPtr, Secs + p.ISI);
              % show disk & return current time
        Screen('Flip', windowPtr, t0 + p.stimDuration);
              % turn off the disk after p.stimDuration secs
        DrawFormattedText(windowPtr, refInstruction, …
              'center', 'center', 1);
```

Display 4.4 (continued)

```
        Screen('Flip', windowPtr);
              % show reference instruction text
        key = WaitTill({'space' 'esc'});
              % wait for SPACE response
        if strcmp(key, 'esc'), break; end
              % if response is <escape> , then stop experiment
        Secs = Screen('Flip', windowPtr);
              % turn off text by flipping to background image
    end

    Screen('FillOval', windowPtr, grayLevels(rec(i, 2)), ...
          stimRect); % make stimulus disk
    t0 = Screen('Flip', windowPtr, Secs + p.ISI);
        % show disk & return current time
    Screen('Flip', windowPtr, t0 + p.stimDuration);
        % turn off the disk after p.stimDuration

    num = GetEchoNumber(windowPtr, trialInstruction, ...
          p.ScreenRect(3) / 2 - 200, p.ScreenRect(4) / 2, 1,...
          p.ScreenBackground, -1);
        % read a number response from the keyboard while
        % displaying the trial instruction
    Secs = Screen('Flip', windowPtr);
        % turn off trial instruction
    if ~isempty(num), rec(i, 3) = num; end
        % check the response, and record it
end

p.finish = datestr(now); % record the finish time
save MagnitudeEstimation_rst.mat rec p;      % save the results
```

to understand the relationship between physical manipulations of the stimulus and the internal scale. This experiment measures the effect of variations in luminance. The experiment consists of a total of 140 trials. In each trial, the subject is shown a 3° radius disk at one of seven possible luminance levels (1, 11, 21, 31, 41, 51, and 61 cd/m^2). The disk is displayed for 200 ms. Observers are asked to assign a numerical value to the perceived intensity using the computer keyboard. At the beginning and after every 14 trials (two presentations of each of the seven luminance levels), the subject is presented a patch with luminance set at 25 cd/m^2, whose perceived intensity is defined as 10, to provide an anchor for the rating scale. The same set of luminance levels are tested 20 times in random order. The observer's rating in each trial is saved. The average rating of a particular luminance level provides an estimate of its perceived intensity (figure 4.3).

The topic of magnitude estimation is treated in section 7.2.1 of chapter 7, where we consider the theoretical assumptions and interpretation of magnitude estimation scaling.

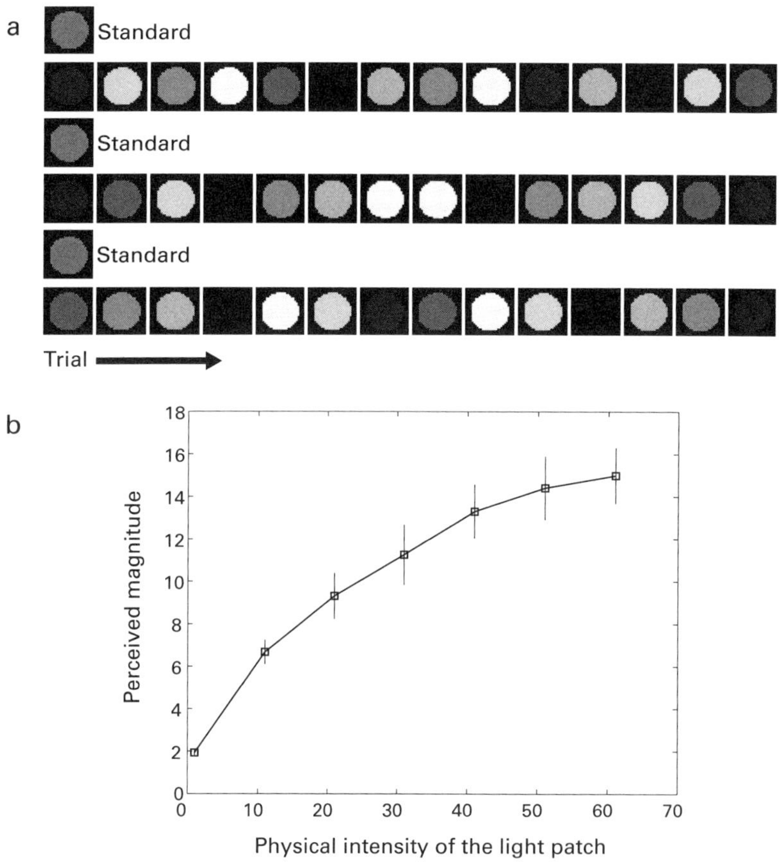

Figure 4.3
Magnitude estimation of perceived light intensity. (a) Trial sequence in the magnitude estimation experiment. (b) Estimated magnitudes of patches with different light intensity.

4.4.2 Contrast Psychometric Function

The program in display 4.5 implements the contrast psychometric function experiment described in section 2.2 of chapter 2. Contrast psychometric functions provide basic data about detectability of a stimulus and allow the estimation of a contrast threshold. This experiment measures the contrast psychometric function for detecting a 3° radius, 1 c/d sine-wave grating. Seven sine-wave grating contrasts, ranging from invisible (0.25%) to clearly visible (4%), are selected as contrast conditions for the method of constant stimuli (see section 8.2.3 of chapter 8). Each stimulus contrast is tested 100 times in a two-interval forced-choice task (see section 8.4.1 of chapter 8), for a total of 700 trials in the entire experiment. On each trial, a contrast is chosen at random, and the sine-wave grating is presented in one of two temporal intervals for 100 ms, with mean gray in the other interval.

Display 4.5

```
%%% Program ContrastPsychometricFunction.m
function ContrastPsychometricFunction

%% Experimental Module

% Specify general experimental parameters
nContrast = 7;      % number of contrast levels in experiment
repeats = 100;      % number of trials to repeat for each
contrast
nTrials = repeats * nContrast;
                    % compute total number of trials
p.randSeed = ClockRandSeed;
                    % use clock to set random number generator
keys = {'1' '2' 'esc'};  % specify response keys for the
                    % two intervals and to break
% Specify the stimulus
p.stimSize = 6;     % image diameter in visual degree
p.sf = 1;           % spatial frequency in cycle/degree
conRange = [0.0025 0.04];
                    % lowest and highest tested contrasts
p.stimDuration = 0.1;  % stimulus duration in seconds
p.interval = 1;     % seconds between two intervals of a trial
p.ISI = 0.5;        % seconds between trials

% Compute stimulus parameters
ppd = pi/180 * p.ScreenDistance / p.ScreenHeight * ...
      p.ScreenRect(4);
                    % compute pixels per degree from
                    % p.ScreenDistance and p.ScreenHeight
m = round(p.stimSize * ppd /2) * 2; % stimulus size
                                    % in pixels
fixRect = CenterRect([0 0 1 1] * 8, p.ScreenRect);
                    % define an 8 x 8 pixel fixation
% create another fixation stimulus of horizontal and
% vertical cross hairs outside the area of the sine wave
% pattern. Crosshairs will be displayed during display
% intervals

fixLen = 32; % length of fixation in pixels
fixXY = [[-1 0 0 1] * fixLen + [-1 -1 1 1 ] * m/2 + ...
     p.ScreenRect(3)/2  [1 1 1 1]*p.ScreenRect(3)/2; ...
     [1 1 1 1]*p.ScreenRect(4)/2 [-1 0 0 1] *fixLen + ...
     [-1 -1 1 1 ] * m/2 + p.ScreenRect(4)/2];

p.contrasts = logspace(log10(conRange(1)), ...
          log10(conRange(2)), nContrast);
```

Display 4.5 (continued)

```
      % use logspace function to choose nContrast contrasts
      % (7) at log intervals between the minimum and
      % maximum specified contrasts
sf = p.sf / ppd; % compute spatial frequency in
                 % cycles per pixel
tex = CreateProceduralSineGrating(windowPtr, m, m, [[1 1 1]…
      *p.ScreenBackground 0], m/2, 0.5);
      % "CreateProceduralSineGrating" is a psychtoolbox
      % function to create a procedural texture for drawing
      % sine grating stimulus patches on the GPU

% Initialize a table to set up experimental conditions
p.recLabel = {'trialIndex' 'contrastIndex' 'whichInterval'…
        'respCorrect' 'respTime'};
      % labels the columns of the data recording array rec
rec = nan(nTrials, length(p.recLabel));
      % matrix rec is initialized as an nTrials x 5 of NaN
rec(:, 1) = 1 : nTrials;
      % initialize trial numbers from 1 to nTrials
contrastIndex = repmat(1 : nContrast, repeats, 1);
      % use repmat to cycle thru contrast #s
intervalIndex = ones(size(contrastIndex)) * 2;
      % first set all trials to 2nd interval
intervalIndex(1 : repeats / 2, :) = 1;
      % change first half to 1
[rec(:, 2) ind] = Shuffle(contrastIndex(:));
      % shuffle contrast indexes to randomize
rec(:, 3) = intervalIndex(ind);
      % shuffle interval indexes in the same order

% Prioritize display to optimize display timing
Priority(MaxPriority(windowPtr));

% Start experiment with instructions
str = sprintf('Press 1 or 2 for the first or second …
      interval.\n\nPress SPACE to start.');
DrawFormattedText(windowPtr, str, 'center', 'center', 1);
      % Draw Instruction text string centered in window
Screen('Flip', windowPtr);
      % flip the text image into active buffer
WaitTill('space'); % wait till space bar is pressed
Screen('FillOval', windowPtr, 0, fixRect);
      % create a black fixation dot
Secs = Screen('Flip', windowPtr);
      % flip the fixation image into active buffer
p.start = datestr(now); % record start time
```

Display 4.5 (continued)

```
% Run nTrials trials
for i = 1 : nTrials
    con = p.contrasts(rec(i, 2));
        % use contrast index from rec to set contrast
        % for this trial
    flipSecs = Secs + p.ISI + [0 p.interval];
        % define the start time of the two intervals

    for j = 1 : 2 % two intervals
        if rec(i, 3) == j
                % draw the grating in the interval defined
                % in rec(i,3) only
              Screen('DrawTexture', windowPtr, tex, [], [], …
                    0, [], [], [], [], [], [180 sf con 0]);
        end     % draw the sine grating with phase 180,
                % spatial frequency, and contrast
        Screen('DrawLines', windowPtr, fixXY, 3, 0.3);
                % add the fixation crosshairs
        t0 = Screen('Flip', windowPtr, flipSecs(j));
                % show the stimulus and return the time
        Screen('Flip', windowPtr, t0 + p.stimDuration);
                % turn off the stimulus by flipping to
                % background image after p.stimDuration secs
    end

    Screen('FillOval', windowPtr, 0, fixRect);
          % draw the smaller centered fixation
    Screen('Flip', windowPtr, t0 + 0.25 + p.stimDuration);
          % show small fixation briefly to cue the
          % observer to respond with the interval

    [key Secs] = WaitTill(keys);
          % wait for response, return key and response time
    if iscellstr(key), key = key{1}; end
          % take the first response in case of multiple
          % key presses
    if strcmp(key, 'esc'), break; end
          % check if response is <escape> to stop experiment
    rec(i, 4 : 5) = [str2double(key)==rec(i, 3) Secs-t0];
          % record correctness and respTime
    if rec(i, 4), Beeper; end % beep if correct
end

p.finish = datestr(now); % record finish time
save ContrastPsychometricFunction_rst.mat rec p;
          % save the results
```

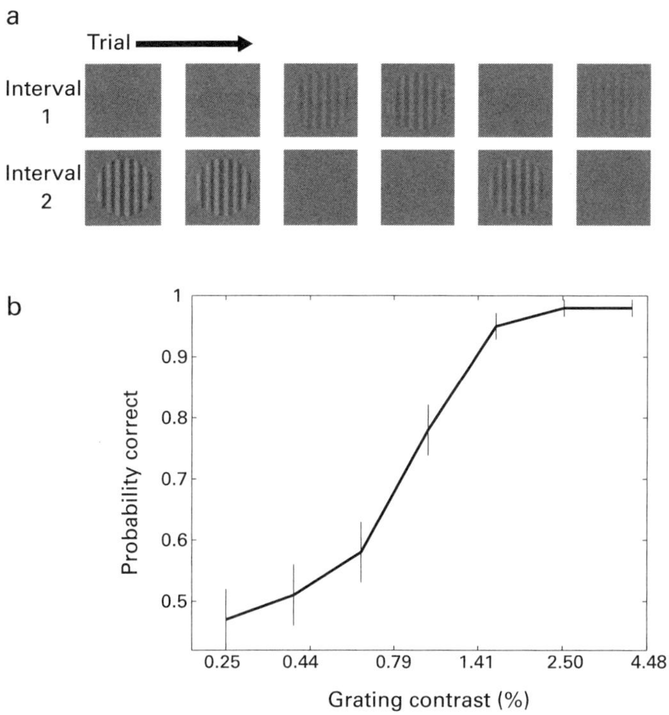

Figure 4.4
Measuring the contrast psychometric function with the method of constant stimuli and two-interval forced-choice (2IFC). (a) Illustration of the 2IFC paradigm for several trials. (b) The measured contrast psychometric function.

The two intervals are each "marked" by the presentation of fixation crosshairs, with onsets of the intervals separated by 1000 ms. The observer reports the interval most likely to contain the grating. The computer records the stimulus condition and the observer's response on each trial. At the end of the 700 trials of the experiment, the computer tabulates the accuracy with which the correct interval was selected by the observer at each contrast. The function that relates the accuracy of performance to the stimulus contrast is the psychometric function (figure 4.4b).

Figure 4.4a shows a sample test sequence over several trials that randomly vary the contrast of the grating and the interval in which it appears (first or second). Table 4.1 shows a tabulation of hypothetical data from the experiment. Figure 4.4b graphs the psychometric function. The error bars for each observed proportion are estimated from the variance of the binomial distribution, or $(pq)/N$, where p is the percent correct, q is the percent incorrect, and N is the number of trials . The data from psychometric function experiments of this kind are analyzed in detail in section 10.4.2 of chapter 10. In that

Table 4.1
A sample psychometric function

Contrast (%)	0.25	0.40	0.63	1.00	1.59	2.52	4.00
Number correct	52	53	59	74	95	97	98
Number incorrect	48	47	41	26	5	3	2

section, the data are modeled with a function, and a threshold contrast is estimated for a particular accuracy level such as 75% correct.

4.4.3 Contrast Sensitivity Function

The program in display 4.6 implements a contrast sensitivity function experiment such as that described in section 2.3 of chapter 2. The contrast sensitivity function is one of the most important functions in visual psychophysics. It describes the sensitivity, or limits of detectability, for stimuli of different spatial frequencies. In this experiment, we measure contrast thresholds for detecting sine-wave gratings at nine different spatial frequencies (0.125, 0.25, 0.5, 1, 2, 4, 8, 16, 32 cycles per degree or c/d) using the method of constant stimuli to measure a psychometric function for each spatial frequency. The method of constant stimuli uses seven preselected contrast levels, ranging from invisible to highly visible for each tested spatial frequency. There are a total of 9 (spatial frequency) × 7 (contrasts) = 63 conditions in this experiment. Each condition is tested 100 times, for a total of 6300 trials. The experiment is broken into 10 sessions. Each session tests each stimulus condition 10 times, leading to 100 trials per condition overall. On each trial, the computer randomly selects a particular sine grating pattern and a particular contrast level and presents it to the subject in one of two randomly chosen temporal intervals for 100 ms, with mean gray in the other interval (see figure 4.5a). Again, the two intervals are marked by crosshairs with the interval onsets separated by 1000 ms. The observer reports the interval that most likely contains the grating. The computer records the stimulus condition and the response on each trial. After the experiment is completed, the accuracy of selecting the correct interval in each condition is tabulated. For each spatial frequency condition, a psychometric function is constructed (figure 4.5b), and the contrast threshold is computed from the best fitting descriptive function. The contrast sensitivity function illustrates the relationship between sensitivity (defined as 1/threshold) and the spatial frequency of the sine-wave grating (figure 4.6).

Figure 4.5a shows a sample test sequence over several trials. Examination of this sequence shows random selection of the spatial frequency condition, the contrast condition, and whether the grating appears in the first or the second interval. Figure 4.5b graphs the psychometric functions for each of the nine spatial frequencies. The error bars for each observed proportion are estimated from the variance of the binomial distribution, or $(pq)/N$. The hypothetical data from each psychometric function were analyzed by methods that are described in detail in section 10.4.2 of chapter 10, and the contrast threshold at 75%

Display 4.6

```
%%% Program ContrastSensitivityFunction.m
function ContrastSensitivityFunction

%% Experimental Module

% specify the experimental parameters
nContrast = 7; % number of contrast levels
nSF = 9;       % number of spatial frequencies
repeats = 10;  % number of trials to repeat for each condition
nTrials = repeats * nContrast * nSF;
               % total number of trials
keys = {'1' '2' 'esc'};   % keys to respond for which
                          % interval, and to break
p.randSeed = ClockRandSeed;
        % use clock to set random number generator

% Specify the stimulus
p.stimSize = 6;             % image diameter in visual degree
conRange = [0.0025 0.04];   % lowest and highest tested contrast
sfRange = [0.125 32];       % lowest and highest tested sf in
                            % cycles/deg
p.stimDuration = 0.1;       % stimulus duration in seconds
p.interval = 1; % seconds between two intervals of a trial
p.ISI = 0.5;                % seconds between trials

% Compute stimulus parameters
ppd = pi/180 * p.ScreenDistance / p.ScreenHeight * ...
       p.ScreenRect(4); % compute pixels per degree from
                        % p.ScreenDistance andp.ScreenHeight
m = round(p.stimSize * ppd /2) * 2; % stimulus size in pixels
fixRect = CenterRect([0 0 1 1] * 8, p.ScreenRect);
       % define an 8 x 8 pixel fixation

% create another fixation stimulus of horizontal and vertical
cross hairs outside the area of the sine wave pattern
fixLen = 32; % length of fixation in pixels
fixXY = [ [-1 0 0 1]*fixLen + [-1 -1 1 1 ]*m/2 + ...
       p.ScreenRect(3)/2 [1 1 1 1] * p.ScreenRect(3)/2; ...
       [1 1 1 1]*p.ScreenRect(4)/2 [-1 0 0 1] *fixLen + ...
       [-1 -1 1 1 ] * m / 2 + p.ScreenRect(4) / 2];

p.contrasts = logspace(log10(conRange(1)), ...
              log10(conRange(2)), nContrast);
p.SFs = logspace(log10(sfRange(1)), log10(sfRange(2)), nSF);
SFs = p.SFs / ppd;
        % compute spatial frequency in cycles per pixel
```

Display 4.6 (continued)

```
% Initialize a table to set up experimental conditions
p.recLabel = {'trialIndex' 'contrastIndex' 'SFIndex' …
   'whichInterval' 'respCorrect' 'respTime'};
        % labels the columns of the data recording array rec
rec = nan(nTrials, length(p.recLabel));
        % matrix rec is initialized as an nTrials x 6
        % matrix of NaN
rec(:, 1) = 1 : nTrials;
        % initialize trial numbers from 1 to nTrials
contrastIndex = repmat(1 : nContrast, [nSF 1 repeats]);
SFIndex = repmat((1 : nSF)', [1 nContrast repeats]);
intervalIndex = ones(nSF, nContrast, repeats) * 2;
        % first set all trials to 2nd interval
intervalIndex(:, :, 1 : repeats / 2) = 1;
        % change fisrt half to 1
[rec(:, 2) ind] = Shuffle(contrastIndex(:));
        % shuffle contrast indexes to randomize
rec(:, 3) = SFIndex(ind); % shuffle SF indexes to randomize
rec(:, 4) = intervalIndex(ind);
        % shuffle intervals to randomize

% Prioritize display to optimize display timing
Priority(MaxPriority(windowPtr));

% Start experiment with instructions
str = sprintf('Press 1 or 2 for first or second ' …
       'interval.\n\nPress SPACE to start.');
DrawFormattedText(windowPtr, str, 'center', 'center', 1);
        % Draw instruction text string centered in window
Screen('Flip', windowPtr);
        % flip the text image into active buffer
Beeper;
WaitTill('space');          % wait for space bar
Screen('FillOval', windowPtr, 0, fixRect);
        % create a black fixation box
Secs = Screen('Flip', windowPtr);
        % flip the fixation image into active buffer

% procedural sinewavegrating allows us to change its
% parameters very quickly
tex = CreateProceduralSineGrating(windowPtr, m, m, …
        [1 1 1 0] * 0.5, m/2, 0.5);
        % "CreateProceduralSineGrating" is a psychtoolbox
        % function to create a procedural texture for drawing
        % sine grating stimulus patches on the GPU

% Run nTrials trials
```

Display 4.6 (continued)

```
for i = 1 : nTrials
    con = p.contrasts(rec(i, 2));
        % use contrast index from rec to set contrast
        % for this trial
    sf = SFs(rec(i, 3));
        % use spatial frequency index from rec to set spatial
        % frequency for this trial
    flipSecs = Secs + p.ISI + [0 p.interval];
        % define the start time of the two intervals

    for j = 1 : 2
        if rec(i, 4) == j   % draw the grating in the interval
                            % defined in rec(i,4) only
            Screen('DrawTexture', windowPtr, tex, [], [], ...
                  0, [], [], [], [], [], [180 sf con 0]);
        end    % draw the sine grating with phase 180,
               % spatial frequency, and contrast
        Screen('DrawLines', windowPtr, fixXY, 3, 0);
               % add the fixation crosshairs
        t0 = Screen('Flip', windowPtr, flipSecs(j));
               % show the stimulus and return the time
        Screen('Flip', windowPtr, t0 + p.stimDuration);
               % turn off the stimulus by flipping to
               % background image after p.stimDuration secs
    end

    Screen('FillOval', windowPtr, 0, fixRect);
       % draw the smaller centered fixation
    Screen('Flip', windowPtr);

    [key Secs] = WaitTill(keys);
       % wait for response, return key and response time
    if iscellstr(key), key = key{1}; end
       % take the first response in case of multiple key
       % presses
    if strcmp(key, 'esc'), break; end
       % check if response is <escape> to stop experiment
    rec(i, 5) = str2double(key) == rec(i, 4);
       % record correctness
    rec(i, 6) = Secs - t0; % record respTime
    if rec(i, 5), Beeper; end % beep if correct

    Screen('FillOval', windowPtr, 0, fixRect); % draw fixation
    Screen('Flip', windowPtr);
end

save ContrastSensitivityFunction_rst rec p; % save the results
```

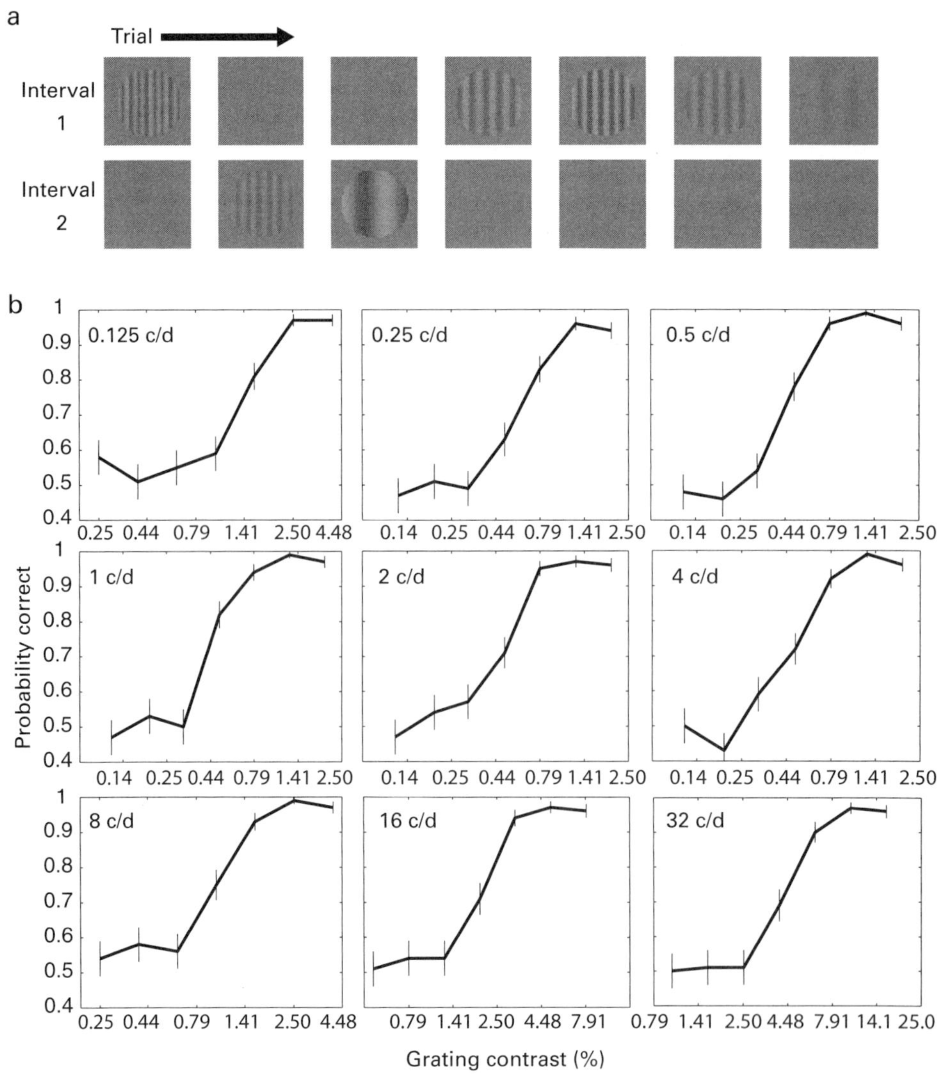

Figure 4.5
Measuring a contrast sensitivity function. (a) Sample trial sequence. (b) Psychometric functions at nine spatial frequencies.

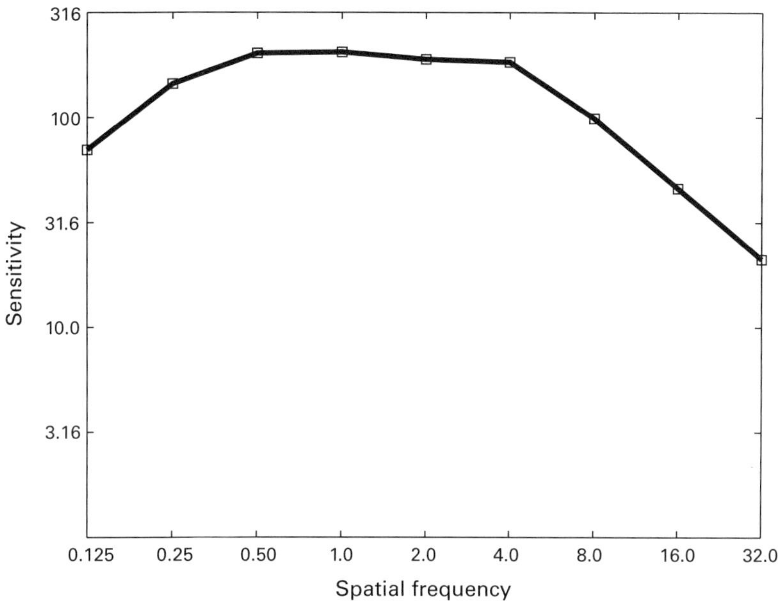

Figure 4.6
Contrast sensitivity function.

Table 4.2
A sample contrast sensitivity function

Spatial frequency (c/d)	0.125	0.25	0.50	1.0	2.0	4.0	8.0	16.0	32.0
Threshold (%)	1.43	0.69	0.49	0.49	0.53	0.54	1.01	2.17	4.74
Sensitivity	70.1	144.7	202.2	204.6	189.0	183.5	98.7	46.0	21.1

correct was estimated for each spatial frequency. These thresholds and the corresponding sensitivity—or 1/threshold—are listed in table 4.2. The sensitivity estimates are graphed in figure 4.6 as the measured contrast sensitivity function. There is an extensive treatment of how to model contrast sensitivity functions in section 10.4.3 of chapter 10.

4.4.4 Rapid Serial Visual Presentation

Rapid serial visual presentation (RSVP) has been a major tool in the study of reading, language processing, attention, and visual memory. In RSVP a sequence of images—letters, words, pictures—is presented at rates between 1 per second up to 25 per second (40 ms per item). Often, the letters, words, or pictures appear in the same location at the fovea,[6,7] but sometimes they are displayed in adjacent locations to mimic time-control of reading[8] and other times perhaps at many locations simultaneously to evaluate uptake of information across the visual field.[9–11] RSVP paradigms have been influential in the development of theories in many of these areas.

Display 4.7 shows a program for a simplified condition of an attention paradigm by Reeves and Sperling[9] and Weichselgartner and Sperling.[7] The purpose of the experiment is to measure the temporal properties of attention leading to storage and report. This program considers the simple example of a single location—sometimes called a single stream—of rapidly presented letters. The task of the observer is to report the first four items following a report cue, which in this case is a square surrounding one of the letters. Opening an attention window prior to the cue would flood a memory store with irrelevant items, so observers must try to open an attention window as soon as they see the square cue. The items are processed through this attention window, and then the most salient four items are selected by the observer for report. In typical data, each reported item is drawn from a temporal position either just before, simultaneous with, or just after the cue. While the observer is trying to report the four items simultaneous with the cue and following it, in fact the distribution of the locations of reported items broadly follows the interval after the cue, with a maximum at some typical delay. In this paradigm, the data are analyzed in a very clever way to estimate the temporal properties of the attention window, which like a gate opens to its full extent and then closes again.[10]

On each trial, a string of 23 letters is displayed in random order and appear in the same location on the screen. The letters are chosen from a set of 23 letters (excluding I, Q, and V because of visual similarity with J, O, and U, respectively). The display rate shows one new letter every 150 ms, consisting of display duration near 50 ms (here three frames), while the rest of the interval is blank. The report cue appears at one of the temporal positions from 6 to 15 equally often. Observers are asked to report the first four letters starting with the one that is simultaneous with the cue. In each trial, the observer produces an ordered report string of four items (i.e., the observer might report H X O E). There are 200 trials.

Figure 4.7a shows a sample sequence in temporal order and the surrounding box that is the report cue. Each letter is translated to a position score based on its position in the RSVP stream that was presented and in turn converted to positions relative to the cue. A position code for letters relative to the ordinal position of the target is shown below each letter. The target letter, which is simultaneous with the cue, is labeled 0 while letters before the cue have negative position scores, and letters after the cue have positive position scores. For example, a reported 4-tuple <H X O E> has position scores of <+2 +4 +1 +3> relative to the report cue. These reports can be scored in several ways. First, the frequencies of report (regardless of report order) are computed by summing the number of reports at each relative temporal position –2, –1, 0, 1, 2, ... Second, the reports can be scored for systematic report orders usually called *iBj* scores. This refers to the probability of reporting the letter displayed in temporal position *i* before the letter displayed in temporal positon

Display 4.7

```
%%% Program RSVP.m
function RSVP

%% Experimental Module

% setup general experimental parameters
nTrials = 10;                   % total number of trials
keys = [cellstr(char(65:90)')' {'delete' 'backspace' …
       'enter' 'return'}];
p.randSeed = ClockRandSeed;     % use the clock to set random
                                % number generator

% Specify the stimulus
p.textSize = 2.5;    % letter size in degrees of visual angle
p.letters = 'ABCDEFGHJKLMNOPRSTUWXYZ';
                     % subset of upper case letters
p.stimDuration = 0.05;    % duration of each letter in seconds
p.blankDuration = 0.1;    % duration of the blank interval
between letters in seconds
p.fixDuration = 0.15;     % fixation duration in seconds
p.iCueRange =[6 15];      % temporal range of the cue frame in
                          % letter screen counts
p.textFont = 'Consolas';  % select a fixed width font for
                          % stimulus display

% Compute stimulus parameters
halfInvt = 1 / p.ScreenFrameRate / 2;
stimDur = round(p.stimDuration * p.ScreenFrameRate) …
          / p.ScreenFrameRate - halfInvt;
blankDur = round(p.blankDuration * p.ScreenFrameRate) …
          / p.ScreenFrameRate - halfInvt;
ppd = pi / 180 * p.ScreenDistance / p.ScreenHeight …
          * p.ScreenRect(4);   % pixels/degree
textSize = round(p.textSize * ppd);
[xc yc] = RectCenter(p.ScreenRect);
fixLen = round(textSize / 4); % fixation size in pixels
fixXY = [[0 0 -1 1] * fixLen + xc;
         [-1 1 0 0] * fixLen + yc] ; % fixation xy
Screen('TextFont', windowPtr, p.textFont);
Screen('TextSize', windowPtr, textSize);
txtSz = Screen('TextBounds', windowPtr, 'A');
x1 = xc - txtSz(3) / 2; y1 = yc - txtSz(4) / 2; % letter xy
txtSz(3) = txtSz(3) * 1.5; % expand a bit to allow more room
cueRect = CenterRect(txtSz, p.ScreenRect);

Screen('TextSize', windowPtr, 48);
txtSz = Screen('TextBounds', windowPtr, 'ABCDEF');
```

Display 4.7 (continued)

```
feedbackRect = CenterRect(txtSz, p.ScreenRect);
feedbackRect(3) = feedbackRect(1) + txtSz(3) / 6;
x2 = feedbackRect(1); y2 = feedbackRect(2);
nLetters = length(p.letters);
p.seq = repmat(p.letters', 1, nTrials);
p.seq = char(Shuffle(p.seq));  % shuffle letter stream for
                               % each trial to randomize

% Initialize a table to set up experimental conditions
p.recLabel = {'trialIndex' 'iCue' 'respTime' 'ans1' …
                'ans2' 'ans3' 'ans4'};
rec = nan(nTrials, length(p.recLabel));
        % matrix rec is nTrials x 7 of NaN
rec(:, 1) = 1 : nTrials;
        % assign trial numbers from 1 to nTrials
rec(:, 2) = randi(p.iCueRange, nTrials, 1);
        % cue index for each trial

% Prioritize display to optimize display timing
Priority(MaxPriority(windowPtr));

% Start experiment with instructions
str = ['Input 4 letters starting from the cued letter.\n\n'…
       'Press Backspace or Delete to remove the last ' …
       'input.\n\n' 'Press Enter or Return to finish your ' …
       'input.\n\n\n' 'Press SPACE to start.'];
Screen('TextSize', windowPtr, 36);
Screen('TextFont', windowPtr, 'Times');
DrawFormattedText(windowPtr, str, 'center', 'center', 1);
        % Draw Instruction text string centered in window
Screen('Flip', windowPtr);
        % flip the text image into active buffer
WaitTill('space');  % wait till space bar is pressed
Screen('Flip', windowPtr);  % turn off instruction
p.start = datestr(now); % record start time

% Run nTrials trials
for i = 1 : nTrials
    Screen('DrawLines', windowPtr, fixXY, 3, 1);
    t0 = Screen('Flip', windowPtr); % show fixation
    t0 = Screen('Flip', windowPtr, t0 + p.fixDuration);
        % turn off fixation

    Screen('TextSize', windowPtr, textSize);
    Screen('TextFont', windowPtr, p.textFont);
```

Display 4.7 (continued)

```
    for j = 1 : nLetters
        if j == rec(i, 2)
            Screen('FrameRect', windowPtr, 1, cueRect, 3);
        end
        Screen('DrawText', windowPtr, p.seq(j, i), x1, y1,...
              1);
        t0 = Screen('Flip', windowPtr, t0 + blankDur);
             % letter on
        t0 = Screen('Flip', windowPtr, t0 + stimDur);
             % letter off
    end
    Screen('TextSize', windowPtr, 48);
    Screen('TextFont', windowPtr, 'Times');
    DrawFormattedText(windowPtr, 'Please input the 4 ...
          letters', 'center', 'center', 1);
    t0 = Screen('Flip', windowPtr); % show prompt string

    Screen('TextFont', windowPtr, p.textFont);
    ans4 = '';
    while 1   % collect 4 responses from keyboard and
              % allow changes until return
        KbReleaseWait;
        [key Secs] = WaitTill(keys);
        if iscellstr(key), key = key{1}; end

        if any(strcmp(key, {'enter' 'return'})) % done
            if length(ans4) >= 4, break; end
        elseif any(strcmp(key, {'delete' 'backspace'}))
            if ~isempty(ans4), ans4(end) = []; end
        else
            ans4 = [ans4 upper(key)];
              % append input as upper case to feedback string
        end

        Screen('DrawText', windowPtr, ans4, x2, y2 - 80, 1);
        Screen('Flip', windowPtr); % show input
    end
    rec(i, 3) = Secs - t0; % record response time
    rec(i, 4 : 7) = ans4(1 : 4); % record the first 4 input

    Screen('DrawText', windowPtr, ans4, x2, y2 - 80, 1);
    Screen('DrawText', windowPtr, p.seq((0 : 5) + ...
          rec(i, 2), i)', x2, y2, 1);
    Screen('FrameRect', windowPtr, 1, feedbackRect, 1);
```

Display 4.7 (continued)

```
    Screen('TextFont', windowPtr, 'Times');
    DrawFormattedText(windowPtr, 'Press SPACE to proceed', ...
        'center', y2 + 120, 1);
    Screen('Flip', windowPtr); % show instruction
    if strcmp(WaitTill({'space' 'esc'}), 'esc'), break; end
end

p.finish = datestr(now); % record start time

save RSVP_rst rec p;    % save the results
```

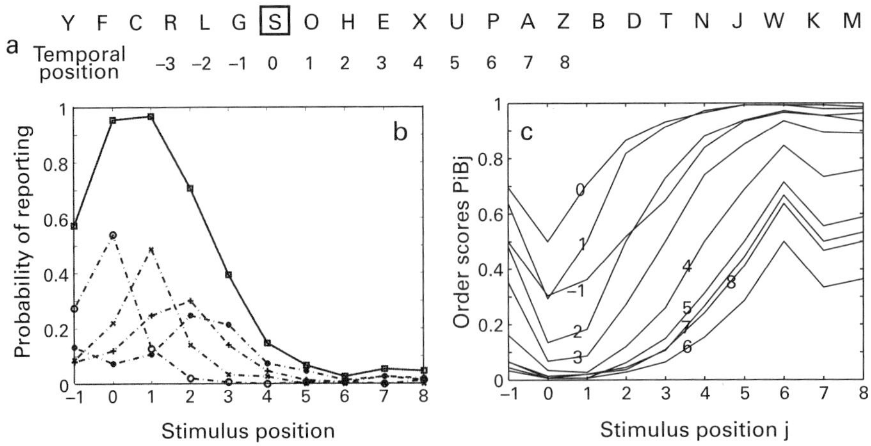

Figure 4.7
Measuring an attention reaction time with an RSVP experiment. (a) A sample display sequence. (b) Item scores for each individual response position. (The probability P(i) of a letter from stimulus position i appearing in response j). (c) The proportion P_{iBj} of trials in which letters from stimulus position i are reported earlier in the response than those from position j, as a function P_{iBj}, with the curve parameter i.

j, which is defined as 0.5 when $i = j$. Both kinds of scoring lead to highly systematic patterns in the data that reveal something about the window of attention.

Figure 4.7b shows the item scores pooled over all response positions (solid line) and for each position in the report order (dashed lines). The figure shows typical results for medium display speeds. The observer does a reasonably good job of reporting the correct items, although the order is somewhat blurred. Figure 4.7c shows the results of the *iBj* analysis. For example, the curve marked 0 shows that the item in the 0 position, simultaneous with the cue, is reported first more often than any other position, and nearly perfectly before all letters in position 3 or greater. The fact that the curves are largely "laminar"—layered like laminated surfaces—reflects a consistent ordering even when items are not reported in the correct order. These highly ordered data have been interpreted as reflecting the operation of attention triggered by the cue and the reporting of items in proportion to their perceived strength rather than explicitly coded order (see ref. 9).

4.4.5 Movie from a Video Source

The last example of RSVP showed how to construct and display a sequence—a "movie"—of images selected and programmed by the investigator. In other experimental settings, however, the stimulus may not be a programmed image sequence but rather an image sequence taken from a natural video source—an actual "movie" of some event. Display 4.8 shows a program that takes a native video source and displays a subset of the movie in a controlled way. In this example, the program reads in a movie in one of the standard movie formats (e.g., *avi*, *mov*, *mpg*, *mp4*), and specifies the clip to display by its starting and ending times, as well as other aspects of the presentation such as the size (relative to the original), the volume of audio, and the frame rate of playback. The display program shows several of these actions within `PlayMovie` as a programmed function of Psychtoolbox.

For example, an experimenter may want to show a brief movie clip at a smaller size and slightly faster rate without sound to imitate some aspects of old silent films. Using other routines in section 3.3.8 of chapter 3, one could add grainy noise to the images to achieve an even stronger impression of the old movie. Display 4.9 shows different calls to `PlayMovie`. `PlayMovie('introhigh.mpg', [ ], [ ], [ ], 1, 1)` shows a movie in its original size, speed, and sound volume, and the call `PlayMovie('introhigh.mpg', [ ], [ ], 0.5, 2, 0)` shows the same movie in half its size, twice the speed, and no sound. The file "introhigh.mpg" (available on the web page for the book) contains a brief video clip.

The movie program introduces some new Psychtoolbox function calls for reading and displaying movies. Movies are played by flipping the buffer from one image to the next. `Screen('OpenMovie',…)` opens a multimedia file (images and sound) named `movieFile` for playback in an open window pointed to by `windowPtr`. The call in the program `[movie dur fps width height] = Screen('OpenMovie', windowPtr, movieFile)` returns the pointer `movie` as well as the duration, the frames per second, and the width and height in pixels.

A call to `Screen('SetMovieTimeIndex', movie, fromTime)` is used here to set the movie to start at `fromTime` in the event that the first image is not at the beginning of the movie.

The function `Screen('PlayMovie', movie, rate, 0, soundVolume)` starts playing the movie pointed to by `movie`. The rate defaults to normal speed. Negative numbers reverse the movie. The `soundVolume` is set between 0 and 1 relative to full volume, and the variable `0` sets the loop variable not to repeat. A call to `Screen('PlayMovie',…)` that sets the rate to 0 stops playing the movie.

Display 4.8

```
%%% Program PlayMovie
function PlayMovie(movieFile, fromTime, toTime, movieSize, …
       rate, soundVolume)
% Play movie with sound. Quicktime is needed for this to work.
% Set fromTime and toTime to choose a clip inside a movie
% clip. movieSize, default 1, controls the ratio between
% displayed and original movie size. rate, default 1, controls
% the playback speed. soundVolume (0~1) controls the sound
% volume. Press ESC to stop playing.

if nargin < 1 || isempty(movieFile)
     [movieFile pathName] = uigetfile('*.mpeg;*.mpg;*.avi; …
            *.mov', 'Select Movie File');
     if movieFile == 0, return; end
     movieFile = fullfile(pathName, movieFile);
end

% Control clip time, size and sound volume
% Set default values if these inputs are not in the argument
% list
if nargin < 2 || isempty(fromTime), fromTime = 0; end
if nargin < 3 || isempty(toTime), toTime = inf; end
if nargin < 4 || isempty(movieSize), movieSize = 1; end
        % 1 for original
if nargin < 5 || isempty(rate), rate = 1; end
        % 1 for original
if nargin < 6 || isempty(soundVolume), soundVolume = 1; end
        % 0 to 1

% Open window & movie
if exist('onCleanup', 'class'), oC_Obj = …
      onCleanup(@()sca); end
        % close pre-existing PsychToolbox screens
whichScreen = max(Screen('screens'));
[windowPtr ScreenRect] = Screen('OpenWindow', whichScreen, 0);
[movie dur fps width height] = Screen('OpenMovie', …
     windowPtr, movieFile);
Screen('SetMovieTimeIndex', movie, fromTime);
        % set starting texture image
Screen('PlayMovie', movie, rate, 0, soundVolume);
        % start to buffer movie

% Deal with movie size
movierect = movieSize * [width height];
ratio = max(movierect ./ ScreenRect(3:4));
```

Display 4.8 (continued)

```
if ratio > 1, movierect = movierect / ratio; end
movierect = CenterRect([0 0 movierect], ScreenRect);

t = Screen('Flip', windowPtr);
toTime = toTime - fromTime + t;
        % compute stop time relative to computer time

% Play movie till toTime, or ESC is pressed,
% or the end of the movie
while t < toTime
    tex = Screen('GetMovieImage', windowPtr, movie);
        % get a frame
    if tex <= 0 || ~isempty(ReadKey('esc')), break; end
    Screen('DrawTexture', windowPtr, tex, [], movierect);
    t = Screen('Flip', windowPtr);
    Screen('Close', tex);
end

%% System Reinstatement Module

Screen('PlayMovie', movie, 0); % stop playing
Screen('CloseMovie', movie);   % close the movie
sca;
```

Display 4.9

```
>> PlayMovie('introhigh.mpg', [], [], [], 1, 1) ;
>> PlayMovie('introhigh.mpg', [], [], 0.5, 2, 0);
```

`Screen('CloseMovie', movie)` releases the pointer `movie` and all OpenGL resources associated with it.

The Psychtoolbox functions for handling movies that are briefly described here are explained in more detail by the help function. The reader should check the help function for the definitions of additional parameters and options.

4.4.6 Retinotopy

The next example program is somewhat more complicated. It is included here to illustrate an important function for data collection in visual fMRI. The program shows how to present a series of displays of the sort that has been widely used to measure the retinotopy in early visual cortex in human brain imaging. Many regions of visual cortex have a retinotopic organization in which different regions of the visual field are represented in the brain in a systematic way. The purpose of the retinotopy display sequences is to measure those voxels of the fMRI—and the corresponding group of neurons—that respond to each point in the visual field. A given point on the cortical surface represents neurons from several cortical layers whose receptive fields center on the same point in visual space. In retinotopic organizations, adjacent points on the cortical surface correspond to adjacent points in the visual field. The purpose of retinotopy experiments is to use a sequence of displays to find the correspondence between stimulation in a part of the visual field and activity in a cortical area in fMRI.

Wedges and rings made of flickering radial color checkerboard patterns are widely used to identify retinotopic visual areas of each observer.[12,13] The relevant display sequences over time stimulate different regions of visual space with flickering stimuli. Figure 4.8a shows a series of black and white checkerboard wedges that cycle radially through stimulating different regions of the visual field, like positions of the hand of an old-fashioned analog clock. Figure 4.8b shows a series of black and white checkerboard rings that cycle through stimulating regions of the visual field of different eccentricity starting from fixation. The aspect ratio of the dark and light checkered regions at different eccentricities is computed by setting their (radial) height equal to their (tangential) mid-width. For the wedge displays, a cycle of 32 images is shown over a period of 32 s, 1 s each. During the 1 s of each image, the checkerboards reverse polarity (change which is black and which is white) at 7.5 Hz. This flicker drives neural activity in visual cortex at that set of locations in the visual field. The wedge has a radius of 8.5° in degrees of visual angle and a width of one-eighth of a disk or 45° orientation and is divided into four subsectors. Every second, the image is rotated counterclockwise one-fourth of its width, so the wedge sweeps the whole visual field in 32 s. The wedge sequence is shown eight times continuously. Each ring was made of two checkers in the radial direction. The rings expand from the center of the display at the speed of one checker box per second. The entire cycle took 20 s. The ring displays are also shown a total of eight times continuously. During both the wedge and the ring display cycles, a central fixation square changed from black to red or vice versa in randomly chosen intervals between 5 and 15 s.

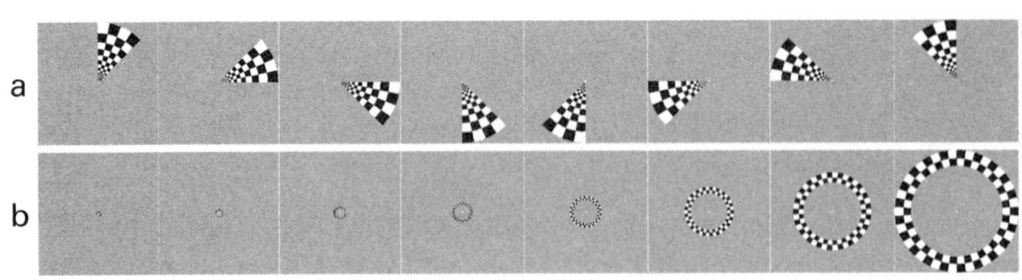

Figure 4.8
Image sequences for defining fMRI retinotopy. (a) Display sequence of the wedge. (b) Display sequence of the ring.

For an fMRI retinotopy session, observers are asked to maintain fixation and to press a key as soon as the fixation region changes color; 1 indicates a white fixation and 2 indicates a red fixation. Image sequences of the rotating wedges or expanding rings are shown while the observer maintains fixation. An fMRI analysis correlates the activity in different voxels representing a location on the cortical surface with the stimulation in a region of the visual field from the wedge and ring flickering displays.

To run an actual fMRI experiment, the experimental computer that generates the visual images and collects the responses would need to receive triggers from the MRI system. For example, a Siemens scanner sends trigger signals that are represented as equivalent to pressing "5" on the keyboard of the experimental computer. Additionally, behavioral responses would likely not be collected on a keyboard, but rather would make use of an fMRI-compatible response collection apparatus.

The retinotopy programs introduce the use of several Psychtoolbox function calls for creating image textures on the GPU as functions under the main function `Screen`. The call to `Screen('MakeTexture',…)` serves a basic function of converting a two-dimensional (2D) or three-dimensional (3D) matrix into an OpenGL texture (a display on the GPU) and returning a pointer that can then be used by various other functions to specify the texture. The form of the call is `textureIndex = Screen('MakeTexture', WindowIndex, imageMatrix)`. See the Psychtoolbox help function for other options. The 2D image(s) containing the black and white circular checkerboards are first set up with calls to `'MakeTexture'` in display 4.10.

Once the images are converted to OpenGL textures, we need to specify the content of the textures. In the case of the wedge program in display 4.10, the arcs of the circular checkerboard were drawn with `Screen('FrameArc', …)`. This function draws an arc (here, essentially one of the black or white checks in the circular checkerboard display) by specifying a starting angle, an angle for the arc, and a length or "pen width." Angles are measured clockwise from vertical. This function is called as `Screen('FrameArc', windowPtr, color, [rect], startAngle, arcAngle, penWidth)`. There are a number of other optional variables that can be looked up using the Psychtoolbox help function.

In the wedge experimental program, first the entire visible wedge of `texture1` is drawn as white by calling `Screen('FrameArc',tex(1), 1, [], 0, p.wedgeDeg, nPixels)`. The `'FrameArc'` subfunction is also subsequently used to paint each black check in the wedge, with the color set to 0, the location specified by `rect`, the initial angle specified by `ang0`, the angle to be swept by the arc of `dA`, and the penwidth (or radial length) of `nPixels` (see display 4.10).

Another function used to "draw" the textures once they are created is `Screen('DrawTexture',…)`. To display the image, this function draws a texture pointed to by a texture pointer, here `tex(i)`, into an (active) screen window pointed to by a window pointer, here `windowPtr`.

Display 4.10

```
%%% Program Wedge.m
function Wedge

%% Experimental Module

% Specify general experiment parameters
keys = {'1' '2' 'esc'};        % response keys for fixation
                               % change and to break
p.randSeed = ClockRandSeed;    % use clock to set random number
                               % generator

% Specify the stimulus
p.radius = 8.5;         % radius in degrees of visual angle
p.innerRad = 0.2;       % room for fixation in degrees of visual
                        %  angle
p.startAng = 0;         % starting angle of wedge
p.wedgeDeg = 45;        % Wedge width in degrees of visual angle
p.fixColor = 0;         % fixation color set to black
p.period = 32;          % seconds of one full rotation
p.repeats = 8;          % total number of full rotations of
                        % the wedge
p.lag = 12;             % seconds of blank screen before and after
                        % the 8 repeats of wedge stimuli
p.tf = 7.5;             % flicker frequency in Hz
p.TR = 1;               % duration of each position, set equal
                        % to TR for fMRI
p.fixDur = [5 15];      % set up range of time intervals between
                        % fixation color changes

% Compute stimulus parameters
ppd = pi/180 * p.ScreenDistance / p.ScreenHeight * …
      p.ScreenRect(4); % pixels/degree
m = round(p.radius * ppd * 2);    % stimulus size in pixels
fixSz = round(p.innerRad * ppd);  % radius of fixation circle
                                  % in pixels
fixRect = CenterRect([0 0 1 1] * fixSz, p.ScreenRect);
      % position fixation on the screen
dA = p.wedgeDeg / 4; % width of each of 4 sub-sectors of
                     % the wedge in degrees
% Compute radial height of each subsector as the tangent of
% its half-width in deg
tanAng = tand(dA / 2);
ratio = (1 - tanAng) / (1 + tanAng);
nRings = floor(log(p.innerRad / p.radius) / log(ratio));
```

Display 4.10 (continued)

```
radius = p.radius * ppd * ratio.^ (0 : nRings-1);
        % The call to Screen('MakeTexture', ) converts a 2D
        % image to a OpenGL texture. one will be the
        % black/white negative of the other, used for flicker
img = ones(m) * p.ScreenBackground;
tex(1) = Screen('MakeTexture', windowPtr, img, 0 ,0, 2);
tex(2) = Screen('MakeTexture', windowPtr, img, 0 ,0, 2);
nPixels = diff(radius([nRings 1]));
        % Fill the whole wedge with white
Screen('FrameArc', tex(1), 1, [], 0, p.wedgeDeg, nPixels);
Screen('FrameArc', tex(2), 1, [], 0, p.wedgeDeg, nPixels);
        % Screen('FrameArc, ) draws an arc in a texture with
        % starting angle, angle of arc and width of line

for i = 1 : nRings-1   % Fill every other sector with black
                       % to make checker pattern
    rect = CenterRect([0 0 2 2] * radius(i), [0 0 m m]);
    nPixels = diff(radius([i+1 i]));
    for j = [0 2]      % every other sector
        for iTex = 1 : 2
            ang0 = (j + mod(i + iTex, 2)) * dA;
            Screen('FrameArc', tex(iTex), 0, rect, ang0, ...
                  dA, nPixels);
        end
    end
end

% Set up onset times of fixation color changes
p.recLabel = {'fixChangeSecs' 'respTime'};
         % labels of columns of condition/data rec
secs = p.period * p.repeats; % total seconds to run
durs = rand(round(secs / p.fixDur(1)), 1); % [0 1]
durs = durs * diff(p.fixDur) + p.fixDur(1); % [5 15]
durs = cumsum(durs);
durs(durs >= secs) = [];
rec = nan(length(durs), length(p.recLabel));
rec(:, 1) = durs; % record onset times of fixation
                  % color changes

% Prioritize display to optimize display timing
Priority(MaxPriority(windowPtr));

% Start experiment with instructions
```

Display 4.10 (continued)

```
str = sprintf(['Please fixate at the center of the ' ...
      'screen. \n\n' 'Press 1 for white fixation or 2 for ' ...
      'red fixation.\n\nPress 5 to start.']);
DrawFormattedText(windowPtr, str, 'center', 'center', 1);
      % Draw Instruction text string centered in window
Screen('Flip', windowPtr);
      % flip the text image into active frame buffer
WaitTill('5');      % wait till 5 (trigger to start the display)
                    % is pressed

Screen('FillOval', windowPtr, 1, fixRect); % draw fixation
startSecs = Screen('Flip', windowPtr);
      % flip the fixation image into active buffer
p.start = datestr(now);     % record start time

t0 = startSecs + p.lag;     % planned onset of wedge
durs = [durs; inf] + t0;    % absolute time to change fixation
iFix = 1;                   % index for fixation color: 1 or 2
                            % for white and red fixation
fixColor = [1 1 1; 1 0 0]; % set up color tables for white and
                            % red fixation colors
i = 1;                      % index to count fixation changes
dA = 360 / p.period;        % rotation in degrees per second
endSecs = t0 + p.period * p.repeats;
                            % end time of wedge rotations
key = ReadKey; key = ReadKey;
      % ReadKey twice to speed up subsequent calls to it.
validResp = 1;              % preset a valid response flag to 1
WaitTill(t0 - 0.02); % wait for start time for wedge display

Screen('FillOval', windowPtr, 1, fixRect);
vbl = Screen('Flip', windowPtr, t0 - 1 / p.ScreenFrameRate);
t0 = vbl;
while vbl < endSecs  % show the wedge stimulus with fixation
                     % changes and collect responses and
                     % record the response times
    iTex = (mod(vbl - t0, 1 / p.tf) < 0.5 / p.tf) + 1;
    ang = p.startAng + floor((vbl - t0) / p.TR) * dA;
    if vbl > durs(i)
        iFix = 3 - iFix; % change fixation color between red
                         % and white
        t1 = vbl + 1 / p.ScreenFrameRate;
                         % fixation change time
        i = i + 1;
    end
```

Display 4.10 (continued)

```
    Screen('DrawTexture', windowPtr, tex(iTex), [], [], ang);
    Screen('FillOval', windowPtr, fixColor(iFix, :), fixRect);
    vbl = Screen('Flip', windowPtr);

    key = ReadKey(keys);
    if strcmp(key, 'esc'), break; end
        % stop the dispay if 'esc' is pressed
    if isempty(key)
        validResp = 1;
        % reset response flag, the next response will be valid
    elseif validResp
        validResp = 0;
        rec(i, 2) = vbl - t1; % record response time
    end
end

Screen('FillOval', windowPtr, fixColor(iFix, :), fixRect);
Screen('Flip', windowPtr);

WaitTill(endSecs +p.lag, 'esc');% wait for p.lag after wedge
p.finish = datestr(now);         % record finish time

save Wedge_rst.mat rec p;        % save the results
```

Figure 4.9 (plate 6) shows the results of a retinotopy experiment using the wedge and ring stimuli from displays 4.10 and 4.11. The fMRI responses to different parts of the visual field are shown overlaid on a flattened representaton of the visual cortex. Lines are drawn to label different visual cortical regions, such as V1, V2, V3, and V4. The retinotopy is based on the organizational principles of the visual cortex: The borders between the early visual areas coincide with the representations of the vertical and horizontal meridians in the visual field, and neighboring areas contain mirror-reversed maps of the visual space.[12,13]

4.5 Summary

In this chapter, we have shown how to use "movies" for controlled visual presentations. Movies are simply sequences of images that are displayed—typically but not necessarily rapidly—that may or may not be accompanied by audio. We started with the detailed presentation of the anatomy of a simple experiment using movies programmed in MATLAB with Psychtoolbox.[1,2] The chapter provides samples of implementations of representative stimuli and tasks. These include cases where the images of the movies are constructed

Display 4.11

```
%%% Program Ring.m
function Ring

%% Experimental Module

% set up general experimental parameters
keys = {'1' '2' 'esc'};      % response keys for fixation
                             % change and to break
p.randSeed = ClockRandSeed;  % use clock to set random number
                             % generator

% Specify the stimulus
p.radius = 8.5;         % radius in degrees of visual angle
p.innerRad = 0.2;       % room for fixation in degrees of
                        % visual angle
p.fixColor = 0;         % set fixation color to black
p.period = 20;          % seconds of one cycle of expanding rings
p.repeats = 8;          % number of expansion cycles
p.lag = 12;             % seconds before and after the repeated
                        % expansion stimuli
p.tf = 7.5;             % flicker frequency in Hz
p.TR = 1;               % duration of each image location in
                        % expansion, set = TR for fMRI
p.fixDur = [5 15];      % set up range of time intervals between
                        % fixation color changes

% Compute stimulus parameters
ppd = pi/180 * p.ScreenDistance / p.ScreenHeight * …
      p.ScreenRect(4); % pixels/degree
m = round(p.radius * ppd * 2);   % stimulus size in pixels
fixSz = round(p.innerRad * ppd);  % fixation size in pixels
fixRect = CenterRect([0 0 1 1] * fixSz, p.ScreenRect);
% Compute radii of all rings, so the black/white check is
% approximately square
nRings = p.period / p.TR;
radius = logspace(log10(p.radius), log10(p.innerRad), …
         nRings+2);
radius = radius * ppd;
ratio = radius(2) / radius(1);
dA = 2 * atand((1 - ratio) / (1 + ratio));
        % sub-wedge width in deg
nSectors = round(180 / dA) * 2;
dA = 360 / nSectors;

% The call to Screen('MakeTexture', ) converts a 2D image to a
```

Display 4.11 (continued)

```
% OpenGL texture. One will be the black/white negative of the
% other, used for flicker
img = ones(m) * p.ScreenBackground;
        % Create 2 textures with background color
tex(1) = Screen('MakeTexture', windowPtr, img, 0 ,0, 2);
tex(2) = Screen('MakeTexture', windowPtr, img, 0 ,0, 2);

rect = CenterRect([0 0 2 2] * radius(1), [0 0 m m]);
        % Color the whole annulus white
nPixels = diff(radius([3 1]));
Screen('FrameArc', tex(1), 1, rect, 0, 360, nPixels);
Screen('FrameArc', tex(2), 1, rect, 0, 360, nPixels);

for i = 1 : 2    % add black to make checks
    rect = CenterRect([0 0 2 2] * radius(i), [0 0 m m]);
    nPixels = diff(radius([i+1 i]));
    for j = 0 : 2 : nSectors-1 % every other sector
        for iTex = 1 : 2
            ang0 = (j + mod(i + iTex, 2)) * dA;
            Screen('FrameArc', tex(iTex), 0, rect, ang0, ...
                   dA, nPixels);
        end
    end
end

% Set up onset times of fixation color changes
p.recLabel = {'fixChangeSecs' 'respTime'};
secs = p.period * p.repeats; % total seconds to run
durs = rand(round(secs / p.fixDur(1)), 1);  % [0 1]
durs = durs * diff(p.fixDur) + p.fixDur(1); % [5 15]
durs = cumsum(durs);
durs(durs >= secs) = [];
rec = nan(length(durs), length(p.recLabel));
rec(:, 1) = durs;  % record onset times of fixation
                   % color changes

% Prioritize display to optimize display timing
Priority(MaxPriority(windowPtr));

% Start experiment with instructions
str = sprintf(['Please fixate at the center of the '...
      'screen.\n\n' 'Press 1 for white or 2 for red.\n\n' ...
      'Press 5 to start.']);
```

Display 4.11 (continued)

```
DrawFormattedText(windowPtr, str, 'center', 'center', 1);
        % Draw Instruction text string centered in window
Screen('Flip', windowPtr);
        % flip the text image into active buffer
WaitTill('5');
        % wait till 5 (trigger to start display) is pressed

Screen('FillOval', windowPtr, 1, fixRect); % draw fixation
startSecs = Screen('Flip', windowPtr);
        % flip the fixation image into active buffer
p.start = datestr(now);                   % record start time

t0 = startSecs + p.lag;  % planned onset of wedge
durs = [durs; inf] + t0; % absolute time to change fixation
iFix = 1;                % index for fixation color:
                         % 1 or 2 for white and red fixation
fixColor = [1 1 1; 1 0 0]; % set color tables for
                           % white and red
i = 1;                   % index for fixation change
endSecs = t0 + p.period * p.repeats;
        % end time of cycle of rings
key = ReadKey; key = ReadKey;
        % ReadKey twice to speed up subsequent calls to it.
validResp = 1;           % set valid response flag to 1
WaitTill(t0 - 0.02);     % wait for wedge start time

Screen('FillOval', windowPtr, 1, fixRect);
vbl = Screen('Flip', windowPtr, t0 - 1 / p.ScreenFrameRate);
t0 = vbl;
while vbl < endSecs  % show the expanding ring stimulus with
                     % fixation changes and collect responses
                     % and record the response times
    iTex = (mod(vbl - t0, 1 / p.tf) < 0.5 / p.tf) + 1;
    iRing = nRings - floor(mod(vbl - t0, p.period) / p.TR);
                     % expanding
    % iRing = 1 + floor(mod(vbl - t0, p.period) / p.TR);
                     % shrinking
    rect = CenterRect([0 0 2 2] * radius(iRing), ...
           p.ScreenRect);
    if vbl > durs(i)
        iFix = 3 - iFix;  % change fixation color between
                          % red and white
        t1 = vbl + 1 / p.ScreenFrameRate;
                          % fixation change time
        i = i + 1;
    end
```

Display 4.11 (continued)

```
        Screen('DrawTexture', windowPtr, tex(iTex), [], rect);
        Screen('FillOval', windowPtr, fixColor(iFix, :), fixRect);
        vbl = Screen('Flip', windowPtr);

        key = ReadKey(keys);
        if strcmp(key, 'esc'), break; end % to stop
        if isempty(key)
            validResp = 1;  % key released, next response
                            % will be valid
        elseif validResp
            validResp = 0;  % debouncing
            rec(i, 2) = vbl - t1;   % record response time
        end
end

Screen('FillOval', windowPtr, fixColor(iFix, :), fixRect);
Screen('Flip', windowPtr);

WaitTill(endSecs +p.lag, 'esc'); % wait for p.lag after wedge
p.finish = datestr(now);         % record finish time

save Ring_rst.mat rec p;         % save the results
```

either ahead of time or on the fly by the experimenter and also an example of presenting modified clips from native movies. Combined with some of the image processing examples in chapter 3, these simple examples provide the basis of the expertise that allows the reader to generate a wide range of interesting experiments and demonstrations and to control their timing and display.

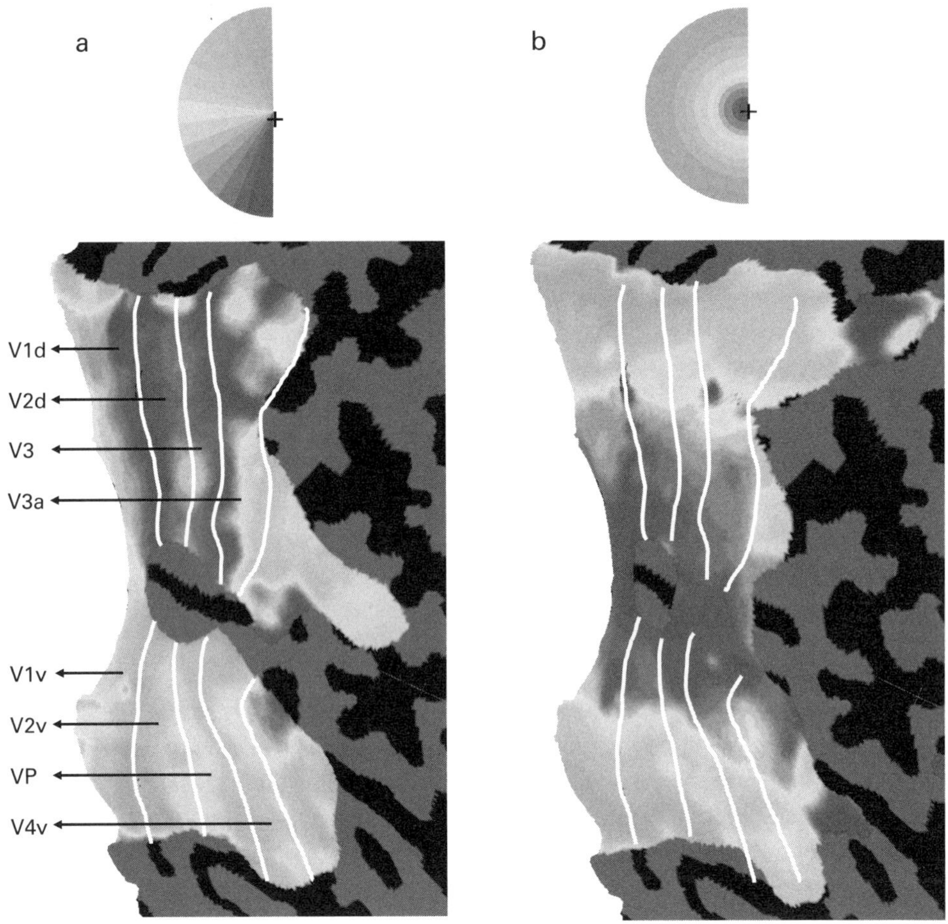

Figure 4.9 (plate 6)
Flattened representations of the visual cortex labeled with retinotopic regions corresponding to different visual areas from an fMRI retinotopy experiment. (a) Retinotopy from the flickering wedge displays. (b) Retinotopy from the flickering ring displays. The activity in different cortical regions is color coded to the positions in the wedges and rings respectively. Retinotopy data from Li, Lu, Tjan, Dosher, and Chu (2008).[14]

References

1. Pelli DG. 1997. The VideoToolbox software for visual psychophysics: Transforming numbers into movies. *Spat Vis* 10(4): 437–442.

2. Brainard DH. 1997. The psychophysics toolbox. *Spat Vis* 10(4): 433–436.

3. Psychtoolbox-3. [computer program] Available at: http://psychtoolbox.org/.

4. Fechner G. *Elemente der psychophysik*. Leipzig: Breitkopf & Härtel; 1860.

5. Stevens SS. 1946. On the theory of scales of measurement. *Science* 103(2684): 677–680.

6. Chun MM, Potter MC. 1995. A two-stage model for multiple target detection in rapid serial visual presentation. *J Exp Psychol Hum Percept Perform* 21(1): 109–127.

7. Weichselgartner E, Sperling G. 1987. Dynamics of automatic and controlled visual attention. *Science* 238(4828): 778–780.

8. Holmes V, Arwas R, Garrett M. 1977. Prior context and the perception of lexically ambiguous sentences. *Mem Cognit* 5(1): 103–110.

9. Reeves A, Sperling G. 1986. Attention gating in short-term visual memory. *Psychol Rev* 93(2): 180–206.

10. Sperling G, Weichselgartner E. 1995. Episodic theory of the dynamics of spatial attention. *Psychol Rev* 102(3): 503–532.

11. Shih SI, Sperling G. 2002. Measuring and modeling the trajectory of visual spatial attention. *Psychol Rev* 109(2): 260–305.

12. Engel SA, Glover GH, Wandell BA. 1997. Retinotopic organization in human visual cortex and the spatial precision of functional MRI. *Cereb Cortex* 7(2): 181–192.

13. Sereno M, Dale A, Reppas J, Kwong K, Belliveau J, Brady T, Rosen B, Tootell R. 1995. Borders of multiple visual areas in humans revealed by functional magnetic resonance imaging. *Science* 268(5212): 889–893.

14. Li X, Lu Z-L, Tjan BS, Dosher B, Chu W. 2008. BOLD fMRI contrast response functions identify multiple mechanisms of attention in early visual areas. *Proc Natl Acad Sci USA,* 105(16): 6202–6207.

5 Visual Displays

The evaluation, calibration, and choice of visual displays are critical to the success of the psychophysics enterprise. In this chapter, we show how to evaluate and calibrate visual displays and summarize the qualities of some current display technologies. A number of simple sample programs for calibration and evaluation of displays are provided. Obviously, the specifics of particular devices at any given time will be subject to change as technologies emerge in the marketplace. However, the goals and principles of evaluation and calibration will be applicable to any new device.

5.1 Evaluating Visual Displays

The quality of the visual display is critical in visual psychophysics. The purpose of visual psychophysics is to measure the behavioral response of the observer to a visual stimulus with known physical properties of size, luminance, timing, and so forth. A characterization of the human (or animal) observer requires the experimenter to design a particular visual stimulus and to then generate that stimulus in a visual display. To understand exactly what has been presented, we must either know or measure the properties of the display. Consider several examples. To test differences in response to a pattern in different orientations, we must ensure that the luminance and contrast properties of the stimulus shown to the observer are equivalent at different orientations. To measure responses to color stimuli, we must control the color information provided in the display. To measure the perception of motion, the experimenter must also control the temporal characteristics of the displays. Assessment of visual memory persistence requires that we can attribute persistence to the viewer and not to persistence on the display.

Any visual display device should be evaluated for the following properties: spatial resolution, refresh rate, pixel depth, color and luminance homogeneity, geometric distortions, temporal response function, pixel independence, and display synchronization.[1–7]

Even though the nominal spatial resolution of a display is specified by the device design, the homogeneity and independence of the values of pixels in the display may be imperfect. Similarly, the refresh rate provided by the manufacturer may not fully reflect the exact temporal properties of the display. There may be persistence of the physical stimulus or delays in the onset of images because of various possible conversions between the graphics card and the display device. The display luminance has an unknown physical range and a nonlinear transformation from the programmed image values specified in a bitmap to physical values on the display device. This nonlinear transformation needs to be measured. In some cases, even monochrome display devices have a characteristic color shift away from gray; that is, the specific phosphors of a cathode ray tube (CRT) device may have a distinctive color.

Measurement devices are required for some kinds of evaluation. A precise ruler is used to measure the spatial properties in relation to a physical standard. A colorimeter is used to measure the spectral composition and intensity of displays. Photodiodes, an oscilloscope, and a video splitter are used to measure the persistence and delay of displayed images.

These properties of displays and how to evaluate them will be discussed in the next sections.

5.1.1 Spatial Resolution, Pixel Depth, and Refresh Rate

Generally, the manufacturers of display devices provide specifications for spatial resolution, refresh rate, and pixel depth. Many current devices offer a selection of possible sets of these properties. For example, you may choose a monochrome (gray) mode, or different spatial resolution, or one of a number of display refresh rates. The spatial resolution consists of available sets of rectangular pixel sizes assigned to the viewing surface, expressed as the number of pixels per width and height of the display (i.e., 1024 × 768). Often, there will be several possible refresh rates (i.e., from 60 to 200 Hz). The pixel depth refers to the number of independent bits of resolution in each display channel (i.e., 8 bits in each color gun). Generally, a display can be set in a monochrome mode with 256 gray levels or true color mode with millions of colors.

Psychtoolbox[8] provides a function called `ScreenTest` that evaluates and reports properties of the display setup. For current graphics cards, the report includes the version of the OpenGL-Renderer, the amount of VRAM and texture memory, the number of display devices associated with the setup, and the refresh rate of each display device (display 5.1).

OpenGL stands for Open Graphics Library, a set of standards for two-dimensional (2D) and three-dimensional (3D) graphics programs.[9] Renderer refers to processes of creating an image from a standardized model or description. A very simple description might be a rectangular bitmap, and a more complicated case might be a model of a 3D object. The surfaces of objects have patterns—called *textures*. The texture may be a regular pattern such as tiles on a wall, or it may be a random pattern such as a pattern of rocks on the ground or fur on an animal. In Psychtoolbox[8] we use the term *texture* to describe any 2D pattern in graphics memory.

Display 5.1

```
***** ScreenTest: Testing Screen 0 *****
PTB-INFO: This is Psychtoolbox-3 for Apple OS X, under Matlab
PTB-INFO: OpenGL-Renderer is NVIDIA Corporation :: NVIDIA
     GeForce 9400M OpenGL Engine :: 2.1 NVIDIA-1.6.36
PTB-INFO: Renderer has 256 MB of VRAM and a maximum 237 MB of
     texture memory.
PTB-INFO: VBL startline = 800 , VBL Endline = 800
PTB-INFO: Measured monitor refresh interval from beamposition
     = 16.605090 ms [60.222498 Hz].
PTB-INFO: Will use beamposition query for accurate Flip time
     stamping.
PTB-INFO: Measured monitor refresh interval from VBLsync =
     16.605377 ms [60.221456 Hz]. (50 valid samples taken,
     stddev=0.065004 ms.)
PTB-INFO: Small deviations between reported values are normal
     and no reason to worry.
PTB-INFO: Using OpenGL GL_TEXTURE_RECTANGLE_EXT extension for
     efficient high-performance texture mapping …
***** ScreenTest: Done With Screen 0 *****
```

VRAM stands for video random access memory, which is available on the graphics card, where models and textures may be stored leading up to display.

The refresh rate, as described earlier, is the rate at which contents of the display buffer are "refreshed" or shown on the display device. This sets the fastest rate at which new images can be presented to an observer. The duration of an image display is the time to complete an integer number of refresh counts.

Many experimental setups have several displays. For example, one display might control MATLAB[10] and the display of messages or information to the experimenter, while another is used as the testing display, which presents images to an observer. `ScreenTest` tests and generates a report for each visual display device.

Display 5.1 shows the output of `ScreenTest` for one of the screens on one of our systems. The `ScreenTest` report begins with the version of MATLAB and other licensing information (omitted here). It then reports the specific manufacturer and version of the OpenGL-Renderer associated with the graphics card. `ScreenTest` reports 256 MB of VRAM of which 237 is usable as texture memory.

The other information concerns the refresh rate and timing of the display. `ScreenTest` measures the refresh rate multiple times and reports the mean and standard deviation. In this example, the nominal refresh rate for Screen 0 was 60 Hz, which corresponds to about 16.67 ms per frame (1000 ms/60 refreshes per second). `ScreenTest` reports two empirical measurements of the refresh rate by slightly different methods: 60.222498 Hz and 16.605090 ms per frame and 60.221456 Hz and 16.605377 ms per frame. One is based

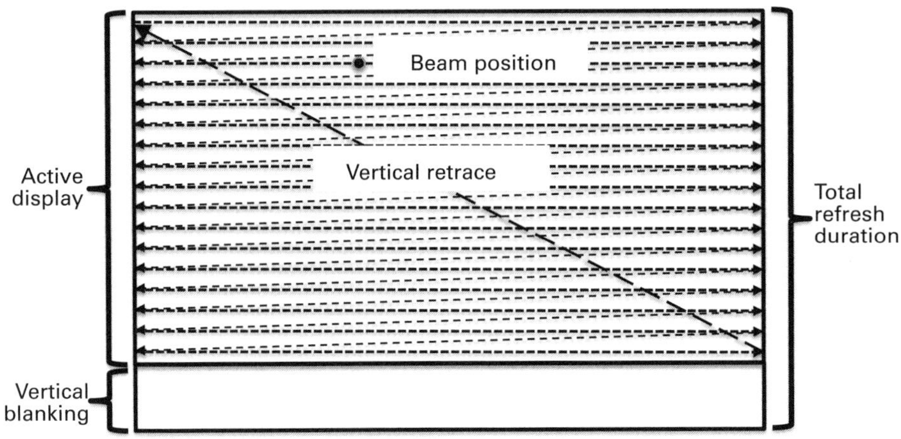

Figure 5.1
Illustration of refresh intervals and vertical blanking in visual displays.

on the vertical blanking interval, and the other is based on the beam position (figure 5.1). Both are within measurement tolerance of the nominal rate. The definitions of horizontal and vertical blanking intervals are explained next.

Most of the terminology and the standards for video and television were developed around the physical constraints of the CRT (figure 5.1). Originally, an *electron beam* activated a phosphor on the surface of the CRT in a conventional order, from top left to bottom right in rows.[11] Very brief *horizontal blanking intervals* allow time for the beam to shift from the end of one row to the beginning of the next, and a longer *vertical blanking interval*[12] (VBI) allows the beam to reset back to the top left of each display. The image on the screen is unchanged during vertical blanking intervals. The beam position refers to the location of the electronic beam in this pattern of display, which is associated with a rather accurate measure of the time during this display interval. In early applications without buffered graphics memory, the VBI might be 15% of the total refresh interval in order to allow new information to be transferred into the frame buffer. For example, of the 16.67 ms per frame at 60 Hz, the VBI might be about 2.5 ms. With modern technology, the physical requirement for a long VBI is no longer relevant—but the conventions are retained for compatibility to television and other video standards, and the VBI is now used to send other kinds of data (i.e., subtitles for the deaf in television) or to carry out graphics computations.

Psychtoolbox Version 3 uses a double buffering system with an onscreen display buffer and an offscreen *back buffer*.[8] Switching the pointer to the next display as the two are "flipped" is almost instantaneous, usually taking less than a microsecond. Psychtoolbox can track both the beam position and the flag (`VBLsync`) that marks the beginning of the

VBI, and it reports measurements of the refresh rate based on both of these. The screen being tested in display 5.1 was set for a spatial resolution of 1280×800, so reporting that the start and end line of the VBI occurred at line 800 indicates that it occurred before starting the next display.

`ScreenTest` is provided by Psychtoolbox to assist in evaluating visual displays, however there are many aspects of visual displays that `ScreenTest` does not handle. In this chapter, we discuss many other calibration and evaluation procedures and implement them in a series of programs. These programs are relatively straightforward—and so we provide fewer comments. They serve as additional sample programs that may be helpful for students of Psychtoolbox.

5.1.2 Luminance Homogeneity

An image provided to a graphics card should have identical physical intensities no matter where it appears on the display system—reflecting luminance homogeneity across the display. In practice, luminance homogeneity is imperfect on many display devices.[3] To evaluate luminance homogeneity, we set the entire screen to a specified image value, usually white (1.0, 1.0, 1.0), and use a photometer or colorimeter to measure the luminance values at the center and around an imaginary circle at different radii in the display (display 5.2). Often, for example almost always in CRT displays, the measured luminance values at the extremes or near the edges of the display are reduced relative to the center. In this case, you can measure luminance values until finding a central region of adequate homogeneity to allow the planned experimental tests.

Many visual tasks manipulate contrast rather than luminance. In these cases, an inhomogeneity in luminance intensity up to 10% may be acceptable for most experimental purposes.

5.1.3 Brightness and Contrast

One aspect of display devices, generally referred to as brightness, evaluates the maximum luminance that the device delivers. A secondary property, the contrast ratio, is the ratio of the maximum luminance (white) to the minimum luminance (black). The contrast ratio defines the dynamic range of the display device in luminance. For a device with low contrast, the issue often is the ability to display near-black values. In general, CRT displays are of high contrast ratio, greater than 1500 to 1. The best liquid crystal displays (LCDs) have contrast ratios of 500 to 1 and may be as poor as 200 to 1.

5.1.4 Geometric Distortions

Display devices may introduce geometric distortions as well as luminance inhomogeneities. A geometric distortion occurs when the image on the display deviates from the

Display 5.2

```
%%% Program LuminanceHomogeneity.m
% Press ESC to exit.
function LuminanceHomogeneity

%% Display Setup Module

% Define display parameters
whichScreen = max(Screen('screens'));
p.ScreenGamma = 2; % from monitor calibration
p.ScreenBackground = 1;

% Open display window and hide the mouse cursor
if exist('onCleanup', 'class'), ...
      oC_Obj = onCleanup(@()sca), end
         % close pre-exiting PTB Screen window
PsychImaging('PrepareConfiguration');
PsychImaging('AddTask', 'General', ...
      'FloatingPoint32BitIfPossible');
         % set up a 32-bit framebuffer
PsychImaging('AddTask', 'General', ...
      'NormalizedHighresColorRange');
PsychImaging('AddTask', 'FinalFormatting', ...
      'DisplayColorCorrection', 'SimpleGamma');
[windowPtr p.ScreenRect] = PsychImaging('OpenWindow', ...
       whichScreen, p.ScreenBackground);
PsychColorCorrection('SetEncodingGamma', windowPtr, ...
       1 / p.ScreenGamma);
HideCursor;
Screen('Flip', windowPtr);

WaitTill('esc'); % wait till ESC is pressed

%% System Reinstatement Module
Screen('CloseAll'); % close window and textures
```

intended shape.[13] For example, a square may appear as a rectangle or a trapezoid or may show some other warping of the shape. One simple way to test this is to display a large square of known pixel size on the display device and use a good ruler to measure the horizontal and vertical size of the square (see display 5.3 for the program). Many display devices have parameters that can be adjusted to determine the overall size and shape (aspect ratio and straight-line properties) of the display. Geometric distortion could be measured in multiple regions of the display, or at least in the regions of the display that will be used in the experiment.

Display 5.3

```
%%% Program GeometricDistortions.m
function GeometricDistortions

%% Experimental Module
% Specify the stimulus
p.stimSize = 256; % stimulus size in pixels

% Compute stimulus parameters
rect = CenterRect([0 0 1 1] * p.stimSize, p.ScreenRect);

% Show the square in the center of the screen
Screen('FillRect', windowPtr, 1, rect);
Screen('Flip', windowPtr);
WaitTill('esc'); % wait till ESC is pressed
```

5.1.5 Pixel Independence

Another important property of the display system is pixel independence.[1] This means that the intensity value of a pixel does not depend upon the intensity values of neighboring pixels. Failures of pixel independence can introduce artifactual or unintended contrast differences between the same image content displayed in different orientations. Each pixel should ideally be displayed as a square wave that corresponds to the physical location of the pixel, with sharp onset and offset just at the pixel location. Because square waves have many frequency components, this requires very high temporal bandwidth for graphics devices—greater than 200 MHz for a monochrome display.

On a typical CRT, however, adjacent pixels on a horizontal line are painted close together in time, from left to right on a row, while adjacent pixels on a vertical line are painted farther apart in time (see figure 5.1). If individual pixels do not achieve a perfect square-wave form (i.e., are blurred over space horizontally) and several white pixels happen to be displayed close together on a horizontal line, the spread will alter the net intensity through nonlinear properties of the phosphor on the CRT. Often, this creates a situation where horizontal lines appear brighter than vertical lines.

Pixel independence is essential for isometry—rotational symmetry—of patterns shown at different angles or orientations. If pixel independence is not achieved, the same image shown at different orientations will generate physically different luminance values in corresponding image regions. If you cannot achieve pixel independence on the display, this limits the stimuli that can be compared without artifact to those that are orientation symmetric. Pixel independence is also critical in cases where luminance intensity patterns with identical intensity histograms (i.e., different noise samples) differ for different orientations.

A classic test of pixel independence is to compare the mean luminance of a horizontal square-wave grating with a 2-pixel period to that of a 2-pixel period vertical square-wave grating or to a checkerboard pattern with a 2-pixel × 2-pixel check size. The pixel independence of the display device in the same region can be measured with a photometer that tests whether the three kinds of patterns presented in succession over time generate an equal amount of luminance. You can also get a strong visual appearance of any failures of independence by simultaneously placing these three kinds of patterns in different regions in space, in which case they are often visible by the eye, as seen in figure 5.2. See display 5.4 for the corresponding program.

One tactic for working around this to achieve pixel independence is to modify images by blanking out every other pixel, thereby limiting the overlap of spatial blur along horizontal lines. However, this reduces the effective resolution of the content of the images.

5.1.6 Temporal Response

The temporal response refers to the exact time-course of the activation of a pixel or small pixel patch in the display. The relevant factors are the actual onset relative to the intended onset in time, and the duration of the activation and/or the entire shape of the temporal response function of activation. Sometimes, display devices perform conversions of the signal provided by graphics cards, and these conversions may cause either delays or changes in the temporal profile of the displayed image.[14] Many display devices convert analog signals to digital signals. Also, if the display device is not set at one of its native modes, it performs spatial transformations on the intended images.

The time course of activation of a pixel or pixel cluster is determined by the physical properties of the light emitting or transmitting materials. Different CRT devices may use distinct types of phosphors, each of which has a characteristic color and time course of persistence.[3] LCD displays are notorious for image intensity persistence.[15] In this case, programming a positive intensity in two adjacent frames in time yields a higher luminance value than the sum of the two, similar to the interactions of pixels in space described earlier. For a typical CRT, the time course includes a rise time of 1–3 ms and a decay period of 1–100 ms, depending upon the phosphor. For commonly used phosphors such as P4, the decay time is on the order of 10 ms. For the typical LCD, in contrast, both the rise time and the decay time are often more than 20 ms.

The temporal properties of displays are usually measured in the laboratory with a photodiode, a video splitter, and an oscilloscope (figure 5.3). The experimenter displays a pixel patch that is turned on and off periodically, and the pattern of luminance is measured by the photodiode and oscilloscope. A program to do this is shown in display 5.5. The temporal shapes of 1, 2, 3, and 6 refresh-cycle activations that appeared on the screen of the oscilloscope for a CRT display and an LCD display are shown in figure 5.4. A display device with a good temporal response function will have the entire time course of activa-

Display 5.4

```
%%% Program PixelIndependence.m
function PixelIndependence(pixels)

%% Experimental Module

% Specify the stimulus
if nargin < 1, pixels = 1; end     % pixels of half period
p.stimSize = 256;   % a quarter of stimulus size in pixels

% Compute stimulus parameters
n = round(p.stimSize / pixels);
sz = n * pixels;   % real number of pixels
[xc yc] = RectCenter(p.ScreenRect);

% Compute and make texture for grating
img = zeros(n);
img(1 : 2 : end, :) = 1;
img = Expand(img, pixels);
grating = Screen('MakeTexture', windowPtr, img, 0, 0, 2);

% Compute and make texture for the checker board
img = zeros(n);
img(1 : 2: end, 1 : 2 : end) = 1;
img(2 : 2: end, 2 : 2 : end) = 1;
img = Expand(img, pixels);
checker = Screen('MakeTexture', windowPtr, img, 0, 0, 2);

% Draw grating and check
Screen('DrawTexture', windowPtr, grating, [], ...
       [xc-sz yc-sz xc yc]);
Screen('DrawTexture', windowPtr, grating, [], ...
       [xc+1 yc+1 xc+sz+1 yc+sz+1], 90);
Screen('DrawTexture', windowPtr, checker, [], ...
       [xc+1 yc-sz xc+sz+1 yc], 90);
Screen('DrawTexture', windowPtr, checker, [], ...
       [xc-sz yc+1 xc yc+sz+1]);
Screen('Flip', windowPtr);

WaitTill('esc');    % wait till ESC is pressed
```

Figure 5.2
Test pattern for pixel independence showing a vertical square-wave grating (upper left), a horizontal square-wave grating (lower right), and two checkerboards.

Display 5.5

```
%%% Program TemporalResponse.m
function TemporalResponse (nFrames)

vbl=Screen('Flip', windowPtr);

%% Experimental Module

% Specify the stimulus
if nargin < 1, nFrames = 1; end   % number of frames
                                  % of the square
p.stimSize = [100 100];   % horizontal and vertical size
                          % in pixels
p.interval = 0.2;         % seconds between two flashes

% Compute stimulus parameters
p.ScreenFrameRate = FrameRate(windowPtr);
rect = CenterRect([0 0 p.stimSize], p.ScreenRect);
flipWait = (nFrames - 0.5) / p.ScreenFrameRate;

% Show the square at the specified location and with
% specified interval
while 1
    key = WaitTill(vbl + p.interval, 'esc');
        % wait till next flash
    if ~isempty(key), break; end

    Screen('FillRect', windowPtr, 1, rect);
    vbl = Screen('Flip', windowPtr); % turn on flash
    vbl = Screen('Flip', windowPtr, vbl + flipWait);
        % turn off after nFrames
end
```

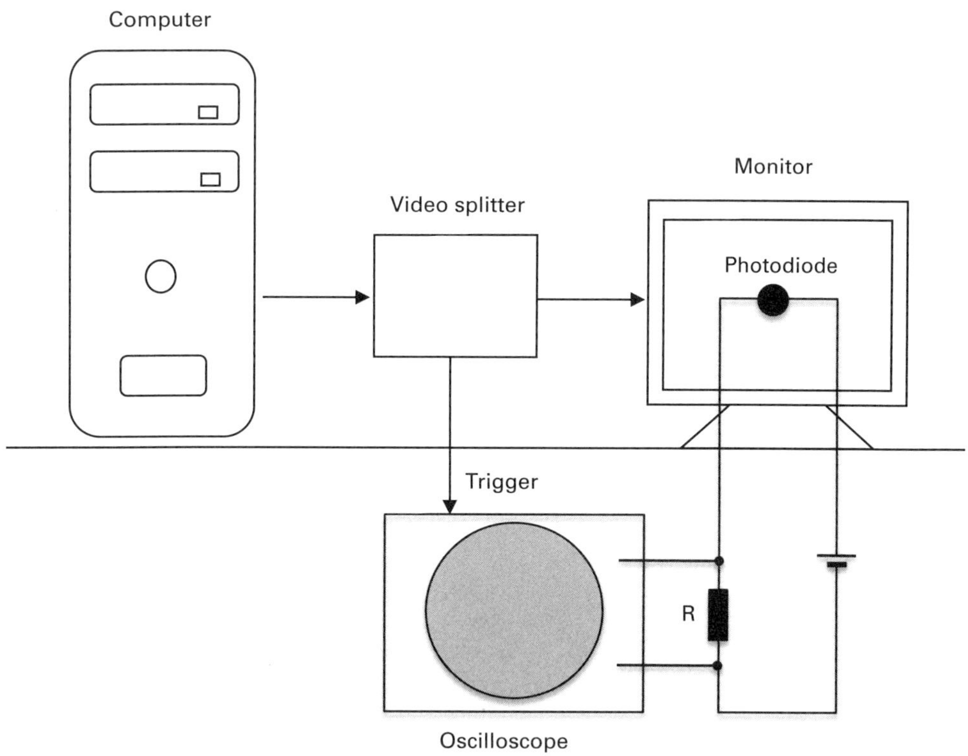

Figure 5.3
Circuit diagram for testing temporal properties of visual displays.

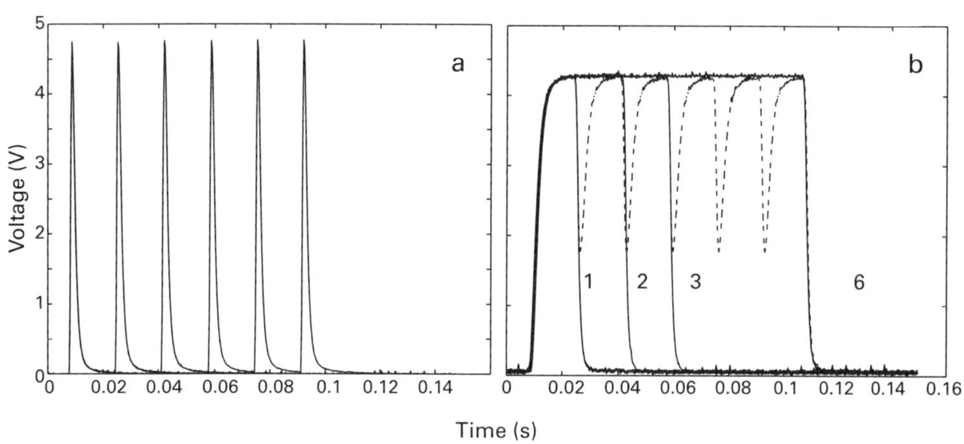

Figure 5.4
Temporal response functions of (a) a CRT display and (b) an LCD as measured by an oscilloscope. Responses to 1, 2, 3, and 6 consecutive refreshes are shown. For the CRT, the trace of each refresh is the same in all conditions. For the LCD, the traces of 2, 3, and 6 refreshes (solid curves) are different from the predictions of the combinations of multiple single refreshes (dotted curves); multiple refreshes smooth out the response functions.

tion, including the rising from and return back to baseline, within the interval of the frame rate. The shape should be consistent across repetitions.

The purpose of the video splitter in the setup in figure 5.3 is to provide a simultaneous signal to the monitor, which will be measured by the photodiode as one input to the oscilloscope, and a direct signal to the oscilloscope in order to assess whether the image is actually presented at the programmed time. To assess a delay requires the comparison of the actual onset of the image patch compared to the onset of the video signal. Another method is to use the video signal from the graphics card via a special-purpose device, such as a video switch device,[16] to trigger the oscilloscope and measure the onset of the photodiode response relative to the trigger.

If the video signal of the graphics card is not directly available, it is possible to use the onset of a pattern on a CRT display—which generally has extremely small delays—to calibrate the delays of other non-CRT devices. The output of the graphics card run through a video splitter may be fed to the CRT and, for example, an LCD. A photodiode measuring the CRT triggers the oscilloscope, and the photodiode measurement of the LCD evaluates the time course in relation to that onset.

5.1.7 Synchronization with the Frame Rate

We discuss two primary aspects of synchronization. The first is synchronizing image frames to the display refresh rate. The second involves the synchronization of images in different color planes or channels.

The time it takes to display an image physically must be less than the refresh interval (see the discussion of refresh rate in section 5.1.1).[12] There is a period of time in which no new information is displayed on the monitor or physical display device, called the VBI. Changing the contents of the graphics buffer being used to display an image—which occurs in the computer or graphics memory—needs to be synchronized with the refresh rate of the monitor.

Older display technologies used single-buffer systems, so changing the contents of the frame buffer needed to occur entirely during the VBI in order that an intact image or picture could be displayed during the next refresh cycle. If the entire image is not read into the active frame buffer at the appropriate time, then the frame buffer may contain parts of the old image and parts of the new one, which may create an impression of image "tearing" (figure 5.5). Newer graphics cards and display technologies often use double-buffered systems where the active frame buffer is changed instantaneously, or "flipped." In this case, the flip still needs to be synchronized with the VBI. This is accomplished in Psychtoolbox by the call `vbl=Screen('Flip', windowPtr)` in our sample programs. With double-buffered systems and the right programs, it may be possible to change the second "back buffer" during the entire refresh interval while the active buffer is being displayed, and so double-buffered systems allow for the transfer or computation of larger or more complicated images.

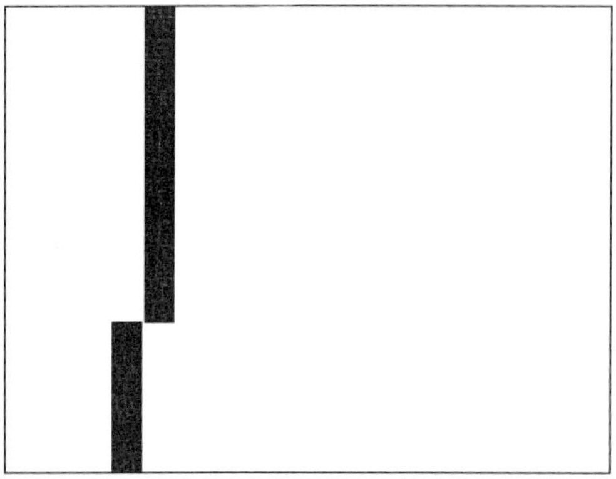

Figure 5.5
Image tearing due to synchronization failure. A single left-to-right moving bar that expands the entire height of the screen is broken due to synchronization failure.

Failures of refresh synchronization can be visualized by programming a black vertical bar on a large image with white background that moves to a new location on each frame (display 5.6). With good synchronization—that is, the transfer of the entire image including the background from the computer memory to the graphics memory is fast enough and occurs in synchrony with the vertical refresh interval—the bar appears to move across the screen intact. With problematic synchronization, the bar may appear to be incomplete or broken. If the display synchronization of a large display fails, it may be corrected by limiting the size of the images displayed to determine the maximum image size that can be correctly synchronized. For graphics cards operating in double-buffered mode, this is less likely to be an issue.

The second important calibration involves the synchronization of images in different color planes or color channels.[2] CRTs ordinarily have three color-channels (R, G, B). In an idealized case, images in the three color-channels would be active at the same time, leading to a fused net color image. One way to see this is to generate a yellow bar ([RGB] = [255 255 0]) that turns on the red and green guns. By programming a yellow vertical bar that moves horizontally across the screen, temporal lags in the color guns may cause the perception of separate red and green bars rather than the single fused yellow bar. Display 5.7 shows a program that displays a moving yellow bar. Some display devices, such as single-chip digital light processing (DLP) displays, often have issues with color synchronization and show visible color "ghosting."

5.2 Current Visual Display Technologies

Most current experimental test setups for visual psychophysics use raster displays, such as the CRT, which paint light in a series of rows. Historically, visual psychophysics used

Display 5.6

```
%%% Program: SynchronizationWithFrameRate.m
function SynchronizationWithFrameRate(pixelsPerFrame)

%% Experimental Module

% Specify the stimulus
p.barwidth = 10; % in pixels
if nargin < 1, pixelsPerFrame = p.barwidth; end
        % shift barwidth pixels per frame

% Make a whole screen texture
img = ones(p.ScreenRect([4 3]));
tex = Screen('MakeTexture', windowPtr, img, 0, 0, 2);

offset = 0;
tEnd = vbl + 30;

% Show the moving bar
while vbl < tEnd
    Screen('DrawTexture', windowPtr, tex);
    Screen('FillRect', windowPtr, 0, [offset 0 offset + ...
           p.barwidth p.ScreenRect(4)]);
    vbl = Screen('Flip', windowPtr);

    offset = offset + pixelsPerFrame;
    if offset + p.barwidth > p.ScreenRect(3), ...
       offset = 0; end
end
```

other display devices to present stimuli, such as tachistoscopes, projectors with shutters and optics, or point-plot or vector-plot displays. Tachistoscopes or slide projectors present either printed materials or slides for controlled times by creating a brief illumination or by briefly opening a shutter. Point-plot or vector-plot display devices activate phosphors only in specific screen locations determined either as a set of individual illuminated points or as a sequence of points along a vector, leaving all other locations inactive. A raster scan display device, as described earlier, is based on television technology, in which an electron beam is moved across the screen, painting pixels from left to right and from top to bottom, one row at a time. Raster displays use either a non-interlaced mode in which each row is painted in turn (see section 5.1.1) or an interlaced mode in which first the odd rows are painted as one image frame followed by even rows in a separate image frame. The interlaced mode is still used in some broadcast television. Interlacing was designed to com-

Display 5.7

```
%%% Program DriftingYellowBar.m
function DriftingYellowBar(pixelsPerFrame)

%% Experimental Module

% Specify the stimulus
p.barWidth = 4; % in pixels
sz = 128; % bar length in pixels
if nargin < 1, pixelsPerFrame = p.barWidth; end
        % shift pixels per frame

% Compute stimulus parameters
[xc yc] = RectCenter(p.ScreenRect);
xy = [[1 1 1 1] * xc; [[-32 -3]-sz/2 -sz/2  sz/2] + yc];

% Show the moving bar
while 1
    Screen('DrawLines', windowPtr, xy, p.barWidth, ...
           [1 1 0]);
    Screen('Flip', windowPtr);
    if ReadKey('esc'), break; end
    xy(1, 3 : 4) = xy(1, 3 : 4) + pixelsPerFrame;
    if xy(1, 3) > xc + sz / 2, xy(1, 3 : 4) = ...
       xy(1, 3 : 4) - sz; end
end
```

pensate for earlier devices with slow transfer of display information to mitigate the appearance that an image appears over an extended time. Display devices currently in use, including most monitors, projectors, and goggles, are raster devices using a non-interlace mode that paints all rows in order.

In this section, we present a brief review of the current state of each display technology and the advantages and disadvantages of each. Attributes of each display type are summarized in table 5.1, followed by a section on each. This table deals with most of the major display technologies used in the laboratory today. We provide this section for individuals who may be setting up a psychophysics laboratory or choosing a new display technology for an existing laboratory. Other readers may choose to skip to section 5.3 to address general issues of calibration and testing.

5.2.1 Monitors: CRT, LCD, LED, OLED

Cathode ray tube[3] (CRT) monitors use a fluorescent screen, activated by a movable electron beam within a vacuum tube, to create images made of light emitted from the fluorescent

Table 5.1
Summary of typical properties of display technologies

Type	Technology	Analog/ digital	Luminance homogeneity	Contrast	Color Depth	Geometric Distortion	Temporal Response	Pixel Independence	Color Synch
Monitor	CRT	A	M	1500:1	E	VG	E	P–M	E
	LCD	D	G	500:1	G	E	P	G–E	E
	LED	D	VG	800:1	VG	E	P	G–E	E
	OLED	D	VG	1000:1	E	E	E	E	E
Projector	LCD	D	G	500:1	G	VG	P	E	E
	DLP	D	VG	2000:1	VG	E	E	E	P
	3-chip DLP	D	VG	2000:1	E	E	E	E	E
Goggles	LCD	D	G	500:1	G	E	P	E	E

A, analog; D, digital; E, excellent; G, good; M, moderate; P, poor; VG, very good.

screen. The chemical compositions of the phosphors painted on the fluorescent screen determine the brightness, color, and the temporal response of the pixels. Color CRTs use three different phosphors packed together around each pixel that correspond to "red," "green," and "blue." Painting an RGB triplet in each pixel with three separate but temporally synchronized electron guns portrays different colors.

CRTs provide good quality displays because they have relatively high numbers of pixels, a range of native (built-in programmable) pixel resolutions, have relatively rapid response times, and achieve good control and quality of colors and color balance. They also are less sensitive to off-axis viewing angles so that the screen can be viewed from a range of side angles and still be seen with little loss of contrast. CRTs are analog devices and often have some issues with homogeneity. Brightness is adequate, and contrast is usually very high. Most phosphors have excellent temporal responses with rapid rise times and relatively short persistence. Some, however, do not. Geometric distortions in flat-panel CRTs can be adjusted to be minimal. Pixel independence can be achieved with special-purpose, high-bandwidth monitors.

A liquid crystal display[15] (LCD) is a display that is based on the light-modulating properties of liquid crystals. Liquid crystals do not emit light directly. LCDs consist of pixels filled with liquid crystals in front of a light source or light reflector, called the *backlight* where the current state of the crystals modulate or filter the light available to the viewer to create either color or monochrome images.

At the current time, there are four major versions of LCD technology that differ in the details of arrangement of the liquid crystals. These are the twisted nematic (TN), in-plane switching (IPS), multi-domain vertical alignment (MVA), and pattern vertical alignment (PVA) technologies. LCDs are lightweight, energy efficient, and are now the dominant display method in computers. However, as devices for visual psychophysics, they have many limitations:

1. The pixel response time for many LCD displays is more than 20 ms. In comparison, the refresh period for a 60-Hz monitor is 16.7 ms. In rapid, dynamic displays, the long pixel response time manifests as low-contrast ghosting, in which a previous image appears as a low-contrast but visible component of the current image.

2. To make things worse, the pixel response time depends on the brightness of the pixel, so the pixel response time is not a constant, but differs depending upon the image and the brightness/contrast settings of the monitor.

3. Another problem for LCD displays in visual testing is the dependence on viewing angle. Pictures look different when seen from different viewing angles, whereas there is relatively little sensitivity to viewing angle in CRTs. Although this problem has been moderated in more recent LCD displays, it may still be significant for purposes of visual testing.

4. Because the LCD matrix is a passive device that only modulates the output of a backlight unit, and because the LCD matrix is not fully opaque, some light is seen even in a

"black" state, showing not a deep black color, but only dark gray. For this reason, LCDs often have a limited contrast ratio of 200–300 to 1.

5. Contrary to the claims of some manufacturers, many LCDs are limited in the number of colors they can depict by a 6-bit resolution in each of three color-channels.

6. In many cases, the backlight is not strictly uniform across the display, and so neither is the LCD modulated light.

On the positive side, LCD displays have excellent geometric properties and good pixel independence. For this reason, LCDs may be a good option in studies with static or slowly varying stimuli in which the details of the brightness and contrast are unimportant. These conditions may hold in many cognitive or memory experiments, for example. Additionally, special-purpose LCD technology for visual psychophysics is under development.[17]

Light emitting diode[18] (LED) displays are essentially LCDs that use semiconductor light sources in backlighting rather than the fluorescent backlights more often used in LCDs. The LCD technology is used in both to filter the light that appears on the display panel. Although sharing many of the disadvantages of the conventional LCD, the LEDs improve spatial uniformity across the display surface. Certain LED displays also can reduce the backlighting source in dark regions to achieve a darker black, and so increase the achievable contrast ratio. If RGB LEDs are used, then this may also improve the representation of color relative to the standard LCD.

Another emerging display contender is the organic light emitting diode[19] (OLED) display. It uses organic carbon–based compounds that emit red, green, and blue lights when stimulated by electric current, and so emits light rather than filtering a backlight. It uses much less power and can be made extremely thin. It has a rapid pixel response time, improves color representation over the LCD, and achieves a high contrast ratio. The current technology has several issues. It has limited pixel resolution per inch, is expensive, and has a limited life expectancy. Although not yet ready for general use, the OLED technology is interesting and may become a contender for visual psychophysics in the future.

5.2.2 Projectors: LCD, DLP, Three-Chip DLP

Projectors are used as preferred displays in some circumstances. For example, they are often used when large-scale displays are desirable, for group demonstrations, and in functional imaging situations. The most commonly used projection technologies are LCD and DLP devices.

The display quality of LCD projectors shares many characteristics with that of LCD monitors, including good geometric properties, but also shares the weaknesses in timing and color representations.

Digital light processing[20] (DLP) projectors are optical semiconductor systems that operate as sophisticated light switches. The physical pixel plane, or x–y plane, inside the DLP is populated with millions of hinge-mounted microscopic mirrors that control the amount of the light emitted for each pixel. The amount of light is controlled by switching each mirror on—toward the projection surface—or off extremely rapidly, and the relative amount of on to off in a rapid time cycle controls the gray scale up to 1024 gray levels.

Filtering white light through red, green, and blue filters operating in rapid sequence creates image colors. The single-chip DLP projection systems create many millions of colors by mixing different proportions of on and off mirror states of this colored light rapidly alternated over time. In a single-chip DLP, only one color filter operates at any given instant, so the components of the RGB image cannot occur strictly simultaneously. Another issue with the single-chip DLP technology is that the synchronization of the color filters and the mirrors can cause visible color separation due to slips in synchronization. For example, when a yellow light is programmed, the R and G components may become visible separately (see section 5.1.7).

A three-chip DLP uses three independent DLP chips to reflect light from red, green, and blue light sources to achieve good synchronization and produce about 12 bits per component, leading to excellent depiction of more than 35 trillion colors. However, the three-chip DLP projection systems are quite expensive, and so are only likely to be used in commercial applications such as large theaters or in specialized laboratories.

The contrast ratio of a DLP can be as high as 2000 to 1. It has excellent luminance homogeneity. Color depth or resolution is excellent, especially with three-chip DLPs. Geometric distortion and pixel independence are excellent. Temporal resolution is excellent in three-chip DLP, but may be an issue in single-chip DLP projection systems.

5.2.3 LCD Goggles

LCD goggles have two small LCD monitors with high pixel per inch resolution embedded in the eyepieces. By virtue of being close to the eye, they can present a large visual field in a highly portable format. Often used as part of gaming systems, they may also be used in functional imaging settings, as portable display devices, and to display stereo stimuli as different images may be presented to the two eyes simultaneously. Goggles are sometimes used to display ongoing modified or virtual visual input in real time. The properties of the LCD goggle are essentially identical to those of the LCD monitor. However, goggles can be quite difficult to calibrate. In fact, calibration typically requires the matching by the human eye of images in the goggles and images on well-calibrated monitors.

5.2.4 Three-Dimensional Displays

Three-dimensional displays present two nearly identical 2D images with a horizontal offset to the left and right eyes separately (see section 3.2.6 of chapter 3). The brain combines the two monocular images to generate the perception of 3D depth. Displaying different images to the two eyes is done in four major ways: (1) using goggles to present different images to the two eyes, (2) using optics (e.g., prism glasses, stereoscopes) to present different images to the two eyes, (3) using filters (e.g., complementary color anaglyphs, polarized glasses) or shutter-glasses to present different images to the two eyes, and (4) generating images for the two eyes on the same display device in different locations and displaying them to different eyes through a physical projection scheme.

The diagram in figure 5.6 shows a four-mirror stereoscope,[21] a setup often used in the laboratory. The images for the left and right eyes are presented spatially separated on the display device and delivered to each eye through half-silvered, front-coated 45° mirrors, where they are fused in the percept of the observer. The distance between the centers of the two inner mirrors should match the distance between the two eyes. One way to check this setup is to create a device with two parallel laser pointers that are separated by the same distance as the eyes and that reverse the optical path. When the two pointers are placed in the stereoscope in the position of the eyes, the points projected on the display screen should be centered in the image for each eye.

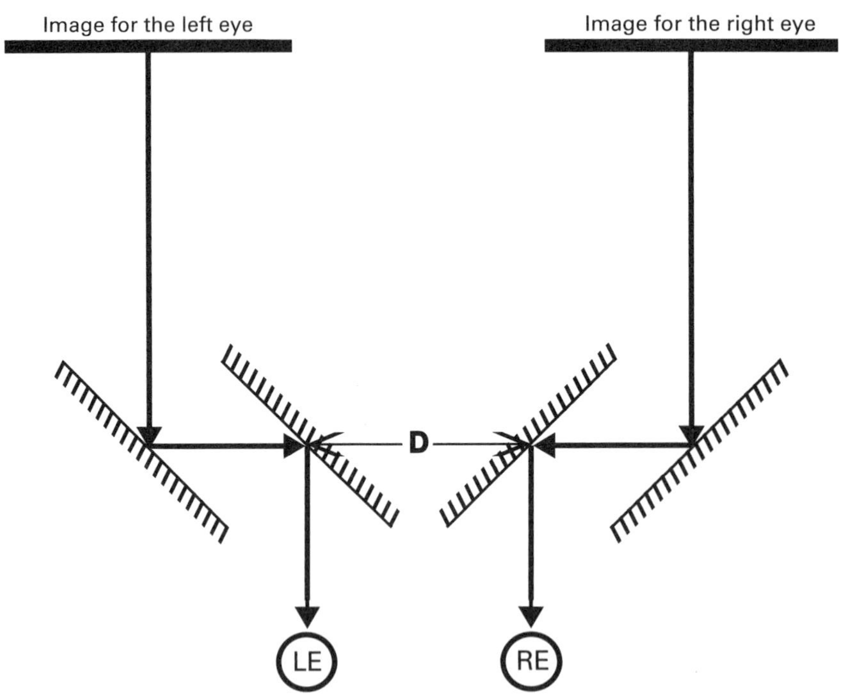

Figure 5.6
A four-mirror stereoscope for displaying separate images to the two eyes.

In addition to considering all the display properties relevant for 2D displays, one major concern for 3D display technologies based on filtering and directional lights is ghosting. In this context, ghosting reflects blending of the two monocular images and so failure to segregate the images to the two eyes. The presence of ghosting may be important in many experiments and should be evaluated.

5.3 Luminance Calibration

The basic input to the visual system is light. All of our knowledge about the properties of the visual system is predicated on careful measurements of the visual response to different patterns of light. It is therefore critical for psychophysical experiments to represent accurately the patterns of light in the stimulus. This requires the measurement of light and color in our display devices.

First, we need to know how to measure light and color. Then, systematic measurements of display devices generate mathematical descriptions of the display through proper calibration. It is then possible to represent faithfully any new visual stimuli based on the known mathematical properties of the display rather than on individual physical measurements of every stimulus.

We focus on the issues and methods of calibration for monitors and projection devices.

5.3.1 Photometric Measures

Luminance is a description of light intensity emitted from a surface area for a particular viewing angle. Often, for visual experiments the viewing angle is assumed to be essentially direct, or perpendicular to the plane of the screen. Measured as candelas per square meter (cd/m^2), luminance characterizes the light intensity received by the eye looking at the surface from the angle of measurement. The maximum luminance from a computer display is typically between 50 and 300 cd/m^2, whereas the maximum luminance for a projection system is typically between 100 and 2000 cd/m^2. In comparison, the sun in direct viewing at noon has a luminance of about 1.6×10^9 cd/m^2.

Luminance is typically measured in the laboratory with a photometer, an instrument that is designed to measure light intensity from surfaces using photoresistors, photodiodes, and/or photomultipliers. To mimic human vision, photometers either assume a filter or allow you to choose from a small number of filters that represent the light sensitivity of the human visual system. These filters focus on the spectral regions or wavelengths of visible light. To measure the luminance of a display, the experimenter arranges to show uniform stimuli on the display device, aims the photometer from the intended viewing angle, and collects measurements. Each photometer is operated in a slightly different way that will be described in instructions from the manufacturer.

5.3.2 Gamma Calibration

The relationship between the programmed intensity values in computer or graphics memory and actual output luminance is almost never linear. Historically, early CRTs had nonlinearities or *Gamma constants* of 2 or a bit higher, rather than 1 (see later). Broadcast television took this into account in their image design, and so modern display devices that could easily be linear typically have purposely introduced Gamma nonlinearities into the displays.

In consequence, if we want to know output luminance exactly, we must first measure the nonlinear relationship between pixel gray level (U) specified in the computer graphics card and output luminance on the display. The procedure for measuring and correcting for this function is called *Gamma correction*. For most displays, the relationship between output luminance and the programmed intensity U (figure 5.7) can be described as

$$L(U) = L_{\min} + (L_{\max} - L_{\min}) \times U^{\gamma}, \tag{5.1}$$

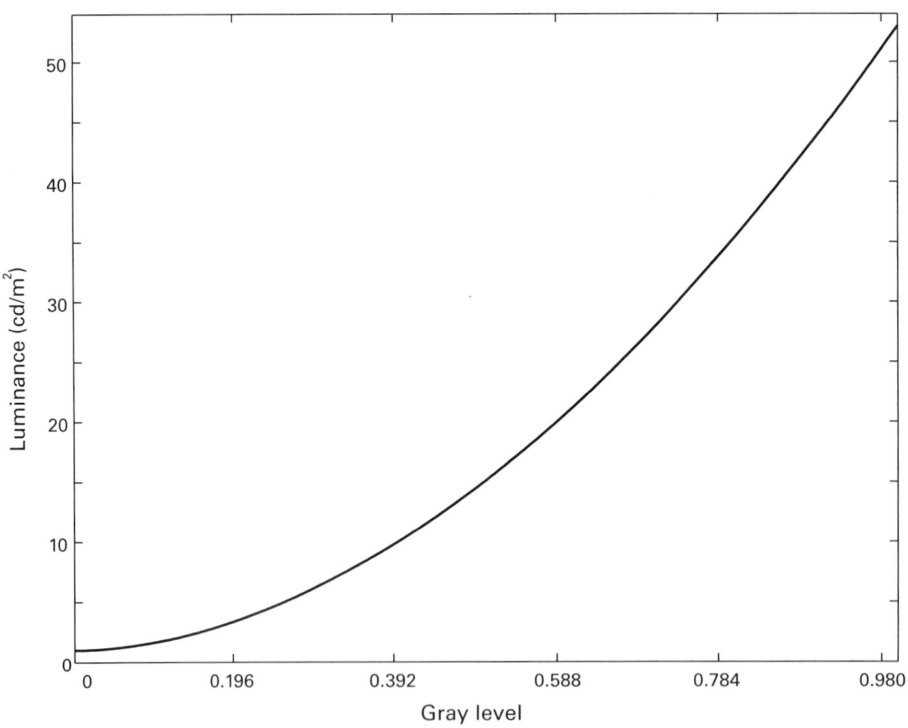

Figure 5.7
A Gamma function with $L_{\min} = 1$ cd/m^2, $L_{\max} = 53$ cd/m^2, and $\gamma = 1.9$.

where L_{min}, L_{max}, and γ are determined either by using photometers or by a combination of psychophysical procedures and photometric measurements.[22,23] The constant γ is called the *Gamma constant*.

Display 5.8 provides a program that allows you to set the RGB value for a displayed square at a starting value, measure the luminance with the photometer, and then increase the value of one or more guns and measure again. The results of many such measurements are fit with equation 5.1 to obtain the characteristic γ of the display. This program for calibration includes calls to functions that set up the display using `PsychImaging`. These calls were treated in the discussion about and comments for display 4.1 in chapter 4.

The human eye is even more sensitive than the photometer in discriminating luminance, so photometric calibration can be augmented with calibration "by eye." We use a bisection matching procedure to match successively different proportional values of output luminance (figure 5.8). The eye combines equal mixtures of maximum and minimum luminance pixels in one mixed intensity patch—through temporal and spatial

Display 5.8

```
%%% Program PhotometricTest.m
function PhotometricTest

% Get frame rate and set screen font
p.ScreenFrameRate = FrameRate(windowPtr);
Screen('TextFont', windowPtr, 'Times');
Screen('TextSize', windowPtr, 24);
%% Experimental Module
%Specify the stimulus
bits = 4;           % determine the minimum step, could be
                    % changed by the user
val = round(0.5 * 2 ^ bits) / 2 ^ bits;
                    % initial gray value, may change
rgb = [1 1 1];      % channels to test; can be changed by
                    % the user using the r, g and b keys
keys = {'down' 'up' 'space' '+' '=' '-' 'r' 'g' 'b' ...
        'f1' 'esc'}; % Define control keys

% Help instruction to change parameters, and to turn
% on/off instruction
hStr=['Up / Down arrow keys: increase / decrease value\n'...
      '+ /  - : incease / decrease bits\n', ...
      'r g or b: turn a channel on / off\n' ...
      'F1: Show / hide this key help information\n', ...
      'ESC: Exit\n'];
```

Display 5.8 (continued)

```
showHelp = 1; % whether to show help text at beginning
showVals = 1; % whether to show values text at beginning

% Show the full screen for test, allow changing parameters
while 1
    Screen('FillRect', windowPtr, val * rgb);
        % draw the value

    texColor = mod(val + 0.45, 1);
        % make sure text is visible
    if showVals
        str = sprintf('Value = %.*f;', …
              ceil(log10(2 ^ bits)), val);
        str = sprintf('%s  Bits = %g;', str, bits);
        str = sprintf('%s  RGB = [%g %g %g]', str, rgb);
        Screen('DrawText', windowPtr, str, 4, 4, texColor);
    end

    if showHelp
        DrawFormattedText(windowPtr, hStr, 4, 64, …
             texColor, [], 0, 0, 1.5);
    end
    Screen('Flip', windowPtr);

    KbReleaseWait; % wait until all keys on the keyboard
                   % are released
    key = WaitTill(keys);      % wait for defined key press

    % Change parameters according to key press
    if     strcmp(key, 'down'),  val = val -2 ^ -bits;
    elseif strcmp(key, 'up'),    val = val +2 ^ -bits;
    elseif strcmp(key, 'space'), showVals = 1 - showVals;
    elseif strcmp(key, '+'),     bits = bits + 1;
    elseif strcmp(key, '='),     bits = bits + 1;
    elseif strcmp(key, '-'),     bits = bits - 1;
    elseif strcmp(key, 'r'),     rgb(1) = 1 - rgb(1);
    elseif strcmp(key, 'g'),     rgb(2) = 1 - rgb(2);
    elseif strcmp(key, 'b'),     rgb(3) = 1 - rgb(3);
    elseif strcmp(key, 'f1'),    showHelp = 1 - showHelp;
    elseif strcmp(key, 'esc'),   break;
    end

    val = round(val * 2 ^ bits) / 2 ^ bits;
    val = max(0, min(1, val));      % within 0 and 1
    bits = max(2, min(16, bits));  % 4 to 16 bits
end
```

blur. This physically defines the halfway point in luminance output. The experimenter then measures which single programmed grayscale level in another patch visually matches this mixture. This halfway point can then in turn be bisected above and below, and so on, until one has measured the function relating programmed intensity to actual output luminance from minimum to maximum. These data are fit with the function in equation 5.1 to describe the relationship. Display 5.9 shows the corresponding visual Gamma calibration program.

Most visual experiments in fact define stimuli in terms of contrast on a background with luminance L_0 halfway between L_{min} and L_{max}, rather than by luminance values. The following equations define the relationship between the contrast $c(U)$ and gray level U (0 to 1) as

$$c(U) = \frac{L(U) - L_0}{L_0} = \frac{L_{max} - L_{min}}{L_{max} + L_{min}}(2U^{\gamma} - 1). \tag{5.2}$$

This equation can be inverted to solve for the programmed grayscale level for a desired contrast:

$$U = \left(\frac{c+1}{2} \times \frac{L_{max} + L_{min}}{L_{max} - L_{min}} \right)^{1/\gamma}. \tag{5.3}$$

Although U can take a continuous value from 0 to 1 in the equation, only a subset of values can actually be displayed because of the color resolution of the display device. The achievable values of U for each device are quantized. For example, for an 8-bit display, only 256 values between 0 and 1 are possible. U must be rounded to the nearest achievable value.

The *Gamma correction* parameter γ in equation 5.3 is a characteristic of a particular display device that should be incorporated into most experimental running programs to "linearize" or correct the output luminance. Failure to calibrate a display device for the Gamma correction can lead to distortions in the luminance profile of stimuli; for example, an incorrect γ could alter an intended sine-wave profile into something closer to a square-wave luminance pattern. The γ of a display device is likely to change when device settings are changed. It also can change with the aging of the device. The display device should be calibrated periodically, and in some very demanding experiments should be calibrated very frequently. Some display devices may need to be turned on for a period of stabilization, usually a few minutes, prior to calibration or experimental running.

5.4 High-Resolution Luminance

The human visual system is highly sensitive. The human eye can sometimes be sensitive to contrasts on the order of 0.1%, or 0.001.[24–26] To display a sine wave with a peak contrast of 0.1% requires a gray-level resolution of at least 0.02%, or 1 part in 5000. This

Display 5.9

```
%%% Program CalibrateGamma.m
function CalibrateGamma (rgb)

% Set screen font
fontsz = round(36 / 1024 * p.ScreenRect(4) / 2) * 2;
Screen('TextFont', windowPtr, 'Times');
Screen('TextSize', windowPtr, fontsz);
% Control keys
keys={'left' 'right' 'up' 'down' 'return' 'enter' 'esc'};
% Instruction 1
str = ['View the display from a distance so that you do '…
       'not see any stripes.\n\n' 'Adjust the luminance' …
       'of the middle part until it matches that of the' …
       'flankers.' …
       '\n\n' 'Press an arrow key to continue.'];
DrawFormattedText(windowPtr, str, 100, 200, 1, 48, 0, 0, 1.5);
Screen('Flip', windowPtr);
WaitTill(keys(1 : 4)); % wait for a key to start

% Instruction 2
str = ['Use Left/right arrow keys for coarse adjustments,\n'…
       'and Up/down keys for fine adjustments.\n' …
       'Press Enter key when you are done'];
DrawFormattedText(windowPtr, str, 100, 200, 255, [], 0, 0,…
       1.5);
Screen('Flip', windowPtr, 0, 1);

%% Experimental Module
% Specify the stimulus
sz = 128; % size of the middle patch in pixels

% Compute and set up the test stimulus
lum = linspace(0, 1, 17);
vol = lum .^ (1 / 2.2);       % initial guess of Gamma is 2.2
hI = [17 17 9 17 13 9 5 17 15 13 11 9 7 5 3];
                              % high value index
lI = [1 9  1 13 9 5  1 15 13 11 9 7 5 3 1]; % low value index
step = [0.3 3] / 255;         % fine and coarse steps
img = zeros(sz, sz * 3);
rect = CenterRect([0 0 1 1] * sz, p.ScreenRect);
tex = zeros(1, 2);

% Loop through 15 steps in a fixed order
for i = 1 : 15
```

Display 5.9 (continued)

```
    high = vol(hI(i));
    low = vol(lI(i));
    ind = (hI(i) + lI(i)) / 2;
    img(:) = high; img(2 : 2 : sz, :) = low;
    tex(1) = Screen('MakeTexture', windowPtr, img, 0, 0, 2);
    tex(2) = Screen('MakeTexture', windowPtr, …
              img([2 : sz 1], :), 0, 0, 2);
    mid = vol(ind);          % start with guessed value

    while 1
        KbReleaseWait;        % avoid continuous change
        while 1
            for j = 1 : 2    % generate flickering flankers
                Screen('DrawTexture', windowPtr, tex(j));
                Screen('FillRect', windowPtr, mid, rect);
                Screen('Flip', windowPtr, 0, 1);
            end
            key = ReadKey(keys);
            if ~isempty(key), break; end
        end
        switch key
            case 'left',     mid = mid - step(2);
            case 'right',    mid = mid + step(2);
            case 'up',       mid = mid + step(1);
            case 'down',     mid = mid - step(1);
            case 'esc',      sca; error('ESC pressed.');
            otherwise % return or enter
                if i == 1
                    guessGamma = log(0.5) / log(mid);
                    vol = lum .^ (1 / guessGamma);
                else
                    vol(ind) = mid;
                end
                Beeper;
                break; % go to next step
        end

        if mid > 1 || mid < 0
            Beeper; % out of range warning
            mid=max(0, min(mid, 1));
        end
    end
end
```

Display 5.9 (continued)

```
% Fit Gamma
costfunc = @(x) sum((lum .^ (1 / x) - vol) .^ 2);
gamma = fminsearch(costfunc, guessGamma);
rsq = 1 - costfunc(gamma) / var(vol, 1) / 17;
% plot the result
figure(3);
x = linspace(0, 1, 256);
plot(x, x .^ gamma, 'r'); hold on;
plot(vol, lum, 'o'); hold off;
xlabel('Normalized Output');
ylabel('Normalized Luminance');
text(0.2, 0.9, sprintf('RGB = [%g %g %g]', rgb));
text(0.2, 0.8, sprintf('Gamma = %.3g', gamma));
text(0.2, 0.7, sprintf('Rsq = %.4g', rsq));
```

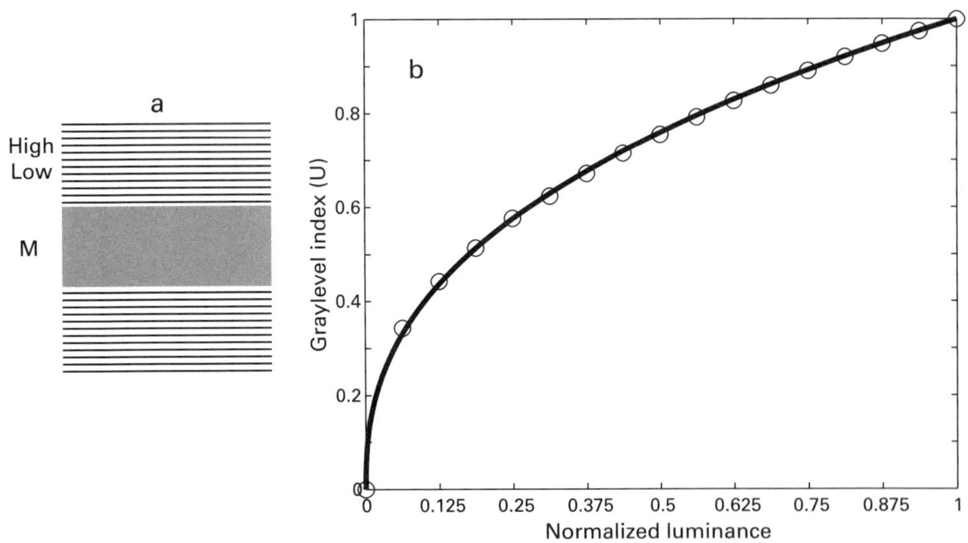

Figure 5.8
Calibration by eye. (a) In the bisection procedure, the single luminance value of M is adjusted to match that of the physical mixture of luminance values L and H. (b) The resulting matching values and Gamma fit for a CRT display. Note that the axes are flipped relative to figure 5.7 because here gray level U is the measured variable.

corresponds to about 12.4 bits of gray-level resolution. However, typical computer graphics cards provide only 8-bit intensity resolution per channel. Certain graphics cards provide 10-bit intensity resolution per channel. Many experiments in visual psychophysics require higher-resolution control of gray or color levels. Several different approaches have been developed to achieve high gray-level resolution in display systems. We review a number of methods and issues in some detail.

For analog display devices such as the CRT, the limiting factor is the bit resolution (pixel depth) of the digital to analog conversion of the video card. A higher-resolution card or a combination of several channels of a lower-resolution card can be used to deliver a net higher resolution to the display device.

For digital display devices such as LCD monitors, achieving high gray-level resolution requires both a high-resolution video card and a high gray-level resolution display, such as a 10- or 12-bit LCD display.

Here, we focus on the input to the display device. Display 5.10 provides a program that uses a number of options to show a luminance ramp that uses high gray-level resolution. It displays a series of rectangles that step up the luminance from the lowest to the highest programmed value in fine steps. The function `PsychImaging` is used to set up and enable display options for testing. Some display options are described in more detail in the following sections. The sample program uses the native capability of the graphics device, such as a 10-bit graphics display. A number of other software and hardware options may be used to improve resolution. We provide syntax in the sample program for these different approaches to high grayscale resolution in the lines that are commented out. Only one option at a time can be implemented in a running program.

5.4.1 Software Methods

Bit stealing is one method to display monochrome images of higher grayscale resolution. Bit stealing adds small increments or decrements ("delta" values) to the different color channels. Usually, monochrome gray levels are displayed on color monitors through setting R, G, and B guns to the same value. Bit stealing[27] creates additional gray levels by adding one quantization level to the value in a subset of the three guns. This effectively adds 2.7 bits of gray-level resolution to achieve 10.7 bits on an 8-bit graphics card, or 12.7 bits on a 10-bit card. Although the RGB values are no longer exactly equal, the color changes are so subtle as to be essentially invisible to the eye.

Noise-bit is an easy-to-implement method that uses random dithering to generate high gray-level resolution.[28] To display a gray-level intermediate to two adjacent quantized luminance levels, the method randomly chooses between those two nearest values with the relative probabilities over time determined by proportional distance between two adjacent quantized gray level values. For example, if the desired gray level is 35% of the way between two adjacent quantized values, the two values are programmed with probabilities of 0.65 and 0.35, respectively.

Display 5.10

```
%%% HighResolutionLuminance.m
function HighResolutionLuminance(rgb)

%% Display Setup Module

% Define display parameters
if nargin < 1, rgb = [1 1 1]; end
       % set the color channels to test
whichScreen = max(Screen('screens'));
p.ScreenGamma = 2;  % from monitor calibration
p.ScreenBackground = 0;

% Open display window and hide the mouse cursor
if exist('onCleanup', 'class'), …
      oC_Obj = onCleanup(@()sca); end
       % close any pre-existing PTB Screen window
PsychImaging('PrepareConfiguration');
PsychImaging('AddTask', 'General', …
      'FloatingPoint32BitIfPossible');
        % set up a 32-bit framebuffer
PsychImaging('AddTask', 'General', …
      'NormalizedHighresColorRange');
% Options for different hardware or software solutions.
% PsychImaging('AddTask', 'General', …
      'EnableNative10BitFramebuffer');
% PsychImaging('AddTask', 'General', …
      'EnablePseudoGrayOutput');
% PsychImaging('AddTask', 'General', …
      'EnableVideoSwitcherSimpleLuminanceOutput', 126.3);
% PsychImaging('AddTask', 'General', …
      'EnableDataPixxM16Output');
% PsychImaging('AddTask', 'General', …
      'EnableBits++Color++Output');
PsychImaging('AddTask', 'FinalFormatting', …
      'DisplayColorCorrection', 'SimpleGamma');
[windowPtr ScreenRect] = PsychImaging('OpenWindow', …
      whichScreen, p.ScreenBackground);
PsychColorCorrection('SetEncodingGamma', windowPtr, …
      1 / p.ScreenGamma);
HideCursor;
% Set screen font
Screen('TextFont', windowPtr, 'Times');
Screen('TextSize', windowPtr, 24);
% Control color channel by alpha blending
Screen('BlendFunction', windowPtr, 'GL_ONE', 'GL_ZERO', …
      [rgb 1]);
```

Display 5.10 (continued)

```
Screen('Flip',windowPtr);

%% Experimental Module

% Make two textures, one 32-bit, the other 8-bit
n = ScreenRect(4) - 100;  % leave 100 pixel for instruction
                          % display
img = linspace(0, 1, n * 4)';
img = [img(1 : n) img((n : -1 : 1) + n) img((1 : n) + …
      2 * n) img((n : -1 : 1) + 3 * n)];
m = floor(ScreenRect(3) / 4);
img = Expand(img, m, 1);
tex(1) = Screen('MakeTexture', windowPtr, img, [], [], 2);
                         % 32-bit
tex(2) = Screen('MakeTexture', windowPtr, uint8(img * …
                255));   % 8-bit
dstRect=[0 100 m*4 ScreenRect(4)];
txt1 = sprintf('ESC to exit. Space to toggle. Gamma:' …
      '%.2g\n\n', p.ScreenGamma);
% Show the ramp
highBit = 0;
while 1
    if highBit, txt2 = ' Maximum bits';
    else txt2 = ' 8 bits';
    end
    % Draw the 32- or 8-bit texture
    Screen('DrawTexture', windowPtr, tex(2 - highBit), …
          [], dstRect, [], 0);
    DrawFormattedText(windowPtr, [txt1 txt2], 30, 0, 1);
                          % instruction
    Screen('Flip', windowPtr);

    % Wait for space or esc
    key = WaitTill({'space' 'esc'});

    if strcmp(key, 'esc')
        break;
    else
        highBit = 1 - highBit; % toggle between the 32-
                               % and 8-bit textures
        KbReleaseWait;
    end
end

%% System Reinstatement Module
sca; % close window and textures,
     % restore color lookup table
```

The method is equivalent to displaying constant luminance intensities plus a certain amount of noise. Psychophysical testing using standard spatiotemporal resolution (60 Hz and 1024 × 768 pixels) has found that the noise-bit method yields the same results for contrast threshold as other methods, and the temporal flicker is not visible.[28] When the spatiotemporal resolution is high enough, this method, combined with the 8-bit 256 luminance levels, is perceptually equivalent to an analog display with a continuous luminance intensity resolution.[28]

5.4.2 Attenuators and High-Resolution Devices

At present, most computer graphics cards have three 8-bit digital to analog converters (DACs), each capable of generating 256 (2^8) quantized voltage levels. If the CRT is capable of displaying all 256 voltages as different luminance levels, the resolution of the display system is 256 levels of red, 256 levels of green, and 256 levels of blue. If we generate monochromatic displays following the common practice of setting the red, green, and blue channels to the same voltage, the system can provide 256 levels of monochromatic grayscale luminance.

In the next few paragraphs, we briefly describe several currently available physical devices that have been developed to present monochrome images with high gray-level resolution.

For a *monochrome attenuator* system, a passive resistor network was developed to attenuate the outputs of two of the three DACs and recombine the attenuated signals into an analog signal to drive monochrome CRTs.[29] The device combines analog outputs of the R (V_R) and B (V_B) channels of the computer graphics card with weights determined by resistors R_1 and R_2:

$$V_+ = \frac{R_2}{R_1 + R_2} V_R + \frac{R_1}{R_1 + R_2} V_B. \tag{5.4}$$

When $R_1 >> R_2$, the attenuated R signal provides the "fine" resolution. This video attenuator is capable of generating video signals with up to 16 bits of gray-level resolution, satisfying the needs of most visual psychophysical experiments.

One issue with this attenuator is that high-quality monochrome CRTs are very hard to find and very expensive; color CRTs are easier to find and less expensive. For this reason, a video switcher that modifies the outputs of conventional computer graphics cards to generate high luminance resolution monochromatic displays on color monitors was developed.[16,30] The design incorporates an attenuator to generate a single-channel high-resolution video signal and then uses amplifiers to duplicate the same signal to drive the three RGB channels of color monitors. This special device design also includes a pushbutton to switch between a normal color mode and a high gray-level resolution monochrome mode as well as a trigger output that can be used to synchronize other equipment (e.g., fMRI recording) to the video signal.

Cambridge Research Systems Ltd. developed BITS++, a high-speed, real-time digital video processor.[31] The "BITS++" technology increases the color resolution from 8 bits to 14 bits per RGB channel through a 14-bit DAC system. It also includes an integrated digital I/O interface, which can be used to synchronize and control external equipment such as eye movement recorders and electrophysiologic amplifiers.

The DATAPixx data acquisition and graphics system[32] provides high-performance video, analog I/O, digital I/O, and audio I/O, all with microsecond synchronization to the video refresh. It provides 16 bits of resolution per channel and synchronization capabilities.

5.5 Color Description and Calibration

The purpose of color calibration is to specify the color of a stimulus in a color space. All color spaces are intrinsically multidimensional. The two major color spaces are the International Color Standards (Commission Internationale de l'Éclairage; CIE) space and the Derrington–Krauskopf–Lennie (DKL) space.[33] The CIE space approximates the excitation in cone types—long, medium, and short wave, or red, green, and blue sensitive. The DKL space represents color opponent systems that code red–green, yellow–blue, and luminance (black–white) responses. DKL is used in many experiments that study isoluminant color perception because it separates color variations from luminance variations.

Colors as presented on display devices are programmed values of the intensities in RGB channels. In this section, we illustrate how to convert programmed RGB values in computer memory into the corresponding descriptions of the stimulus in CIE or DKL spaces; or, conversely, how to translate a CIE or DKL specification of a desired color stimulus into programmable RGB values on a display device. The diagram in figure 5.9 summarizes the steps and equations that are explained in the next several sections.

5.5.1 International Color Standards (CIE)

Human color vision derives from the responses of three types of cones in the retina: long wavelength ("red"), medium wavelength ("green"), and short wavelength ("blue"). Therefore, colors are described in three dimensions. Color stimuli on computers are rendered as (RGB) triplets, red, green, and blue, roughly corresponding to the stimulation of the three types of cones. It is no accident that the design of display devices capitalizes on the principles of human vision.

In 1932, the CIE[34] established a system to specify colors based on human vision (figure 5.10, plate 7). The CIE chromaticity space is organized in two dimensions that incorporate hue and color saturation or color purity—as well as variations in luminance. There is a white point near the center of the chromaticity space. The hue is represented by the angular position relative to the white point, and the color purity corresponds with the radial distance from the white point normalized by the maximum radial distance for that hue.[35]

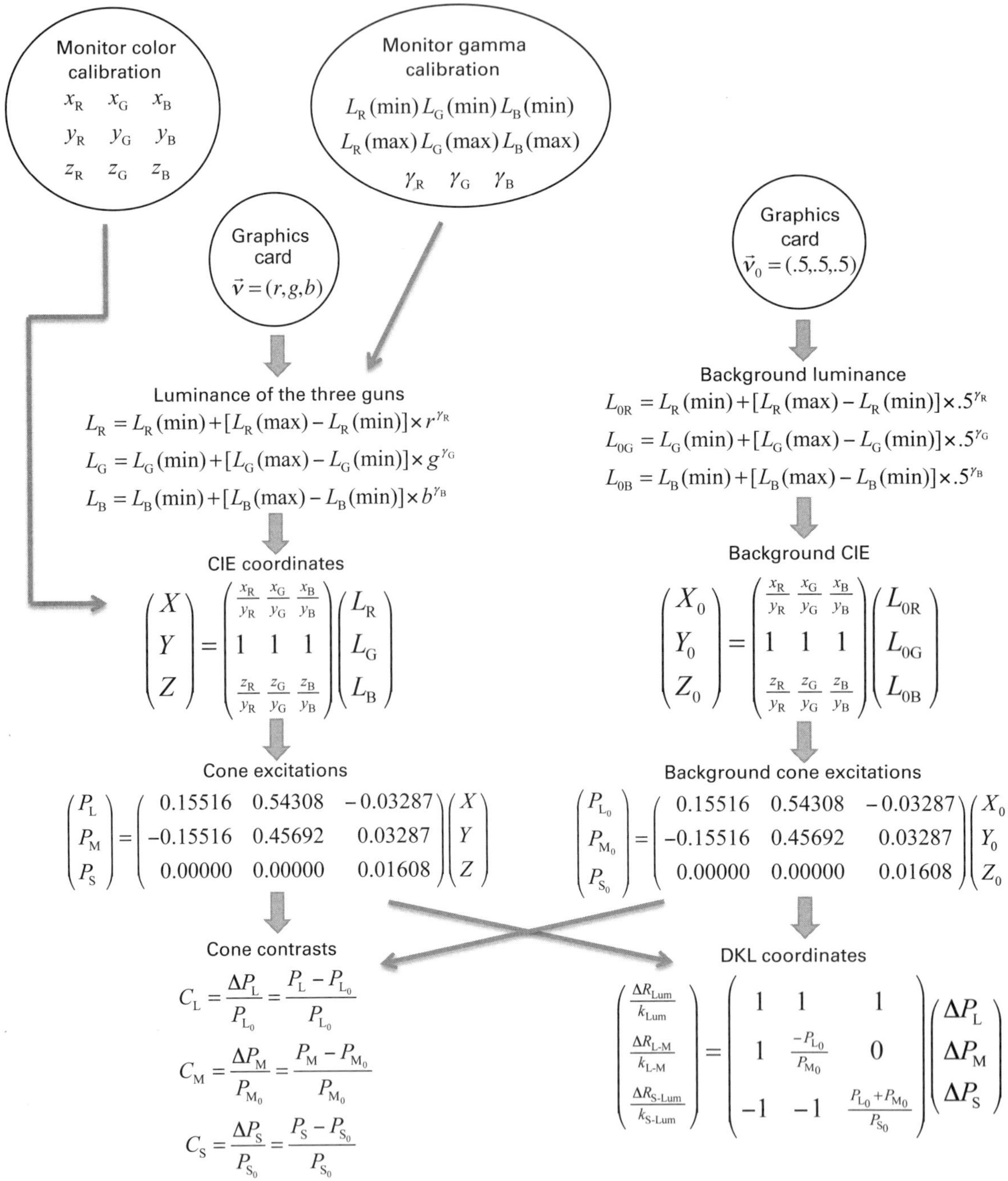

Figure 5.9
Diagram of the components, steps, and equations that convert RGB values from a graphics card into expected cone excitations and coordinates in DKL color space.

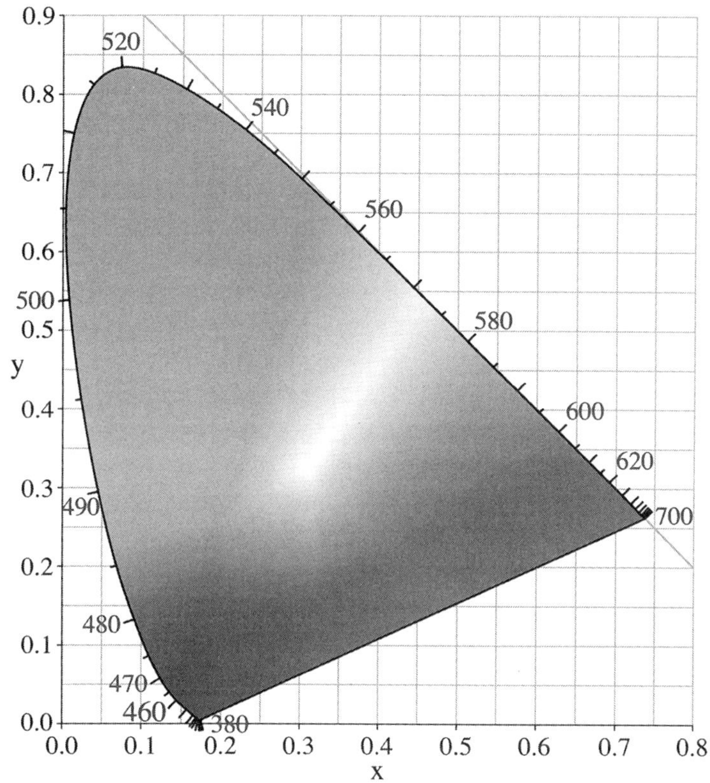

Figure 5.10 (plate 7)
CIE 1931 *xy* chromaticity diagram. The diagram shows the positions of different chromaticities in the CIE space, with positions varying in hue and in saturation or depth of color. The positions of pure wavelength lights are indicated around the edges of the color manifold, and a white point is in the middle.

When the CIE standards were established, spectral sensitivities of the cone photoreceptors in the eye were not known independently but were inferred from behavioral study of perceived colors. Color matching experiments showed that a light of any pure wavelength λ could be perceptually matched by a weighted linear combination of three wavelengths—700.0 nm, 546.1 nm, and 435.8 nm—mixtures of three primary colored lights.[34] Color matching functions graph the weights on each of the three primary colors required for a perceived color match as a function of the pure wavelength to be matched. These functions are analogous with the spectral sensitivity of the three cone types to different wavelengths. The resulting functions for a standard observer are $\bar{x}(\lambda)$, $\bar{y}(\lambda)$, and $\bar{z}(\lambda)$, which represent the weights of the three primary colors. These functions are one description of the sensitivity of the observer to different wavelengths of light; the functions depend on the choice of the three primary light colors. The wavelengths of the three primary colors are approximately equal to the wavelengths that generate the maximum responses from the three cone types in the human eye.

Every physical stimulus has a spectral radiance distribution $W(\lambda)$ that determines the perceived color. Because the percept of any *single* wavelength stimulus can be matched by a linear combination of three selected wavelengths, any *distribution* of wavelengths can also be matched. The CIE space assigns each stimulus to an (*X*, *Y*, *Z*) value. It does this by weighting the different wavelengths in the stimulus by the color matching functions [$\bar{x}(\lambda)$, $\bar{y}(\lambda)$, and $\bar{z}(\lambda)$] of a "standard" CIE observer and correcting by a constant K_{m} called the maximum photopic luminance efficacy.

The equations are

$$\begin{aligned} X &= K_{\mathrm{m}} \int W(\lambda)\bar{x}(\lambda)d\lambda \\ Y &= K_{\mathrm{m}} \int W(\lambda)\bar{y}(\lambda)d\lambda \\ Z &= K_{\mathrm{m}} \int W(\lambda)\bar{z}(\lambda)d\lambda. \end{aligned} \tag{5.5}$$

So, if the color matching functions of the CIE observer, the constant, and the spectral distribution of the image are known, you can compute the color of the stimulus in (*X*, *Y*, *Z*) space. Images with different spectral radiance distributions but the same tri-stimulus descriptions are perceptually color matched for the standard CIE observer.

The XYZ system based on the CIE space combines hue and luminance information. An alternative description of the stimulus transforms the XYZ into normalized values with chromaticity coordinates x and y along with luminance L. Luminance is set as equal to Y.

The transformation from XYZ space to normalized chromaticity space is

$$\begin{aligned} x &= \frac{X}{X+Y+Z} \\ y &= \frac{Y}{X+Y+Z} \\ z &= \frac{Z}{X+Y+Z} = 1 - x - y. \end{aligned} \tag{5.6}$$

The third variable, z, is fully specified by x and y and is usually ignored.

The reverse conversion from chromaticity and luminance (usually as measured by a colorimeter) to XYZ is straightforward:

$$\begin{aligned} X &= \frac{x}{y}L \\ Y &= L \\ Z &= \frac{z}{y}L = \frac{1-x-y}{y}L. \end{aligned} \tag{5.7}$$

5.5.2 Calibrating a Display Device for Color

For a display device such as a color CRT display with three phosphors, each phosphor has a characteristic chromaticity and programmable luminance. For display monitors, the characteristic chromaticity of the R, G, and B guns depends on the phosphors used by the manufacturer. Calibrating a monitor or other display device requires measurement of chromaticity, minimum and maximum luminance, and Gamma correction for each "gun" separately. The characteristic color chromaticity of each gun is almost always approximated by a single (x, y), and any small color shifts over changing luminance levels are often ignored.

The color and intensity response of each gun of a display is calibrated with a colorimeter using methods similar to those of grayscale calibration described in section 5.3. The experimenter displays patches of color of different luminance values using a single RGB gun. The other two guns are set to zero. A colorimeter measurement for the patch returns three values, x, y, and L, of normalized CIE space. The user's manual of the colorimeter will have instructions for how to make measurements—which settings to choose and which buttons to push.

Selected colorimeter measurements of patches of different single-gun luminances yield the minimum, maximum, and Gamma constant γ for each gun. The γ estimate may be refined using psychophysical methods described in section 5.3.2. The chromaticity and luminance of a background color is measured by setting each gun at midpoint luminance, for example, and using the colorimeter to measure the corresponding x, y, and L.

At the end of calibration, the following parameters will have been specified by measurements: $\{x_R, y_R, L_R(\min), L_R(\max), \gamma_R\}$ for the red gun, $\{x_G, y_G, L_G(\min), L_G(\max), \gamma_G\}$ for the green gun, and $\{x_B, y_B, L_B(\min), L_B(\max), \gamma_B\}$ for the blue gun.

5.5.3 Specifying XYZ for a Stimulus on a Display Device

For any color stimulus specified as an RGB triplet, $\vec{v} = (r, g, b)$, on the graphics card, you can compute the corresponding XYZ values given the measured chromaticity coordinates of the R and G and B guns of the display device, (x_R, y_R), (x_G, y_G), and (x_B, y_B), and the luminances, L_R, L_G, and L_B. From these measured values we can derive the corresponding XYZ values for the color stimulus:

$$\begin{pmatrix} X(\vec{v}) \\ Y(\vec{v}) \\ Z(\vec{v}) \end{pmatrix} = \begin{pmatrix} \frac{x_R}{y_R} & \frac{x_G}{y_G} & \frac{x_B}{y_B} \\ 1 & 1 & 1 \\ \frac{z_R}{y_R} & \frac{z_G}{y_G} & \frac{z_B}{y_B} \end{pmatrix} \begin{pmatrix} L_R \\ L_G \\ L_B \end{pmatrix}. \tag{5.8}$$

The values of L_R, L_G, and L_B can be computed from the results of Gamma calibration of each gun.

Conversely, to display a pixel with tri-stimulus values XYZ, the programmed luminance of the red, green, and blue guns can be computed as:

$$\begin{pmatrix} L_R \\ L_G \\ L_B \end{pmatrix} = \begin{pmatrix} \frac{x_R}{y_R} & \frac{x_G}{y_G} & \frac{x_B}{y_B} \\ 1 & 1 & 1 \\ \frac{z_R}{y_R} & \frac{z_G}{y_G} & \frac{z_B}{y_B} \end{pmatrix}^{-1} \begin{pmatrix} X \\ Y \\ Z \end{pmatrix}. \tag{5.9}$$

The corresponding RGB triplet, $\vec{v} = (r, g, b)$, on the graphics card can be computed from the Gamma calibration of each gun.

5.5.4 Cone Excitation and Cone Contrast

Stimuli on a display device have a description in XYZ CIE color space, as discussed in the previous sections. These stimuli also elicit cone excitations in the eye of an observer. This section considers the correspondence between the stimuli presented and the resulting expected cone excitations for a standard observer. Equations translate between programmable RGB values and the corresponding XYZ description and cone excitation, and vice versa.

The expected amount of cone excitation for a standard human observer is computed from XYZ values based on the chromaticity coordinates of the three types of cones. The expected cone excitations for long, medium and short cones, P_L, P_M, and P_S, for XYZ values have been estimated from physiological responses, and follow the translation equation[36]:

$$\begin{pmatrix} P_L \\ P_M \\ P_S \end{pmatrix} = \begin{pmatrix} 0.15516 & 0.54308 & -0.03287 \\ -0.15516 & 0.45692 & 0.03287 \\ 0.00000 & 0.00000 & 0.01608 \end{pmatrix} \begin{pmatrix} X \\ Y \\ Z \end{pmatrix}. \tag{5.10}$$

This equation is based on estimates from humans—different remapping matrices would need to be estimated for other animals from non-human visual experiments.

To specify a display that will produce a given state of expected cone excitation, the equation can be inverted to compute the required XYZ values:

$$\begin{pmatrix} X \\ Y \\ Z \end{pmatrix} = \begin{pmatrix} 0.15516 & 0.54308 & -0.03287 \\ -0.15516 & 0.45692 & 0.03287 \\ 0.00000 & 0.00000 & 0.01608 \end{pmatrix}^{-1} \begin{pmatrix} P_L \\ P_M \\ P_S \end{pmatrix} = \begin{pmatrix} 2.9448 & -3.5001 & 13.1745 \\ 1.0000 & 1.0000 & 0.0000 \\ 0.0000 & 0.0000 & 62.1891 \end{pmatrix} \begin{pmatrix} P_L \\ P_M \\ P_S \end{pmatrix}. \tag{5.11}$$

Combining equations 5.8 and 5.10 allows us to use what we know about the chromaticity and luminance of a display device and what we know about the responses of cones in the visual system to compute expected cone excitations from luminances set for the three guns of the display device:

$$\begin{pmatrix} P_L \\ P_M \\ P_S \end{pmatrix} = \begin{pmatrix} 0.15516 & 0.54308 & -0.03287 \\ -0.15516 & 0.45692 & 0.03287 \\ 0.00000 & 0.00000 & 0.01608 \end{pmatrix} \begin{pmatrix} \frac{x_R}{y_R} & \frac{x_G}{y_G} & \frac{x_B}{y_B} \\ 1 & 1 & 1 \\ \frac{z_R}{y_R} & \frac{z_G}{y_G} & \frac{z_B}{y_B} \end{pmatrix} \begin{pmatrix} L_R \\ L_G \\ L_B \end{pmatrix}. \tag{5.12}$$

By inverting equation 5.12, we can specify the luminance values of each color gun to achieve a given state of expected cone excitation:

$$\begin{pmatrix} L_R \\ L_G \\ L_B \end{pmatrix} = \begin{pmatrix} \frac{x_R}{y_R} & \frac{x_G}{y_G} & \frac{x_B}{y_B} \\ 1 & 1 & 1 \\ \frac{z_R}{y_R} & \frac{z_G}{y_G} & \frac{z_B}{y_B} \end{pmatrix}^{-1} \begin{pmatrix} 2.9448 & -3.5001 & 13.1745 \\ 1.0000 & 1.0000 & 0.0000 \\ 0.0000 & 0.0000 & 62.1891 \end{pmatrix} \begin{pmatrix} P_L \\ P_M \\ P_S \end{pmatrix}. \tag{5.13}$$

Often, the most important aspect of color vision is not the absolute color but instead color contrast relative to the background color. For this reason, color stimuli may also be described in terms of expected cone contrasts:

$$C_L = \frac{\Delta P_L}{P_{L_0}}, C_M = \frac{\Delta P_M}{P_{M_0}}, C_S = \frac{\Delta P_S}{P_{S_0}}. \tag{5.14}$$

For a given background on a display device specified as $\vec{v}_0 = (r_0, g_0, b_0)$ and characterized by L_{0R}, L_{0G}, and L_{0B}, expected cone excitations P_{L_0}, P_{M_0}, and P_{S_0} can be computed from equation 5.12. One can compute luminance increments or decrements of the three different color channels by the following equation:

$$\begin{pmatrix} \Delta L_R \\ \Delta L_G \\ \Delta L_B \end{pmatrix} = \begin{pmatrix} \frac{x_R}{y_R} & \frac{x_G}{y_G} & \frac{x_B}{y_B} \\ 1 & 1 & 1 \\ \frac{z_R}{y_R} & \frac{z_G}{y_G} & \frac{z_B}{y_B} \end{pmatrix}^{-1} \begin{pmatrix} 2.9448 & -3.5001 & 13.1745 \\ 1.0000 & 1.0000 & 0.0000 \\ 0.0000 & 0.0000 & 62.1891 \end{pmatrix} \begin{pmatrix} \Delta P_L \\ \Delta P_M \\ \Delta P_S \end{pmatrix}, \tag{5.15}$$

where $\Delta P_L = C_L P_{L_0}$, $\Delta P_M = C_M P_{M_0}$, $\Delta P_S = C_S P_{S_0}$.[36]

5.5.5 Derrington–Krauskopf–Lennie Color Space

The DKL color space is related to the color opponent model of early visual processing.[33] That is, color vision starts with extraction of different cone signals that are *recoded* by three post-receptor mechanisms: a luminance mechanism and two opponent chromatic mechanisms. This representation is very useful because it separates color from luminance. Stimuli that differ only in color but not in luminance are called *isoluminant* stimuli. Isoluminant stimuli have many interesting properties, and the study of perception at isoluminance is a core aspect of understanding color vision.[37–41]

The DKL color space coordinates or axes, R_{Lum}, R_{L-M}, and R_{S-Lum}, correspond with the color contrasts of three theoretical mechanisms (figure 5.11). These three mechanisms are sometimes labeled luminance, red–green, and yellow–blue opponent channels. The DKL development focuses on responses to the background and deviations from the background, reflecting the fact that DKL space represents color *contrasts* of stimuli.

The DKL coordinates of a stimulus are related to its expected cone excitations in the following way:

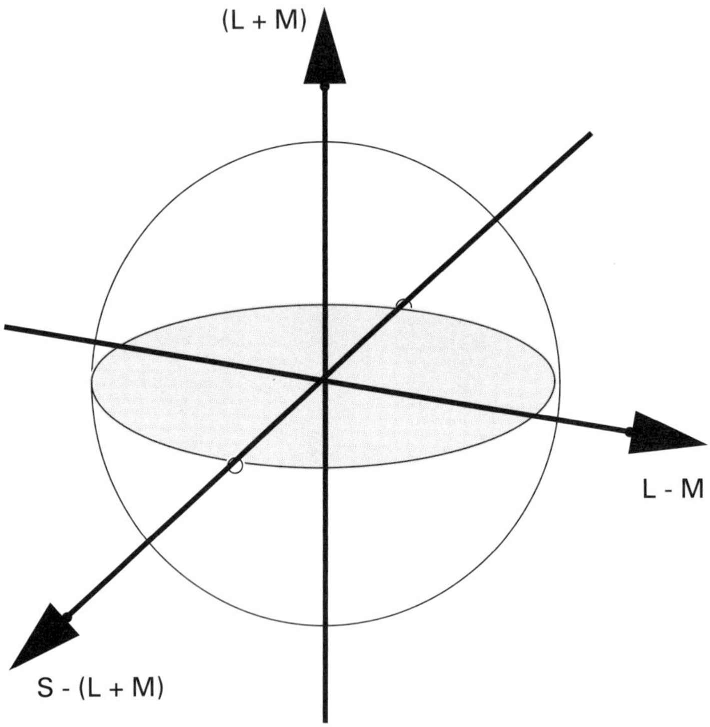

Figure 5.11
The 3D DKL color space. The primary axes are L + M (luminance), L – M (red–green), and S – (L + M) (yellow–blue). Isoluminant stimuli vary within a plane in the space varying [L – M, S – (L + M)] at a constant value of (L + M) luminance.

$$\begin{pmatrix} \frac{\Delta R_{\mathrm{Lum}}}{k_{\mathrm{Lum}}} \\ \frac{\Delta R_{\mathrm{L-M}}}{k_{\mathrm{L-M}}} \\ \frac{\Delta R_{\mathrm{S-Lum}}}{k_{\mathrm{S-Lum}}} \end{pmatrix} = \begin{pmatrix} 1 & 1 & 1 \\ 1 & \frac{-P_{\mathrm{L0}}}{P_{\mathrm{M0}}} & 0 \\ -1 & -1 & \frac{P_{\mathrm{L0}}+P_{\mathrm{M0}}}{P_{\mathrm{S0}}} \end{pmatrix} \begin{pmatrix} \Delta P_{\mathrm{L}} \\ \Delta P_{\mathrm{M}} \\ \Delta P_{\mathrm{S}} \end{pmatrix}. \tag{5.16}$$

Conversely,

$$\begin{pmatrix} \Delta P_{\mathrm{L}} \\ \Delta P_{\mathrm{M}} \\ \Delta P_{\mathrm{S}} \end{pmatrix} = \begin{pmatrix} 1 & 1 & 1 \\ 1 & \frac{-P_{\mathrm{L0}}}{P_{\mathrm{M0}}} & 0 \\ -1 & -1 & \frac{P_{\mathrm{L0}}+P_{\mathrm{M0}}}{P_{\mathrm{S0}}} \end{pmatrix}^{-1} \begin{pmatrix} \frac{\Delta R_{\mathrm{Lum}}}{k_{\mathrm{Lum}}} \\ \frac{\Delta R_{\mathrm{L-M}}}{k_{\mathrm{L-M}}} \\ \frac{\Delta R_{\mathrm{S-Lum}}}{k_{\mathrm{S-Lum}}} \end{pmatrix}. \tag{5.17}$$

The constants k_{Lum}, $k_{\mathrm{L-M}}$, and $k_{\mathrm{S-Lum}}$ in equations 5.16 and 5.17 define or scale the units of response for each dimension in the DKL space. One standard convention is to choose the scaling constants such that, when $(\Delta R_{\mathrm{Lum}}, \Delta R_{\mathrm{L-M}}, \Delta R_{\mathrm{S-Lum}}) = (1, 0, 0), (0, 1, 0)$, and $(0, 0, 1)$, $C = \sqrt{C_{\mathrm{L}}^2 + C_{\mathrm{M}}^2 + C_{\mathrm{S}}^2} = 1.0$.[36]

To summarize, color calibration of your display yields the chromaticity parameters for that device. The expected cone excitations for a chosen background, P_{L0}, P_{M0}, and P_{S0} can be computed from the device-specific chromaticity parameters. From the color calibration constants and the expected cone excitations for a selected background, a system of equations to estimate the scaling constants k_{Lum}, $k_{\mathrm{L-M}}$, and $k_{\mathrm{S-Lum}}$ is generated by setting $(\Delta R_{\mathrm{Lum}}, \Delta R_{\mathrm{L-M}}, \Delta R_{\mathrm{S-Lum}}) = (1, 0, 0), (0, 1, 0)$, and $(0, 0, 1)$. Every color can be specified in the DKL space.

5.5.6 Generating Color Stimuli in DKL Space

Often in color experiments, stimuli are specified in DKL space, and the experimenter needs to convert the coordinates in DKL space into RGB values for their display device. The program shown in display 5.11 takes a desired or target color specified in DKL space coordinates (two color dimensions and a luminance dimension) and computes the corresponding RGB settings. This program takes a DKL specification as the input and—in combination with information about the calibration of the display device—provides the computations that translate this into desired RGB values.

A second program in display 5.12 computes and presents an isoluminant grating that varies only in the red–green (L – M) dimension of the DKL space, while holding the value on the luminance dimension and the yellow–blue dimension constant.

5.6 Isoluminance Calibration

Isoluminant stimuli, important in many studies of color vision, vary only in color without contamination from luminance signals. Calibration in DKL space provides candidate stimuli for isoluminance. The DKL specification of a stimulus depends upon calibration by a photometer using the filter setttings of a standard observer. However, an isoluminant stimulus varying solely in the color dimensions of DKL space is only nominally isoluminant based on photometric measurements and may not be perceptually isoluminant. Random variation in responses at every level of the visual system means that what is isoluminant for one neuron may not be perfectly isoluminant for other neurons. Isoluminance may also depend on characteristics of the pattern, the size, and the location of the stimulus. The unintended luminance signals may help an observer solve a task and so contaminate the study of pure color vision.

Special calibration based on human perceptual tests, sometimes called "higher-order" calibration, is necessary to remove unintended luminance impurities from signal transmission in the visual system.[42,43]

Another area of vision research that requires functional, higher-order calibration is the study of motion perception. According to some theories, human motion perception is

Display 5.11

```
%%% Program: DKL2RGB.m
function [rgb background] = DKL2RGB(dR)
                              % dR is a desired DKL vector

% The following are chromaticity measurements at
% [0.5 0 0], [0 0.5 0] and [0 0 0.5]
x = [0.665 0.338 0.145];  % xR xG xB
y = [0.335 0.642 0.044];  % yR yG yB
L0 = [30 132 10]';        % YR YG YB

gamma = [2.1 2 2];        % monitor gamma for r g b guns
Lmin = 0.5; % Y at rgb = [0 0 0]
Lmax = [60.5 264.5 20.5]; % Y at rgb = [1 0 0], [0 1 0]
                          % and [0 0 1]

z = 1 - x - y;            % zR zG zB

% Combine two 3x3 matrix in Eq 5.12
L2P = [0.15516 0.54308 -0.03287;
      -0.15516 0.45692  0.03287;
       0       0        0.01608] * ...
      [x ./ y;
       1 1 1;
       z ./ y];

P0 = L2P * L0;            % compute background cone
                          % excitation (Eq 5.12)

% Conversion matrix (3x3) in Eq 5.17
DKL2dP = inv([1 1 1;
              1 -P0(1)/P0(2) 0;
             -1 -1 (P0(1)+P0(2))/P0(3)]);

% In Eq 5.17, set [dRlum dRL_M dRS_lum] to [1 0 0], [0 1 0]
% and [0 0 1] respectively. For each one, solve one of the
% constants k based on CL^2 + CM^2 + CS^2 = 1.
kFactor = sqrt(sum((DKL2dP ./ (P0 * [1 1 1])) .^ 2))';

% Now convert DKL contrast dR (3x1) into normalized rgb
dP = DKL2dP * (dR ./ kFactor); % Eq 5.17
dL = inv(L2P) * dP;            % Eq 5.15

c = (1 + dL ./ L0) / 2;  % convert to normalized rgb
                         % contrast
rgb = ((Lmax + Lmin) ./ (Lmax - Lmin) .* c') ...
      .^ (1 ./ gamma); % Eq 5.3
background = ((Lmax + Lmin) ./ (Lmax - Lmin) ...
      .*[0.5 0.5 0.5]) .^ (1 ./ gamma);
```

Display 5.12

```
%%% Program IsoLuminantGrating
function IsoLuminantGrating (L_M_contrast)
if nargin < 1 || isempty(L_M_contrast), ...
      L_M_contrast = 0.2; end
stimSize = 256;      % pixels
sf = 2;              % cycles/image
dkl = zeros(stimSize, 1, 3);
dkl(:, 1, 2) = sin(2*pi*sf* (1 : stimSize)/stimSize)/2;
dkl(:, 1, 2) = dkl(:, 1, 2) * L_M_contrast;
img = dkl; % pre-allocate
for i = 1 : stimSize
    img(i, 1, :) = DKL2RGB(squeeze(dkl(i, 1, :)));
        % B = SQUEEZE(A) returns an array B with the same
        % elements as A but with all the singleton
        % dimensions removed.  A singleton is a dimension
        % such that size(A,dim)==1.
end
img = repmat(img, [1 stimSize 1]);
[~, background] = DKL2RGB([0 0 0]');

imshow(img);
set(gcf, 'color', background);
```

served by three independent motion systems[26,44]: A first-order motion system computes flowfields from moving luminance modulations. A second-order system extracts motion from stimuli with moving feature modulations where the expected luminance is the same everywhere but some features (e.g., texture patterns) move systematically in space and time. And a third-order system detects the movement of regions of "salience." To better reveal the properties of individual motion systems, the experimenter must produce motion displays that stimulate only one motion system. For example, studies on second-order motion must use stimuli that do not have any first-order motion contamination, and studies on third-order motion must use stimuli that do not contain any first- and second-order contaminations.

Flicker photometry and minimum motion methods, described next, have been developed to calibrate isoluminance displays.[45] In these methods, a varying amount of luminance modulation is added to a candidate isoluminant stimulus in order to cancel out the luminance contamination. The new combined stimulus is used to create a flicker or a motion stimulus. The stimulus that produces the minimum amount of perceived flicker or motion is defined as the functionally isoluminant stimulus.

The procedure is illustrated in figure 5.12 in a minimum motion task with a red–green grating that translates in space. A particular amplitude m of luminance modulation is added

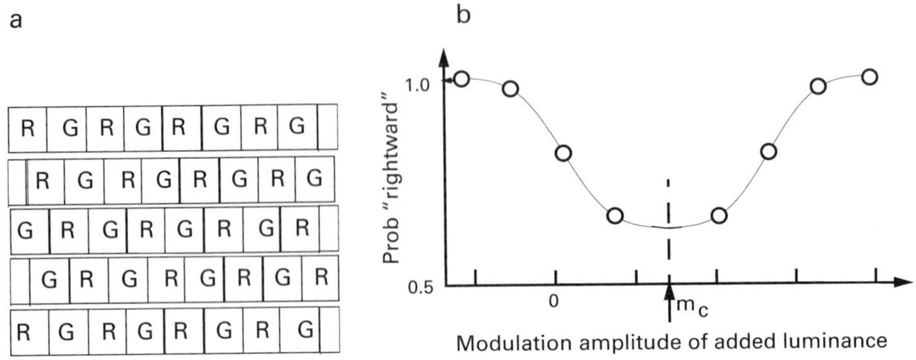

Figure 5.12
The minimum motion procedure to generate functional color isoluminance. (a) A representation of alternating isoluminant red and green (R, G) patches that move rightward in successive frames (top to bottom). (b) A functional isoluminance test adds luminance variation in some phase (i.e., in phase with green) with a luminance modulation that minimizes the perception of rightward motion.

in a particular phase to the candidate isoluminant stimulus. A search is then made for the m that produces minimum motion. This added m is assumed to cancel the contamination component. Usually, only one phase is tested. In the minimum motion method, the two motions, the possible contamination and the added canceling modulation, both move in the same direction.

A more sensitive method to remove luminance contamination is based on sandwich displays in which the experimenter alternates the original image frames with amplifier frames. This procedure we discuss here is based on five-frame calibration displays.

The odd frames are candidate isoluminant red–green sine-wave gratings with 180° phase shifts between them. The even "amplifier frames" are luminance sine-wave gratings also with 180° phase shifts. The phase shift between successive frames is 90° (figure 5.13). There is no motion signal in the odd frames alone or the even frames alone. The perception of motion requires the combination of information in odd and even frames. No consistent motion would be perceived in a color luminance sandwich display if the "isoluminant" red–green frames were truly isoluminant. If there were residual luminance contamination in the "isoluminant" frames so that, for example, the green areas were slightly less luminous than the red areas, motion would be seen in a consistent leftward direction based on luminance contamination in the color frames. If green were more luminous than red, motion would be seen in the rightward direction based on luminance contamination in the color frames. The high-contrast luminance frames in the sandwich displays amplify the luminance contaminants in the nominally isoluminant frames. In the sandwich procedure, luminance is added to the red–green "isoluminant" stimuli to cancel luminance artifacts.

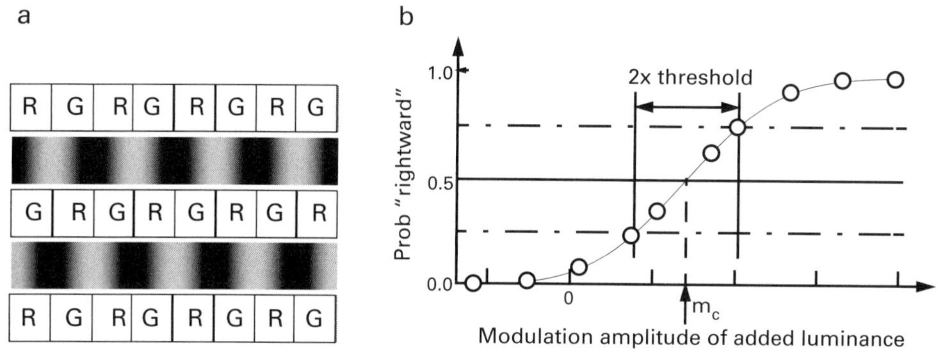

Figure 5.13
An illustration of the sandwich display for isoluminant color calibration. (a) Five frames consisting of odd frames of candidate color isoluminant stimuli and even frames of luminance stimuli. (b) Psychometric function to cancel motion artifacts.

Several examples of higher-order calibration methods for motion displays are described in Lu and Sperling (2001).[42] A program based on the sandwich method is shown in display 5.13 for interested readers.

5.7 Summary

This chapter was designed to be a road map to the evaluation, calibration, and selection of visual displays for the psychophysicist's laboratory. The capabilities and limitations of many of the currently available display technologies were considered. Although new technologies may be developed or the existing technologies may be improved, the same goals and principles of calibration remain. We have provided simple methods to evaluate the geometric distortions of displays and to calibrate monochromatic and color luminances and temporal and persistence properties of the medium. We have explained methods to describe visual stimuli in color spaces and how to translate between different stimulus representations. Knowing exactly which physical stimuli are being presented to observers is the first step in high-quality psychophysical investigation.

Display 5.13

```
%%% Program: RemoveFirstOrderContamination.m
function RemoveFirstOrderContamination

%% Experimental Module

% Specify the stimulus
p.alpha = 0.7 : 0.1 : 1.1;  % set up alpha values
p.radius = 6;               % image radius in visual degree
p.innerRad = 1;
p.contrast1 = 0.1;          % 1st order signal contrast
p.contrast2 = 0.5;          % 2nd order signal contrast
p.noise1 = 0.5;             % 1st order noise contrast
p.noise2 = 1;               % 2nd order noise contrast
p.nFactor = 2;              % noise pixel size
p.sf = 6;                   % cycles per degree
p.fixSize = 0.4;            % fixation size in deg
p.fixColor = 0;             % fixation color
p.tf = 7.5;                 % rotations per second
p.ITI = 0.5;                % seconds between trials
repeats = 16;               % Number of trials to repeat
                            % for each alpha
keys = {'left' 'right' 'esc'}; % keys to respond for
                               % direction, and to break

% Compute stimulus parameters
ppd = pi/180 * p.ScreenDistance / p.ScreenHeight * …
      p.ScreenRect(4);      % pixels per degree
m = round(p.radius * 2 * ppd / p.nFactor) * p.nFactor;
                            % stimulus size in pixels
fixRect=CenterRect([0 0 1 1] * p.fixSize * ppd, …
      p.ScreenRect);
nAlpha = numel(p.alpha);    % number of alpha levels
nTrials = repeats * nAlpha; % total number of trials
nFrames = round(p.ScreenFrameRate / p.tf / 4);
                            % # of refreshes for each image
dt = (nFrames - 0.5) / p.ScreenFrameRate;
                            % time for Flip wait
p.randSeed = ClockRandSeed; % use clock to set random
                            % number generator
[x, y] = meshgrid(linspace(-1, 1, m) * p.radius);
                            % coodinates
[theta, r] = cart2pol(x, y); % transform x,y coordinates to
                            % polar coordinates
clip = r < p.innerRad | r > p.radius; % set up annulus
nPixels = m / p.nFactor;
```

Display 5.13 (continued)

```
bNoise = ones(nPixels) * 0.5;
bNoise(1 : numel(bNoise) / 2) = -0.5;
                              % binary noise -0.5 or 0.5
tex = zeros (1, 5);

% Initialize a table to set up experimental conditions
p.recLabel = {'trialIndex' 'alphaIndex' 'clockwise' ...
      'respCorrect' 'respTime'};
rec = nan(nTrials, length(p.recLabel));
        % matrix rec is nTrials x 5 of NaN
rec(:, 1) = 1 : nTrials;    % count  trial numbers from 1
                              % to nTrials
alphaIndex = repmat(1 : nAlpha, repeats, 1);
clockwise = ones(repeats, nAlpha);
                              % first set all to 1: clockwise
clockwise(1 : repeats / 2, :) = -1;
                         % change first half to -1: counter-cw
[rec(:, 2) ind] = Shuffle(alphaIndex(:));
                              % shuffle alpha index
rec(:, 3) = clockwise(ind); % shuffle clockwise in the same
                              % order

% Prioritize display to optimize display timing
Priority(MaxPriority(windowPtr));

% Start experiment with instructions
str = ['Press left or right arrow keys for\n\n' ...
       'counter-clockwiset or clockwise rotation.\n\n' ...
       'Press SPACE to start.'];
DrawFormattedText(windowPtr, str, 'center', 'center', 1);
% Draw Instruction text string centered in window
Screen('Flip', windowPtr);
       % flip the text image into active buffer
WaitTill('space');         % wait till space bar is pressed
Secs = Screen('Flip', windowPtr);
p.start = datestr(now);    % record start time

% Run nTrials trials
for i = 1 : nTrials
    % Make textures for each frame
    for j = 1 : 5
        bNoise(:) = Shuffle(bNoise(:));
              % new noise image in each frame
        nImg = Expand(bNoise, p.nFactor);
        ang = (j - 1) * pi / 2 / p.sf * rec(i, 3);
```

Display 5.13 (continued)

```
                % phase shift: 90 deg per frame
        wedge = sin(p.sf * (theta + ang)) / 2; % [-0.5 0.5]
        if mod(j, 2) % odd number frames: 2nd order
             img = nImg * p.noise2   ...
                  .* (wedge * p.contrast2 + 0.5);
        else  % even number frames: 1st order
            img = nImg * p.noise1 ...
                 + wedge * p.contrast1;
        end
        ind = img < 0; % find pixels with negative contrast
        img(ind) = img(ind) * p.alpha(rec(i, 2));
                     % apply alpha
        img(clip) = 0;
        tex(j) = Screen('MakeTexture', windowPtr, img + ...
                   p.ScreenBackground, 0, 0, 2);
    end

    [key vbl] = WaitTill(Secs + p.ITI);

    % Display each frame
    for j = 1 : 5
        Screen('DrawTexture', windowPtr, tex(j));
        Screen('FillOval', windowPtr, p.fixColor, fixRect);
        vbl = Screen('Flip', windowPtr, vbl + dt);
    end
    Screen('FillOval', windowPtr, p.fixColor, fixRect);
    Screen('Flip', windowPtr, vbl + dt);
          % turn off last frame
    Screen('Close', tex); % close textures of each trial
    [key Secs] = WaitTill(keys);    % wait till response
    if iscellstr(key), key = key{1}; end
          % take the first in case of multiple keys
    if strcmp(key, 'esc'), break; end % to stop
    rec(i, 4) = strcmp(key, 'right') == (rec(i, 3) == 1);
          % record correctness
    rec(i, 5) = Secs - vbl; % record respTime
end
p.finish = datestr(now); % record finish time
save RemoveFirstOrderContamination_rst rec p;
       % save the results

%% System Reinstatement Module
Priority(0);   % restore priority
sca;           % close window and textures,
               % restore color lookup table
```

Display 5.13 (continued)

```
% Plot the result
pCorrect = zeros(nAlpha, 1);
for i = 1 : nAlpha
    foo = rec(rec(: ,2) == i, 4);
    foo(isnan(foo)) = [];
    pCorrect(i) = sum(foo) / numel(foo);
end
figure(3);
plot(p.alpha, pCorrect * 100, '+-');
axis([0.5 1.1 0 108]);
xlabel('Alpha');
ylabel('Percent Correct')
```

References

1. Pelli DG. 1997. Pixel independence: Measuring spatial interactions on a CRT display. *Spat Vis* 10(4): 443–446.

2. Brainard DH, Pelli DG, Robson T. Display characterization. In: *Encyclopedia of imaging science and technology*. Hoboken, NJ: Wiley; 2002.

3. Bach M, Meigen T, Strasburger H. 1997. Raster-scan cathode-ray tubes for vision research—limits of resolution in space, time and intensity, and some solutions. *Spat Vis* 10(4): 403–414.

4. Packer O, Diller LC, Verweij J, Lee BB, Pokorny J, Williams DR, Dacey DM, Brainard DH. 2001. Characterization and use of a digital light projector for vision research. *Vision Res* 41(4): 427–440.

5. García-Pérez MA, Peli E. 2001. Luminance artifacts of cathode-ray tube displays for vision research. *Spat Vis* 14(2): 201–215.

6. Compton K. 2001. Factors affecting cathode ray tube display performance. *J Digit Imaging* 14(2): 92–106.

7. Zele AJ, Vingrys AJ. 2005. Cathode-ray-tube monitor artefacts in neurophysiology. *J Neurosci Methods* 141(1): 1–7.

8. Psychtoolbox-3. [computer program] Available at: http://psychtoolbox.org/.

9. Segal M, Akeley K. The OpenGL graphics system: A specification. Available at: www.opengl.org/registry/doc/glspec20.20041022.pdf.

10. The MathWorks Inc. MATLAB [computer program]. Natick, MA: MathWorks; 1998.

11. Tyson J, Carmack, C. How computer monitors work. Available at: http://www.howstuffworks.com/monitor.htm.

12. Pelli DG. 1997. The VideoToolbox software for visual psychophysics: Transforming numbers into movies. *Spat Vis* 10(4): 437–442.

13. Samei E, Badano A, Chakraborty D, Compton K, Cornelius C, Corrigan K, et al. 2005. Assessment of display performance for medical imaging systems: Executive summary of AAPM TG18 report. *Med Phys* 32: 1205–1225.

14. Winterbottom MD, Geri GA, Morgan B, Pierce BJ. 2004. An integrated procedure for measuring the spatial and temporal resolution of visual displays. Interservice/Industry Training, Simulation, and Education Conference (I/ITSEC); paper 1855, pp 1–8.

15. Cristaldi DJR, Pennisi S, Pulvirenti F. *Liquid crystal display drivers: Techniques and circuits*. Berlin: Springer-Verlag; 2009.

16. Li X, Lu ZL, Xu P, Jin J, Zhou Y. 2003. Generating high gray-level resolution monochrome displays with conventional computer graphics cards and color monitors. *J Neurosci Methods* 130(1): 9–18.

17. Wang P, Nikolić D. 2011. An LCD monitor with sufficiently precise timing for research in vision. *Front Human Neurosci* 5: 1–10.

18. Okuno Y. Light-emitting diode display: Google Patents, US patent 4298869; 1981.

19. Müllen K, Scherf U. *Organic light emitting devices: Synthesis, properties and applications.* Hoboken, NJ: Wiley; 2006.

20. Hornbeck LJ. 1997. Digital light processing TM for high brightness, high-resolution applications. *Proc SPIE* 3013: 27–40.

21. Wheatstone C. 1852. The Bakerian Lecture—Contributions to the physiology of vision—part the second. On some remarkable, and hitherto unobserved, phenomena of binocular vision (continued). *Philos Trans R Soc Lond* 142: 1–17.

22. Lu ZL, Sperling G. 1999. Second-order reversed phi. *Atten Percept Psychophys* 61(6): 1075–1088.

23. Colombo E, Derrington A. 2001. Visual calibration of CRT monitors. *Displays* 22(3): 87–95.

24. Kelly D. 1979. Motion and vision. II. Stabilized spatio-temporal threshold surface. *JOSA* 69(10): 1340–1349.

25. Burr DC, Ross J. 1982. Contrast sensitivity at high velocities. *Vision Res* 22(4): 479–484.

26. Lu ZL, Sperling G. 1995. The functional architecture of human visual motion perception. *Vision Res* 35(19): 2697–2722.

27. Tyler CW. 1997. Colour bit-stealing to enhance the luminance resolution of digital displays on a single pixel basis. *Spat Vis* 10(4): 369–377.

28. Allard R, Faubert J. 2008. The noisy-bit method for digital displays: Converting a 256 luminance resolution into a continuous resolution. *Behav Res Methods* 40(3): 735–743.

29. Pelli DG, Zhang L. 1991. Accurate control of contrast on microcomputer displays. *Vision Res* 31(7–8): 1337–1350.

30. Li X, Lu ZL. 2012. Enabling high grayscale resolution displays and accurate response time measurements on conventional computers. *J Vis Exp* 60: e3312.

31. BITS++. Available at: http://www.crsltd.com/tools-for-vision-science/visual-stimulation/bits-sharp/.

32. vPixx. Available at: http://www.vpixx.com/.

33. Derrington AM, Krauskopf J, Lennie P. 1984. Chromatic mechanisms in lateral geniculate nucleus of macaque. *J Physiol* 357(1): 241–265.

34. CIE. *Commission Internationale de l'Eclairage proceedings, 1931.* Cambridge, UK: Cambridge University Press; 1932.

35. Kaiser PK, Boynton RM, Swanson WH. *Human color vision*, 2nd ed. Washington, DC: Optical Society of America; 1996.

36. Brainard D. 1996. Cone contrast and opponent modulation color spaces. *Human Color Vision* 2: 563–579.

37. Gregory RL. 1977. Vision with isoluminant colour contrast: 1. A projection technique and observations. *Perception* 6(1): 113–119.

38. D'Zmura M. 1991. Color in visual search. *Vision Res* 31(6): 951–966.

39. Sekiguchi N, Williams DR, Brainard DH. 1993. Efficiency in detection of isoluminant and isochromatic interference fringes. *JOSA A* 10(10): 2118–2133.

40. Lindsey DT, Teller DY. 1990. Motion at isoluminance: Discrimination/detection ratios for moving isoluminant gratings. *Vision Res* 30(11): 1751–1761.

41. Lu ZL, Lesmes LA, Sperling G. 1999. The mechanism of isoluminant chromatic motion perception. *Proc Natl Acad Sci USA* 96(14): 8289–8294.

42. Lu ZL, Sperling G. 2001. Sensitive calibration and measurement procedures based on the amplification principle in motion perception. *Vision Res* 41(18): 2355–2374.

43. Scott-Samuel NE, Georgeson MA. 1999. Does early non-linearity account for second-order motion? *Vision Res* 39(17): 2853–2865.

44. Lu ZL, Sperling G. 2001. Three-systems theory of human visual motion perception: Review and update. *JOSA A* 18(9): 2331–2370.

45. Anstis SM, Cavanagh P. A minimum motion technique for judging equiluminance. In: Mollon JD, Sharpe LT, eds. *Colour vision: Physiology and psychophysics.* London: Academic Press; 1983:155–166.

6 Response Collection

Once the stimulus is generated and displayed on the monitor or other display device, the next step is to measure aspects of the observer's response, which may be as simple as a key-press or as complex as an implicit behavior such as eye movements or physiological responses observed through external devices. In this chapter, we describe the use of many standard methods of collecting responses and consider issues of synchronization between stimulus presentation and response collection or between multiple devices collecting distinct responses. The first two sections describe the collection of discrete responses and the measurement of response times, respectively. The third section provides relevant programs and examples. The fourth section treats the measurement of eye movements, and the fifth section describes the synchronization issues in the collection of physiological data.

6.1 Discrete Responses

6.1.1 Keyboard

The simplest method of collecting a person's response to a visual display is to ask the individual to press a key on the computer keyboard. He or she may be asked to press different keys to indicate different responses. Computer keyboards are inexpensive, do not require any special hardware setup or programming, and naturally work with any computer hardware, operating system, and standard software toolkit.

`KeyboardInput.m` is a program that displays the first character you type on your keyboard and the elapsed time since the onset of the instruction screen. The Experimental module of the program is shown in display 6.1.

6.1.2 Mouse/Touchpad

Most modern computers are equipped with a mouse or touchpad. A mouse is a device that controls the position of a cursor in the two dimensions of a screen or other graphic device. The mouse can be used to select different response categories by clicking a different button or by clicking at a different cursor location. It may also be used to trace the trajectory of the movements of a mouse-controlled cursor on the screen.

Display 6.1

```
%%% Program KeyboardInput.m
function KeyboardInput

%% Experimental Module

% Start experiment with instructions
str = 'Type letters or numbers. Press ESC to exit.';
DrawFormattedText(windowPtr, str, 'center', 'center', 1);
        % Draw Instruction text string centered in window
t0 = Screen('Flip', windowPtr);
        % flip the text image into active buffer

% Show typed letter or number and time since onset of
% instruction
keys = [cellstr([num2str((0:9)'); char(97:122)']); 'esc'];
while 1
    KbReleaseWait; % avoid reading a single press repeatedly
    [key Secs] = WaitTill(keys);    % wait till response
    if strcmp(key, 'esc'), break; end
    if iscellstr(key), key = key{1}; end
        % take the first in case of multiple keys
    str = sprintf('You typed: %s\n\nTime since ' ...
          'instruction onset: %.3f s', key, Secs - t0);
    DrawFormattedText(windowPtr, str, 'center', ...
          'center', 1);
    Screen('Flip', windowPtr);
end
```

`MouseButton.m` (display 6.2) is a sample program that demonstrates how to draw a "button" on the screen and detect whether the location of the button is clicked by the mouse. When you use the mouse to click on the "button," it will darken. When you click and release the "Quit" button on the screen, the program terminates.

`MouseDraw.m` (display 6.3) is a sample program that demonstrates how to draw a curve with the mouse. Press the left button to enter drawing mode, move the mouse to draw, and release the left button to finish drawing. (On a Macintosh, just hold down the single mouse button and draw.) The trace is recorded and shown in a figure.

6.1.3 Joystick

A joystick is an input device consisting of a stick that pivots on a base and reports its angle or direction to the computer. A two-dimensional (2D) joystick is like a mouse—moving the stick left or right signals movements along the *X*-axis and moving it up or down signals movement along the *Y*-axis. A three-dimensional (3D) joystick uses twist (clockwise or counterclockwise) to signal movement along the *Z*-axis. These three axes—*X*, *Y*, and *Z*—correspond with roll, pitch, and yaw in the movements of an aircraft. Some joysticks also have one or more simple on/off switches, called "fire buttons," to be used for other kinds of responses or actions. Some also have haptic feedback capability.

An analog joystick returns an angular measure of the movement in any direction as a continuous value. A digital joystick gives only on/off signals for four different directions and those combinations that are mechanically possible. Most I/O interface cards for PCs have a game control port for a joystick. Modern joysticks mostly use a USB interface for connection to the PC. The current version of Psychtoolbox does not support joysticks.

6.1.4 Touchscreen

A *touchscreen* is an electronic visual display that detects the presence and location of a touch of a finger or a stylus within the display area. It enables a user or observer to interact directly with what is displayed, rather than indirectly with a cursor controlled by a mouse or touchpad. The typical touchscreen detects an event about 10 ms after contact.

The Psychtoolbox treats a touchscreen in the same way as the left mouse button. You can use `MouseButton.m` to test a touchscreen if you have one.

6.1.5 Button Boxes

Button boxes or response boxes have been developed to collect responses on an external device rather than through the built-in keyboard or mouse of the computer. There are three primary reasons to use external devices to collect responses. First, button boxes allow experimenters to develop modified sets of buttons or interfaces that are simpler or more ergonomic than the keyboard, perhaps providing fewer keys or keys of different sizes that may be useful for testing children or older adults. Second, special button boxes may be designed to operate in special circumstances, for example at a distance from the computer or in environments such as MRI machines that require special non-magnetic materials or shielding. Finally, most button boxes have electronic mechanisms that increase the accuracy of timing of responses in relation to stimulus presentation or inputs from other response devices such as those that measure psychophysiological activity.

Button boxes constructed for the first two purposes simply mimic computer keyboards. Button boxes constructed for more precise response time collection often involve specialized hardware and software in order to deliver more accurate timing.

6.2 Response Times

Response times (RTs) provide valuable measures of human performance.[1,2] RT and accuracy together define a performance point in a trade-off between speed and accuracy, the speed–accuracy trade-off.[3–5] There are areas in psychology where RT is treated as the primary measured variable of interest because RT varies while there may only be small variations in accuracy. Defined as the elapsed time between stimulus or task onset and a

Display 6.2

```
%%% Program MouseButton.m
function MouseButton

%% Experimental Module

% Draw a button with "Try Me" on it
buttonColor = [0 1 1];     % button color
tryme = 'Try Me';          % text to label a button
trymeRect = CenterRect([0 0 120 60], p.ScreenRect);
                           % button rectangle
[x y] = RectCenter(trymeRect); % center of the rectangle
textRect = Screen('TextBounds', windowPtr, tryme);
                          % size of text
xtryme = x- textRect(3) / 2;    % x position so the text is
                                % in the rect center
ytryme = y- textRect(4) / 2;
Screen('FillRect', windowPtr, buttonColor, trymeRect);
                          % draw a "button"
Screen('DrawText', windowPtr, tryme, xtryme, ytryme, …
      [1 1 1]);           % button label

% Draw a button with "Quit" on it with different text
% color
Quit = 'Quit';
quitButton = trymeRect - [0 1 0 1] * 200;
[x y] = RectCenter(quitButton);
textRect = Screen('TextBounds', windowPtr, Quit);
xquit = x - textRect(3) / 2;
yquit = y - textRect(4) / 2;
Screen('FillRect', windowPtr, buttonColor, quitButton);
                          % quit "button"
Screen('DrawText', windowPtr, Quit, xquit, yquit, [1 0 0]);

% Show all the drawing. We don't clear buffer, since we will
% over-draw at the same location. Otherwise, we need to
% redraw both buttons for each flip.
Screen('Flip', windowPtr, 0, 1);
while 1
    while 1 % wait till left button press
        [x, y, buttons] = GetMouse(windowPtr);
        if buttons(1), break; end
    end
    if IsInRect(x, y, trymeRect)
                          % is the click on the TryMe button?
        Screen('FillRect', windowPtr, buttonColor/2, …
```

Display 6.2 (continued)

```
                trymeRect);  % darken the button
        Screen('DrawText', windowPtr, tryme, xtryme, …
                ytryme, [1 1 1]);
        Screen('Flip', windowPtr, 0, 1);
        while buttons(1) % wait till button release
            [x, y, buttons] = GetMouse(windowPtr);
        end
        Screen('FillRect', windowPtr, buttonColor, …
                trymeRect);  % restore button color
        Screen('DrawText', windowPtr, tryme, xtryme, …
                ytryme, [1 1 1]);
        Screen('Flip', windowPtr ,0, 1);
    elseif IsInRect(x, y, quitButton)
            % is the click on the Quit button?
        Screen('FillRect', windowPtr, buttonColor/2, …
                quitButton); % darken the button
        Screen('DrawText', windowPtr, Quit, xquit, yquit,…
                [1 0 0]);
        Screen('Flip', windowPtr, 0, 1);
        while buttons(1)
            [x, y, buttons] = GetMouse(windowPtr);
        end
        Screen('FillRect', windowPtr, buttonColor, …
                quitButton); % restore button color
        Screen('DrawText', windowPtr, Quit, xquit, yquit, …
                [1 0 0]);
        Screen('Flip', windowPtr, 0, 1);

        % Exit if mouse on the Quit button and released
        if IsInRect(x, y, quitButton), break; end
    end
end
```

subject's response, RTs have been measured in a large variety of tasks. Although computer keyboards and the mouse are widely used to collect reponses, this often leads to possibly substantial variations and biases in the measurement of RTs. Specialized button boxes must be used to obtain more accurate measurements.[6,7]

6.2.1 Measuring RT with Computer Keyboard/Mouse

Although computer keyboards and mouse devices are frequently used in measuring RTs, the accuracy of these measurements is limited. Regular computer keyboards and mice are optimized for daily use and low manufacturing costs, not for scientific experiments with

Display 6.3

```
%%% Program MouseDraw.m
function MouseDraw

%% Experimental Module

lineWidth = 1; % pixels
% Set the cursor to its initial location
SetMouse(100, 100);
str = 'Press on the left mouse button and move it to '…
      'draw. Release to finish';
Screen('DrawText', windowPtr, str, 50, 50, 1);
Screen('Flip', windowPtr);
% Wait for click and hide the cursor
while 1
   [x0, y0, buttons] = GetMouse(windowPtr);
   if buttons(1), break; end
end
HideCursor;
thePoints = nan(10000,2);     % pre-allocate record of mouse
                              % locations
thePoints(1, :) = [x0 y0];    % first point
i = 2;
while 1
    [x, y, buttons] = GetMouse(windowPtr);
    if ~buttons(1), break;    end
                              % exit when button is released
    if (x ~= x0 || y ~= y0) % make sure mouse is moving
        Screen('DrawLine', windowPtr, 128, x0, y0, x, y,…
               lineWidth);
        Screen('Flip', windowPtr, 0, 1); % update the drawing
                                         % on the display
        thePoints(i, :) = [x y];
        x0 = x;
        y0 = y;
        i = i + 1;
    end
end

% Plot the contour in a Matlab figure
thePoints(i : end, :) = [];
figure(7);
plot(thePoints(:, 1), p.ScreenRect(4) - thePoints(:, 2), …
     'linewidth', lineWidth);
axis equal;
xlim(p.ScreenRect([1 3]));
ylim(p.ScreenRect([2 4]));
```

highly accurate timing. There are a number of sources of timing error in registering a keyboard/mouse response, including mechanical lag in depressing the keys (how far the key has to travel and the pressure required), "debouncing" of multiple key presses (discounting mechanical bouncing of the contact), sequential scanning of key presses, polling of keyboard events, and event handling of the computer operating system.

Cumulatively, the various sources of timing errors in keyboard responses can add up to delays and uncertainties of between 20 and 70 ms. For some tasks, this variability is comparable to or greater than the variability of human RTs in simple detection tasks, which is typically about 20 ms.[2] Importantly, some of the timing variability in keyboard/mouse responses is biased. The bias may make comparisons of RTs between different conditions invalid. In situations in which the possible RT difference is small, the additional variability from the measurement device may make the difference undetectable. An external device, such as a special-purpose RT box, makes RT measurements more accurate.

6.2.2 External RT Devices

An ideal RT measurement device should be (1) compatible with the standard, widely used computer hardware, (2) compatible with the commonly used operating systems, (3) easy to use and control from stimulus presentation software, (4) easy to integrate into common experimental setups, (5) capable of storing the timing of all the response events until it is convenient for the user software to retrieve them, and (6) capable of detecting and time-stamping signals from other external devices.[6]

A number of commercial RT devices are now available. Virtually all of the external RT devices are designed with high-quality keys with good mechanical properties. There are other differences between devices that can be very important depending upon the requirements of a particular application or experiment.

Most or all devices incorporate good key-press debouncing routines. The connections made by depressing mechanical keys may not be stable, but instead may bounce, or have very rapid on/off cycles as the key is depressed. These extremely rapid on/off cycles occur very quickly (on the millisecond scale or less) and clearly cannot reflect intentional movements of the human. Either hardware or software mechanisms are used to discount the rapid bounces as a key is depressed.

One important characteristic that determines the accuracy of RT measurement is the way that the device is connected to the computer. An important factor limiting the measurement of key-press responses with the regular computer keyboard is that keyboard inputs are sensed by scheduled polling or scanning. If the input device can be scanned at 1-ms intervals and other aspects of the device are optimized, then the bias in RT measurement may be as low as 1 ms. However, standard keyboards are usually polled less frequently, typically every 8 ms. This means that if a response actually occurred just after a polling cycle, then the RT is recorded as occurring later than it actually did—introducing an average bias in RT of half the polling interval.

Table 6.1
Properties of some existing button boxes

Device Name	Port Style	Debounce Protocol	Onboard Clock	TTL Inputs	Operating System	Accuracy
Psych Software[a]	Keyboard emulator	+	–	–	Microsoft	
Engineering Solutions[b]	USB keyboard emulator	+	–	–	Microsoft	8-ms delays
Stewart[c]	Parallel port		–	–	Linux	1 ms
PsyScope[d]	USB serial emulator		+			
Cedrus[e]	USB serial emulator		+	+		1 ms
RTBox[f]	USB serial emulator	+	+	+	Any	1 ms

a. PST Serial Response Box from Psychology Software Tools Inc. (http://www.pstnet.com/).
b. Engineering Solutions, Inc. (http://www.response-box.com/tools.shtml).
c. Stewart[9]: A PC parallel port button box.
d. PsyScope Button Box (http://psy.cns.sissa.it/).
e. Cedrus Corporation RB Series Response Pad (http://www.cedrus.com/support/rb_series/).
f. Li et al.[6]: RTBox.

In this section, we consider several RT devices for which specifications are available, summarized in table 6.1. Several commercial RT boxes, such as the device by Empirisoft Corporation,[8] have good mechanical keys but still use a keyboard emulator and the standard keyboard intake drivers, and so do not eliminate the biases of delayed polling cycles. Other devices (see table 6.1) use either standard serial or parallel port communication that—with appropriate software—can be polled frequently enough to reduce the polling bias and allow millisecond accuracy.

Some RT devices use only simple button boxes and solve the various logging and timing problems in software through the use of specially programmed device drivers. One example is the software by Stewart[9] for the Linux operating system that can be used to achieve 1-ms accuracy in RT measurement.

On the other end of the dimension are devices that are very sophisticated and have their own real-time clocks, microprocessors, and ports to take inputs from other devices that allow autonomous action. They use very rapid or continuous polling by the onboard microprocessor of the RT inputs in order to minimize polling lags. The physical response of the keys is debounced either in hardware or software. A microprocessor may keep a sequence of time-stamped RTs and key identities and other events until this information can be communicated to a main computer during a non-time-critical part of the trial.

The more elaborate RT devices also allow other events to trigger inputs that can be time-stamped. If the device and the main computer each have their own real-time clocks and the main computer is controlling, for example, the onset and duration of the display, it is necessary to coordinate or synchronize the timing of the two devices. If we know the time-stamp of a display onset relative to the clock of the RT device and we know the time-stamp of the display onset by the main computer clock, we can infer the relative timing and use this information to coordinate or synchronize the two.

One example of a full-service device is the RTBox.[6,7] It combines many of the good properties of the full-service devices. A software driver is provided to control the RTBox in MATLAB, with Psychtoolbox-3 extensions.[10] Once the RTBox is connected to the host computer through a USB connection, the software driver detects the device and is used to control and use all of its functions.

6.2.3 Timing With and Without External Triggers

Stimulus and response events can be timed either with or without the use of external triggers. We discuss this issue here using the RTBox as an example system. If external triggers are available to signal either an event or the state of the display system, for example, a device such as the RTBox can provide accurate timing without synchronizing with the clocks of the main computer. An external trigger, usually a TTL (transistor–transistor logic) signal, marks a key time point. TTL refers to the delivery of logical 0 or logical 1 response using a standardized voltage (usually 0–0.4 V for 0 and 2.6–5 V for a 1).

Coordinating timing using the RTBox and associated software device drivers involves several steps. This can be illustrated by timing a response relative to a visual or auditory display. After the main computer completes the computation of the display, a function call to the RTBox driver clears the serial buffer to remove any possible irrelevant keypress events that might have occurred previously. Then, a trigger signal is sent to the RTBox by the stimulus generator on the main display computer. An observer responds to the display with a button press. At the end of the trial, the main computer reads the button event and time (t_{button}) and the trigger event and time (t_{trigger}) from the RTBox. The difference in times can be used to compute RT relative to the stimulus onset for that trial, all referenced to the clock in the RTBox rather than the main computer.

In some experimental setups, it may not be possible or convenient to generate accurate external triggers for stimulus onset or other events. In these cases, the event time-stamps of the RTBox must be translated into the time on the host computer clock. This requires synchronization of the clocks of the RTBox and the host computer. Synchronization is accomplished by a call to the RTBox driver before any timing is measured. An offset time is the difference between the main computer clock time and the RTBox clock time, and is used to interpret other times sent from the RTBox.

The RTBox website[11] includes several example programs that show how the RTBox can be used to measure response times with TTL triggers, with a light trigger, and without external triggers.

6.3 Example Experiments

6.3.1 Example RT and Speed–Accuracy Trade-off Experiments—Visual Search

Consider two variants of a sample experiment testing a visual search task.[12] The first variant measures the RT for visual search of displays of different sizes (numbers of objects). The second variant measures the time course of visual search by interrupting the observer after different amounts of processing time and scoring the accuracy of response——a speed–accuracy trade-off (SAT) measure of speed of processing.

This visual search example illustrates the differential difficulty of searching for a C among O's and searching for an O among C's, where the former is easier than the latter.[12,13] This is called a *search asymmetry*. Visual search asymmetry occurs when the speed and/or accuracy of visual search differs depending upon which of the same two items is assigned as the target of the search. Another example is the relative ease of finding a tilted line among vertical lines compared to finding a vertical line among tilted ones.[14] Researchers have suggested that search asymmetries reveal the existence of coded features of primary visual analysis, such as gap detectors or tilt detectors.[13,15–19]

Display 6.4 shows the experimental code for using the computer keyboard to measure the RT and accuracy of responses for visual search for C among O's or O among C's for display sizes of 4, 8, and 12. Display 6.5 shows the same experiment using the RTBox without an external trigger to record RTs.

Observers decide whether a search display contains a target or not and respond yes or no. In the RT variant, RT is measured from the onset of the search display, which remains on the screen until response, and then is erased. Sample displays of different set sizes with target present or absent are shown in figure 6.1, along with corresponding mean RT and accuracy data based on actual data in this experiment.[12]

It takes longer to respond to larger displays for the more difficult O in C searches, whereas RT is nearly unaffected by display size for the C in O searches. This pattern of RT data has often been interpreted as evidence for serial search—evaluation of one item at a time—in the O in C searches, but not in the C in O searches. Further analysis of visual search with an SAT experiment suggests instead that the time course of these two kinds of searches is largely overlapping, with differences in the relative accuracy of performance for a given processing time[12] (see later).

The SAT version of the search experiment manipulates the amount of time spent processing the display and measures the corresponding accuracy.[12] This experiment tests two display sizes, 4 and 12. The visual search display is exposed only briefly (50 ms), and accuracy of target absent or present discrimination is measured at seven different points in time, with delays to an auditory response cue (beep) of 0, 0.05, 0.15, 0.30, 0.50, 1.15, and 1.80 s after display offset. Display 6.6 shows experimental code for the SAT experiment with a 50-ms display duration. It measures the time from the offset of the display to the auditory cue, the RT of the response relative to the auditory cue, and displays the RT as feedback to the observer.

In these SAT paradigms, observers are trained to keep the RTs to the response cue quite short (150–350 ms) so that the experimenter manipulates processing time while measuring the corresponding accuracy of the responses. At very short cue delays, the observer has not had time to process the content of the search display, and the responses are guesses. Performance accuracy increases as observers are allowed more processing time. Sample data from an actual SAT experiment of this design[12] are shown in figure 6.2. The accuracy curves are fit with best-fitting time-course curves, as described in section 10.4.5 of chapter 10.

These sample RT and SAT experiments and the corresponding code to run the experiments provide some of the typical building blocks of coordination of visual displays, presentation and timing of simple auditory cues, and measurement of RTs. These building blocks can be combined to create many RT and SAT experiments.

6.4 Eye Tracking

In many experiments, the point of gaze ("where we are looking") and/or the movement of the eye relative to the head are measured.[20] Eye position is sometimes used as an implicit measure of what an observer is thinking about or where the observer is getting information. It is also the object of study in experiments that investigate the functioning of the eye movement system. How eye movements depend upon physical characteristics of the display or on context, expectancy, or attention is sometimes the topic of study.[21–24]

There is a long history of the measurement of the eye position. Several generations of technology have been used in its measurement. The different methods for measuring eye position reflect a trade-off between precision of measurement and degree of invasiveness of the procedure. For example, hard-core studies of the dynamic system properties, including onset, acceleration, maximum eye speed, slowing of the eye, and so forth, require more accurate measures, and investigators may use more invasive methods of measurement. At the other extreme, investigations of infant looking behavior may use only very rough measures of where an infant looks and for how long in order to provide a completely noninvasive method of measurement.[25,26] Earlier research often fixed the position of the observer's head with a "bite bar"—wax impressions of the teeth attached to a fixed metal bar. More often today, the observer stabilizes the head by placing it against a chin and a forehead rest. Some systems are designed to tolerate modest head movements.

6.4.1 Measuring Eye Position

The most sensitive eye-tracking system uses tight-fitting contact lenses with an embedded mirror or magnetic search coil, and the movement is measured as the eye rotates during eye movements.[20] This method is uncomfortable and is used only in laboratories that are studying the eye movement system itself or with animals. The more popular eye-tracking systems use video-based methods for measuring eye movements.[27] Some also use electric potentials measured with electrodes placed around the eyes (electrooculography, or EOG) to track eye movements.[28]

Display 6.4

```
%%% Program RT.m
function RT (C_in_Os)

%% Experimental Module

% Specify the stimulus
p.stimSize = 0.98;        % letter size in visual degree
p.radius = 4.12;          % radius of the annulus in degree
p.jitter = 4;             % range of location jitter in pixels
p.dispSize = [4 8 12];    % total # of items in the display
repeats = 80;             % number of trials in each
                          % experimental condition
p.fixDuration = 0.25;
p.ITI = 1;                % time interval between trials in
                          % seconds
keys = {'left' 'right' 'esc'};  % response keys for target
                          % absent and present, and to break

% Compute stimulus parameters
ppd = pi/180 * p.ScreenDistance / p.ScreenHeight *
p.ScreenRect(4);          % pixels per degree
m = round(p.stimSize * ppd);  % letter size in pixels
fixXY = [[[-1 1] * m / 2 0 0] + p.ScreenRect(3) / 2;
        [0 0 [-1 1] * m / 2] + p.ScreenRect(4) / 2];
nDispSize = length(p.dispSize);
nTrials = repeats * nDispSize * 2; % total number of trials
p.randSeed = ClockRandSeed;    % use clock to set the seed
                          % for the random number generator
radius = p.radius * ppd;       % radius in pixels
theta = (0 : 360/15 : 359)';   % polar angles of the 15
                               % locations
[x y] = meshgrid(linspace(-1, 1, m));
r = sqrt(x .^ 2  + y .^ 2);
circle = (1 - exp(-((r - 0.85) / 0.15) .^ 4)) *
p.ScreenBackground;          % circle with blurred edge
texB = Screen('MakeTexture', windowPtr, ones(m) * p.
ScreenBackground, 0, 0, 2); % background
texD = Screen('MakeTexture', windowPtr, circle, 0, 0, 2);
                          % distractors
texT = Screen('MakeTexture', windowPtr, circle, 0, 0, 2);
                          % target
tex = texD;
if nargin && C_in_Os, tex = texT; end
Screen('FillRect', tex, p.ScreenBackground, [m/2 m/4 ...
      m m/4*3]);          % 'C' open to right
```

Display 6.4 (continued)

```
% Initialize a table to set up experimental conditions
p.recLabel = {'trialIndex' 'sizeIndex' 'targetPresent'...
      'respCorrect' 'respTime'};
rec = nan(nTrials, length(p.recLabel));
         % matrix rec initialized with NaN
rec(:, 1) = 1 : nTrials;% count the trial number from 1
                        % to nTrials
sizeIndex = repmat(1 : nDispSize, [2 1 repeats]);
targetPresent = zeros(2, nDispSize, repeats);
                        % first set all to 0
targetPresent(:, :, 1 : repeats / 2) = 1;
                        % change first half to 1
[rec(:, 2) ind] = Shuffle(sizeIndex(:));
                        % shuffle size index
rec(:, 3) = targetPresent(ind); % shuffle target presence
                                % index in the same order

% Prioritize display to optimize display timing
Priority(MaxPriority(windowPtr));

% Start experiment with instructions
str = ['Press the left arrow for target absent, and the '...
       'right arrow for target present reponses\n\n' ...
       'Please respond as quickly and accurately as '   ...
       'possible.\n\n''Press SPACE to start.'];
DrawFormattedText(windowPtr, str, 'center', 'center', 1);
         % Draw Instruction text string
Screen('Flip', windowPtr);
         % flip the text image into the active buffer
Beeper;
WaitTill('space');      % wait till space bar is pressed
Secs = Screen('Flip', windowPtr);
p.start = datestr(now); % record start time

% Run nTrials trials
for i = 1 : nTrials
    % parameters for this trial
    sz = p.dispSize(rec(i, 2));
        % display size: total # of items in the display
    angles = theta + rand * 360;
    x = radius * cosd(angles) + p.ScreenRect(3) / 2;
        % center locations of the recs
    y = radius * sind(angles) + p.ScreenRect(4) / 2;
    x = x + (rand(15, 1) - 0.5) * 2 * p.jitter;
        % apply jitter
```

Display 6.4 (continued)

```
    y = y + (rand(15, 1) - 0.5) * 2 * p.jitter;
    rects = CenterRectOnPoint([0 0 m m], x, y);
        % 15 recs
    nLoc = sz + sz / 4 - 1;
    tex(1 : nLoc) = texD; % set up distractors
    tex([5 10 nLoc+1:15]) = texB;
       % 5, 10 and higher ones are set as background
    if rec(i, 3)           % target present
       ind = randi(nLoc);
       if mod(ind, 5) == 0, ind = ind + 1; end
                           % avoid blank location
       tex(ind) = texT;    % draw a target in a target
                           % present trial
    end
    WaitTill(Secs + p.ITI);
    Screen('DrawLines', windowPtr, fixXY, 3, 0); % fixation
    t0 = Screen('Flip', windowPtr, 0, 1);
    Screen('DrawTextures', windowPtr, tex, [], rects');
       % C and O
    t0 = Screen('Flip', windowPtr, t0 + p.fixDuration);
    [key Secs] = WaitTill(keys);   % wait till response
    Screen('Flip', windowPtr);     % remove stimulus
    if iscellstr(key), key = key{1}; end
       % take the first key press in case of multiple
       % key presses
    if strcmp(key, 'esc'), break; end
    rec(i, 4) = strcmp(key, 'right') == rec(i, 3);
       % record response accuracy
    rec(i, 5) = Secs - t0;         % record respTime
    if rec(i, 4), Beeper; end      % beep if correct
end
p.finish = datestr(now);           % record finish time
save RT_rst.mat rec p;             % save the results
```

We discuss only the infrared or near-infrared video-based eye trackers in this section. These video-based eye trackers are noninvasive, provide relatively good measurements for many though not all people, and the simpler ones can be moderate in price.

Video-based eye trackers use a video camera that focuses on one or both eyes to record eye movements. Typically, an infrared light generated by the device is reflected from the eye and sensed by a video camera. The information is then analyzed to extract eye rotations from changes in reflections from the eye.

Some eye trackers infer the orientation of the eye by using the corneal reflection (the first Purkinje image) of the infrared light source created by the front surface of the cornea,

Display 6.5

```
%%% Program RT_RTBox.m
function RT_RTBox (C_in_Os)

%% Experimental Module

% Specify the stimulus
p.stimSize = 0.98;      % letter size in visual degree
p.radius = 4.12;        % radius of annulus in degree
p.jitter = 4;           % range of location jitter in pixels
p.dispSize = [4 8 12]; % display size
repeats = 80;           % number of trials in each
                        % experimental condition
p.fixDuration = 0.25;
p.ITI = 1;              % time interval between trials in
                        % seconds

% Compute stimulus parameters
ppd = pi/180 * p.ScreenDistance / p.ScreenHeight * …
      p.ScreenRect(4); % pixels per degree
m = round(p.stimSize * ppd); % letter size in pixels
fixXY = [[[-1 1] * m / 2 0 0] + p.ScreenRect(3) / 2;
         [0 0 [-1 1] * m / 2] + p.ScreenRect(4) / 2];
nDispSize = length(p.dispSize);
nTrials = repeats * nDispSize * 2; % total number of trials
p.randSeed = ClockRandSeed;  % use clock to set the seed of
                             % the random number generator
radius = p.radius * ppd;     % radius in pixels
theta = (0 : 360/15 : 359)'; % polar angles of the 15
                             % locations
[x y] = meshgrid(linspace(-1, 1, m));
r = sqrt(x .^ 2  + y .^ 2);
circle = (1 - exp(-((r - 0.85) / 0.15) .^ 4)) * …
      p.ScreenBackground; % circle with blurred edge
texB = Screen('MakeTexture', windowPtr, ones(m) * …
      p.ScreenBackground, 0, 0, 2); % background
texD = Screen('MakeTexture', windowPtr, circle, 0, 0, 2);
                        % distractors
texT = Screen('MakeTexture', windowPtr, circle, 0, 0, 2);
                        % target
tex = texD;

if nargin && C_in_Os, tex = texT; end
Screen('FillRect', tex, p.ScreenBackground, [m/2 m/4 …
       m m/4*3]);    % 'C' open to right
% Initialize a table to set up experimental conditions
p.recLabel = {'trialIndex' 'sizeIndex' 'targetPresent'…
```

Display 6.5 (continued)

```
        'respCorrect' 'respTime'};
rec = nan(nTrials, length(p.recLabel));
                          % matrix rec initialized with NaN
rec(:, 1) = 1 : nTrials;
        % count the trial numbers from 1 to nTrials
sizeIndex = repmat(1 : nDispSize, [2 1 repeats]);
targetPresent = zeros(2, nDispSize, repeats);
        % first set all to 0
targetPresent(:, :, 1 : repeats / 2) = 1;
        % change first half to 1
[rec(:, 2) ind] = Shuffle(sizeIndex(:));
        % shuffle size index
rec(:, 3) = targetPresent(ind);
        % shuffle target presence index in the same order

% Prioritize display to optimize display timing
Priority(MaxPriority(windowPtr));

% Start experiment with instructions
str = ['Press the left buttons for target absent and the '...
        'right buttons for target present responses\n\n' ...
        'Please respond as quickly and accurately as '...
        'possible.\n\n' 'Press any button to start.'];
DrawFormattedText(windowPtr, str, 'center', 'center', 1);
        % Draw Instruction text string centered in window
Screen('Flip', windowPtr);
        % flip the text image into active buffer
Beeper;

% Initialize RTBox
RTBox('ButtonNames', {'left' 'left' 'right' 'right'});
        % define the first two buttons as left; the last
        % two as right.
RTBox(inf);               % wait till any button is pressed
Secs = Screen('Flip', windowPtr);
p.start = datestr(now); % record start time

% Run nTrials trials
for i = 1 : nTrials
    % parameters for this trial
    sz = p.dispSize(rec(i, 2));
        % display size: total # of items in the display
    angles = theta + rand * 360;
    x = radius * cosd(angles) + p.ScreenRect(3) / 2;
```

Display 6.5 (continued)

```
        % center of the recs
    y = radius * sind(angles) + p.ScreenRect(4) / 2;
    x = x + (rand(15, 1) - 0.5) * 2 * p.jitter;
        % apply jitter
    y = y + (rand(15, 1) - 0.5) * 2 * p.jitter;
    rects = CenterRectOnPoint([0 0 m m], x, y);
        % 15 recs
    nLoc = sz + sz / 4 - 1;
    tex(1 : nLoc) = texD; % set up distractors
    tex([5 10 nLoc+1:15]) = texB;
        % 5, 10 and higher ones set as background
    if rec(i, 3)      % target present
        ind = randi(nLoc);
        if mod(ind, 5) == 0, ind = ind + 1; end
                        % avoid blank location
        tex(ind) = texT;
    end
    WaitTill(Secs + p.ITI);
    RTBox('clear'); % clear RTBox, sync clocks
    Screen('DrawLines', windowPtr, fixXY, 3, 0); % fixation
    t0 = Screen('Flip', windowPtr, 0, 1);
    Screen('DrawTextures', windowPtr, tex, [], rects');
                                % C and O
    t0 = Screen('Flip', windowPtr, t0 + p.fixDuration);
    [Secs key] = RTBox(inf);   % wait till response
    Screen('Flip', windowPtr); % remove stimulus
    if iscellstr(key), key = key{1}; end
        % take the first in case of multiple key presses
    if strcmp(key, 'esc'), break; end
     rec(i, 4) = strcmp(key, 'right') == rec(i, 3);
        % record correctness
     rec(i, 5) = Secs - t0;    % record respTime
     if rec(i, 4), Beeper; end % beep if correct
end
p.finish = datestr(now);        % record finish time
save RT_RTBox_rst.mat rec p;    % save the results
```

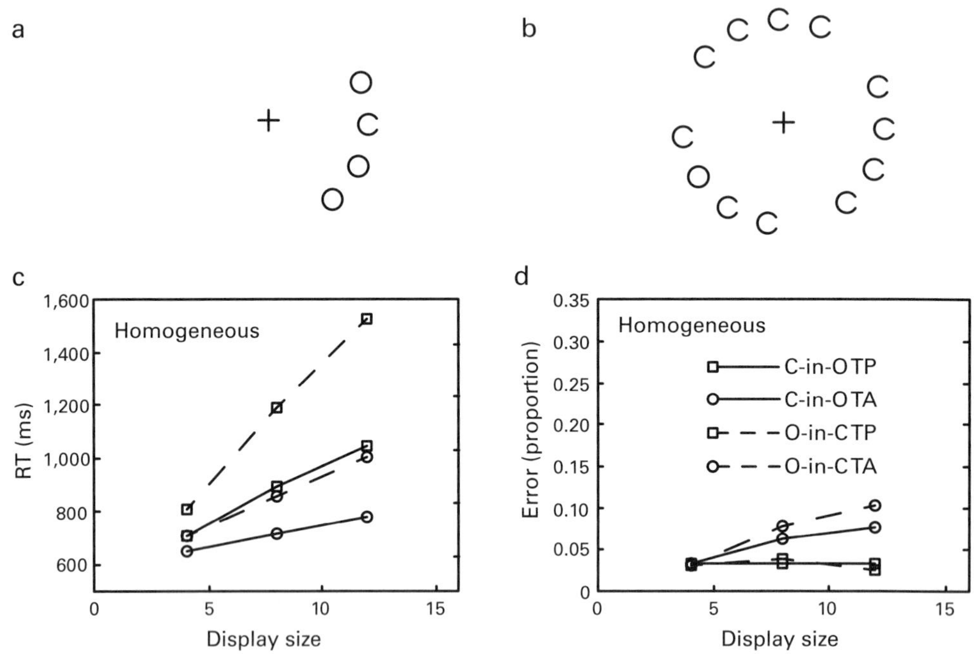

Figure 6.1
Response time study of visual search. (a) This display of size 4 illustrates a search for a C in O's. (b) This sample display of size 12 illustrates a search for an O in homogeneous C's. (c) and (d) Average correct RTs and error rates as a function of display size for search with free viewing of unlimited-time displays (after Dosher, Han, and Lu[12]). TP, target present; TA, target absent.

which acts like a convex mirror, and the center of the pupil as features to track the eye position over time (figure 6.3).[29] An image-processing algorithm looking for a dark circular or elliptical pattern in the video image estimates the location of the pupil and its center. A more sensitive type of eye tracker, the dual-Purkinje eye tracker,[27] uses reflections from the front of the cornea (first Purkinje image) and the back of the lens (fourth Purkinje image) as features to track eye movements.

Eye trackers measure the rotation of the eye relative to a reference frame. Different kinds of systems measure the rotation of the eye relative to different reference frames. A head-mounted system, in which the light source and video camera are mounted on the subject's head, measures the eye-in-head angles. A head-mounted system in which the observer may look around at the external world by moving the head as well as the eyes requires separate measurement of the images experienced by the eye to place where in an image the eye is looking. Goggle systems for measuring eye movements are an example of head-mounted displays where not just the measurement system but also the display system is head-mounted.

Display 6.6

```
%%% Program SAT.m
function SAT (C_in_Os)

%% Experimental Module

% Specify the stimulus
p.stimSize = 0.98;      % letter size in visual degree
p.radius = 4.12;        % radius of annulus in degree
p.jitter = 4;           % range of location jitter in pixels
p.dispSize = [4 12];    % display sizes: total # of items in
                        % a display
p.stimDuration = 0.1;   % letters duration in seconds
p.processTime = [0 0.05 0.15 0.30 0.50 1.15 1.80];
                        % cue delays
repeats = 60;           % Number of trials in each condition
p.fixDuration = 0.25;
p.feedbackDuration = 0.5;
p.ITI = 1;              % time interval between trials in
                        % seconds

% Compute stimulus parameters
ppd = pi/180 * p.ScreenDistance / p.ScreenHeight * …
      p.ScreenRect(4);
m = round(p.stimSize * ppd);   % letter size in pixels
fixXY = [[[-1 1] * m / 2 0 0] + p.ScreenRect(3) / 2;
         [0 0 [-1 1] * m / 2] + p.ScreenRect(4) / 2];
nDispSize = length(p.dispSize);
nProcessTime = length(p.processTime);
nTrials = repeats * nDispSize * nProcessTime * 2;
                        % total number of trials
p.randSeed = ClockRandSeed;  % use clock to set the seed of
                             % the random number generator
radius = p.radius * ppd;     % radius in pixels
theta = (0 : 360/15 : 359)';% polar angles of the 15
                             % locations
stimDur = (round(p.stimDuration * p.ScreenFrameRate) …
      - 0.5) / p.ScreenFrameRate ;
[x y] = meshgrid(linspace(-1, 1, m));
r = sqrt(x .^ 2  + y .^ 2);
circle = (1 - exp(-((r - 0.85) / 0.15) .^ 4)) * …
      p.ScreenBackground;    % circle with blurred edge
texB = Screen('MakeTexture', windowPtr, ones(m) * …
      p.ScreenBackground, 0, 0, 2);  % background
texD = Screen('MakeTexture', windowPtr, circle, 0, 0, 2);
                        % distractors
```

Display 6.6 (continued)

```
texT = Screen('MakeTexture', windowPtr, circle, 0, 0, 2);
                      % target
tex = texD;

if nargin && C_in_Os, tex = texT; end
Screen('FillRect', tex, p.ScreenBackground, [m/2 m/4 …
      m m/4*3]); % C open to right

% Initialize a table to set up experimental conditions
p.recLabel = {'trialIndex' 'sizeIndex' 'processTimeIndex' …
      'targetPresent' 'respCorrect' 'respTime'};
rec = nan(nTrials, length(p.recLabel));
        % matrix rec initialized with NaN
rec(:, 1) = 1 : nTrials;
        % count the trial numbers from 1 to nTrials
sizeIndex = repmat(1 : nDispSize, [2 1 repeats …
      nProcessTime]);
targetPresent = zeros(2, nDispSize, repeats, nProcessTime);
        % first set all to 0
targetPresent(:, :, 1 : repeats / 2, :) = 1;
        % change first half to 1
timeIndex = repmat(1 : nProcessTime, [2 nDispSize …
      repeats 1]);
[rec(:, 2) ind] = Shuffle(sizeIndex(:));
        % shuffle size index
rec(:, 3) = timeIndex(ind);
        % shuffle process time index in the same order
rec(:, 4) = targetPresent(ind);
        % shuffle target presence index in the same order

% Prioritize display to optimize display timing
Priority(MaxPriority(windowPtr));

str = ['Press the left buttons for target absent and the '…
       'right buttons for target present responses\n\n' …
       'Please respond as quickly and accurately as '…
       'possible.\n\n''Press any button to start.'];
DrawFormattedText(windowPtr, str, 'center', 'center', 1);
        % Draw Instruction text string centered in window
Screen('Flip', windowPtr);
        % flip the text image into active buffer
Beeper;

% Initialize RTBox
RTBox('ButtonNames', {'left' 'left' 'right' 'right'});
```

Display 6.6 (continued)

```
        % define the first two buttons as left; the last
        % two as right.
RTBox(inf);             % wait till any button is pressed
Secs = Screen('Flip', windowPtr);
p.start = datestr(now); % record start time

% Run nTrials trials
for i = 1 : nTrials
        % parameters for this trial
    sz = p.dispSize(rec(i, 2));
        % display size: total # of items in the display
    angles = theta + rand * 360;
    x = radius * cosd(angles) + p.ScreenRect(3) / 2;
        % center of rec for item locations
    y = radius * sind(angles) + p.ScreenRect(4) / 2;
    x = x + (rand(15, 1) - 0.5) * 2 * p.jitter;
        % apply jitter
    y = y + (rand(15, 1) - 0.5) * 2 * p.jitter;
    rects = CenterRectOnPoint([0 0 m m], x, y); % 15 rects
    dt = p.processTime(rec(i, 3));
    nLoc = sz + sz / 4 - 1;
    tex(1 : nLoc) = texD;
        % set up distractors
    tex([5 10 nLoc+1:15]) = texB;
        % 5, 10 and higher ones set as background
    if rec(i, 4)      % target present
         ind = randi(nLoc);
        if mod(ind, 5) == 0, ind = ind + 1; end
                    % avoid blank location
        tex(ind) = texT;
    end

    WaitTill(Secs + p.ITI);
    RTBox('clear'); % clear RTBox, sync clocks
    Screen('DrawLines', windowPtr, fixXY, 3, 0); % fixation
    t0 = Screen('Flip', windowPtr, 0, 1);
    Screen('DrawTextures', windowPtr, tex, [], rects');
                    % C and O
    t0 = Screen('Flip', windowPtr, t0 + p.fixDuration);
    t0 = Screen('Flip', windowPtr, t0 + stimDur);
                    % turn off stim
    tCue = WaitSecs('UntilTime', t0 + dt);
    Beeper;         % please double check if there is any
                    % delay introdued by your operating
                    % system
```

Display 6.6 (continued)

```
     RTBox('clear', 0); % clear any response before the tone
     [Secs key] = RTBox(inf);    % wait till response
     if iscellstr(key), key = key{1}; end
         % take the first in case of multiple key presses
     rec(i, 5) = strcmp(key, 'right') == rec(i, 4);
         % record correctness
     rec(i, 6) = Secs(1) - t0; % record respTime
     str = sprintf('%.0f', (Secs(1) - tCue) * 1000);
     DrawFormattedText(windowPtr, str, 'center', 'center', …
         1);
     t0 = Screen('Flip', windowPtr); % show feed back
     Screen('Flip', windowPtr, t0 + p.feedbackDuration);
         % turn off feedback
end
p.finish = datestr(now);   % record finish time
save SAT_rst.mat rec p;   % save the results
```

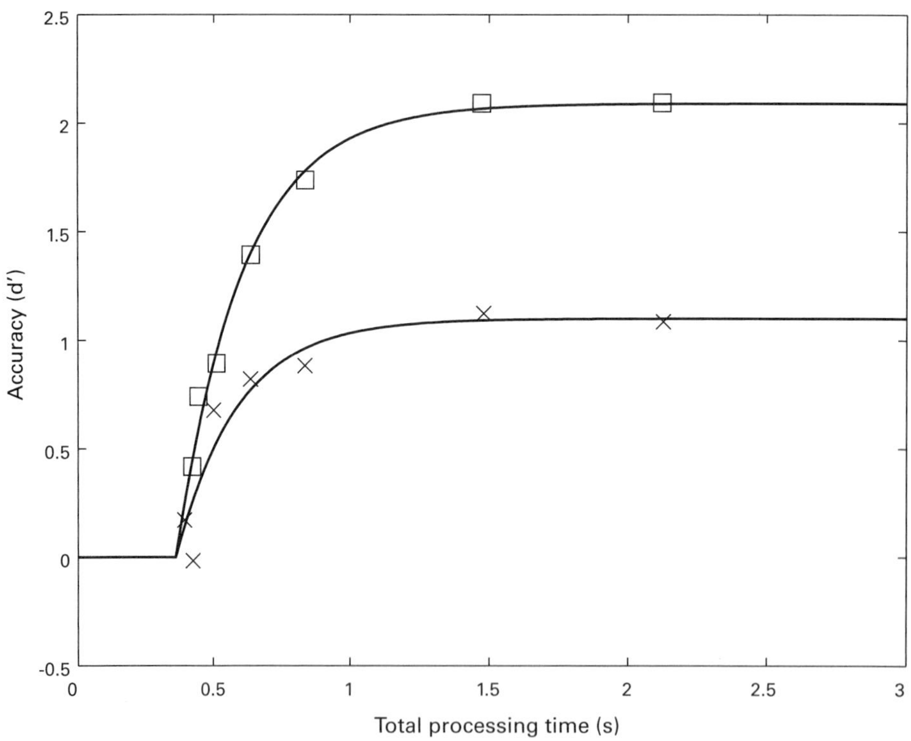

Figure 6.2
Visual search discrimination performance (d′) as a function of total processing time (test onset to response) for display sizes of 4 and 12 (data after Dosher, Han, and Lu[12]).

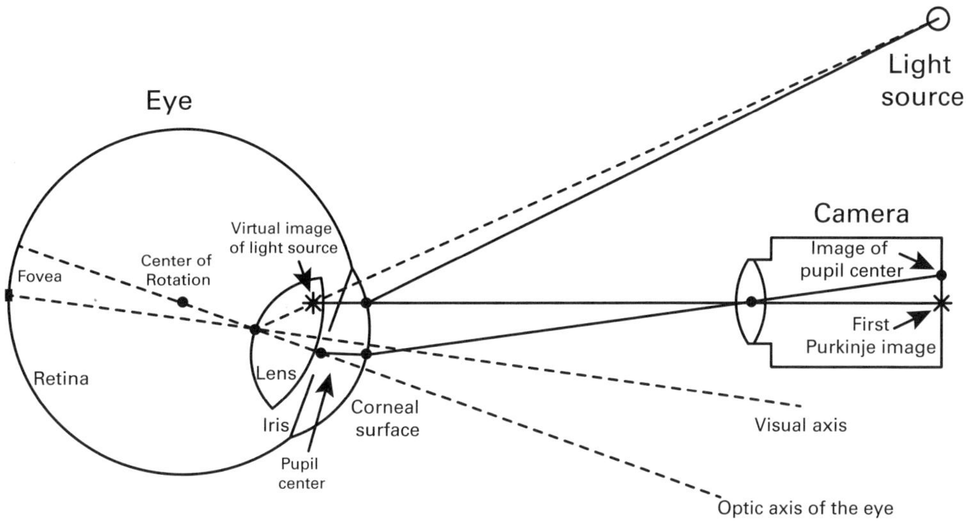

Figure 6.3
Ray-tracing diagram showing schematic representations of the eye, a camera, and a light source.

A table-mounted ("remote") system has a camera mounted on the table and measures gaze angles relative to a physical layout. The head position is fixed (e.g., using a bite bar, or a chin rest and forehead support setup), then eye position and gaze are directly connected. Head direction is subtracted from gaze direction to determine eye-in-head position. Using one camera and one light source, the gaze can be estimated for a stationary head. Using one camera and multiple light sources, the gaze can be estimated with free head movements.

In some experiments, eye trackers monitor eye fixation and track eye movements while the experimenter alters the content depending upon where an observer is looking. These are called *gaze contingent displays*. The host computer of the eye tracker extracts information about the eye position in real time. The information is then communicated to the experimental computer to control the displays. Synchronization of the two computers is very important for accurate timing of eye movements relative to stimulus events.

6.4.2 Calibration of Eye Position

An eye tracker does not measure absolute gaze direction—it can only measure *changes* in gaze direction. To know precisely *where* an observer is looking, a calibration procedure measures the locations relatively to known locations in the external world. The observer is told to look at or fixate a series of points on a display (or in the world), while the eye tracker records the measured value that corresponds to each gaze position. The relative measures corresponding to all locations in the display are interpolated from the calibrated responses to the known locations. The calibration is verified by measuring the relative position for new locations and comparing this with the predicted values for these locations. An accurate and reliable calibration is essential for obtaining valid and repeatable eye movement data properly registered to external references.

Eye movement measurement systems can experience "drift" in their responses over time or with movement of the viewer's head within the measurement frame of the device. For this reason, calibration should be performed periodically. It may be necessary to perform a calibration every dozen trials or so if very precise measurements are required.

For adult humans with healthy eye control systems who are willing and able to stay relatively still, the calibration is relatively straightforward. In contrast, for young children or animals, the situation is more challenging—we cannot simply say exactly where to look during a calibration process and expect these observers to comply. Similarly, there may be measurement issues with certain ocular or other diseases that exhibit variability or bias in the actual control of eye gaze. For such individuals, the experimenter may need to design a more qualitative experiment or use specially designed methods to ensure calibration.

6.4.3 Examples Using the Eyelink Eye-Tracking System

The Eyelink system by SR Research[30] is one of the currently available systems that provides relatively high-quality eye-tracking support for intermediate cost. It is widely used in the psychophysics laboratory. A photograph of the Eyelink-1000 is shown in figure 6.4. This system combines image analysis of pupil position with measured reflection of an infrared light source to compute eye position. The accompanying company-provided soft-

Figure 6.4
Photograph of an Eyelink-1000 eye-tracking system. A chinrest is shown in the foreground attached to two poles. The system of cameras to photograph the eye and light sources are in the small unit placed in front of the monitor. The computer and equipment for eye-movement tracking are shown. Often, a separate computer (not shown) drives the display device and performs timing operations.

ware package provides programs for calibration, measurement, and analysis of eye position. Other functions for the Eyelink eye tracker have been developed and circulated by Psychtoolbox user groups.

Display 6.7 provides the code for `RT_Eyelink.m`, a sample program for tracking eye fixation and generating gaze contingent displays for the Eyelink device. The experiment is the same as the one described in section 6.3.1 except that an Eyelink eye tracker is used to track a subject's eye fixation during each trial. The Eyelink systems use an Ethernet communication protocol between the main computer and the Eyelink computer, with estimated synchronization accuracies of 1–2 ms.

Other commercial eye-tracking systems may use different protocols and rely on different methods of communication between the eye movement and main experimental computers. Subroutines for eye-tracking systems will either be provided by the device developer or by second-source user groups. However, the general principles of operation, calibration, and testing will be similar to the example provided here.

6.5 Physiological Measures

Psychophysical experiments are increasingly combined with physiological measures of the response of the organism to stimuli. The study of general physiological responses such as heart rate or temperature dates back to early in the past century.[31–33] Current fascination with brain science has led to active research using many different measures of brain activity with multiple technologies: *electroencephalography* (EEG),[34–36] *magnetoencephalography* (MEG),[37,38] *functional magnetic resonance imaging* (fMRI),[39–41] or *near-infrared spectroscopy* (NIRS).[42,43] Each of these brain-imaging measures provides a window on general brain activity or the brain response to specific stimuli. Each technology has characteristic levels of spatial localization and temporal precision.

If the research aims to understand how the brain responds to specific stimuli, then the experimental trick is likely to focus on exact synchronization of stimulus displays—whether visual, auditory, tactile—with the measures of brain response. The stimulus presentation computer and the computer that records and stores data from physiological measurement devices have separate clocks. It is critical to synchronize the timing of the separate clocks to align properly stimulus presentation with physiological measures. This is analogous to RT measurement with an external device. Although it might seem that a single cross-synchronization of various computer and device clocks might be sufficient for an entire session, in fact very small differences in clock rates, if they exist, may cumulate over time. Measures of display onset and measures of physiological responses generally must be synchronized at least once per trial to be sure of timing accuracy.

In this section, we briefly discuss several physiological measurements and their basis and some of the issues in incorporating these modalities of measurement into the psychophysical laboratory.

6.5.1 Electroencephalography

EEG measures the electrical activity of the brain through sensors attached to the scalp and a reference sensor attached to another part of the body, such as the ear. The electrical fields of brain activity are measured through the skull and the scalp and reflect the group activity of many active neurons. Many psychophysics and other laboratories include EEG systems.[35]

The level of noise or random variation in the brain response on any given trial is sufficiently high that the stereotypical response to a stimulus and task is usually measured by averaging more than 100 trials synchronized to the stimulus input.[34–36] Individual trial noise can reflect random or slow cyclical variations in intrinsic brain activity or in electromagnetic interference from the environment. The average EEG locked to an event is called an event-related potential (ERP). The temporal precision of the EEG measurement is at a millisecond level and is determined by the sampling rate of the EEG device. Measurable responses of the average ERP may occur within the first few milliseconds following the stimulus and continue for a second or more. Figure 6.5 shows an example of continuous EEG, with several event markers, and an average ERP trace. This shows a stereotypical form of a response to repeated visual stimulus. Experimenters infer properties of the brain response from where an ERP signal is localized in the brain and the shape and time course of the average response to the stimulus.

Source localization of an ERP tries to identify which brain region or regions are the source(s) of the primary ERP response. Localization of brain responses to a specific part of the brain depends upon source-localization modeling. By assuming a net current flow of a certain magnitude at a given source location and applying a model of the conduction properties of the brain, a source-localization model generates a prediction about the distribution of ERP at the scalp where the EEG sensors are placed. Mathematically inverting such models estimates the location of sources of the electrical brain activity from the scalp measurements.[44]

In a typical EEG setup, one computer interacts with the EEG device setup and records the EEG responses while another computer manages visual, auditory, or tactile stimulus display and perhaps collects responses through a button box or on the keyboard. The synchronization of the three sets of activities is the key experimental feature. In almost all setups, synchronization with the stimulus occurs by sending a stimulus onset signal or mark to the EEG computer and including this time-stamp as part of the EEG data stream. This time mark may then be used to mark the beginning of the EEG interval either for online averaging routines or for later offline data processing.

The specifics of the synchronization are dependent upon the setup and the manufacturer of the EEG system. Synchronization signals may be sent via TTL, light or other pulses, or via messages to data acquisition computer cards for the particular EEG system. Details and the specific accompanying function calls are device-specific.

Display 6.7

```
%%% Program RT_Eyelink.m
function RT_Eyelink (C_in_Os)

%% Experimental Module

% Specify the stimulus
p.stimSize = 0.98;            % letter size in visual degree
p.radius = 4.12;              % radius of annulus in degree
p.jitter = 4;                 % range of location jitter in
pixels
p.dispSize = [4 8 12];        % display size: total # of items in
                              % the display
repeats = 80;                 % number of trials in each
                              % experimental condition
p.fixDuration = 0.25;
p.ITI = 1;                    % time interval between trials in
                              % seconds
keys = {'left' 'right' 'esc'}; % absent, present, and to break

% Compute stimulus parameters
ppd = pi/180 * p.ScreenDistance / p.ScreenHeight * …
      p.ScreenRect(4);        % pixels per degree
m = round(p.stimSize * ppd); % letter size in pixels
fixXY = [[[-1 1] * m / 2 0 0] + p.ScreenRect(3) / 2;
         [0 0 [-1 1] * m / 2] + p.ScreenRect(4) / 2];
nDispSize = length(p.dispSize);
nTrials = repeats * nDispSize * 2; % total number of trials
p.randSeed = ClockRandSeed;         % use clock to set random
                                    % number generator
radius = p.radius * ppd;      % radius in pixels
theta = (0 : 360/15 : 359)'; % polar angles of the 15 locations
[x y] = meshgrid(linspace(-1, 1, m));
r = sqrt(x .^ 2  + y .^ 2);
circle = (1 - exp(-((r - 0.85) / 0.15) .^ 4)) * …
      p.ScreenBackground;     % circle with blur edge
texB = Screen('MakeTexture', windowPtr, ones(m) * …
      p.ScreenBackground, 0, 0, 2);  % background
texD = Screen('MakeTexture', windowPtr, circle, 0, 0, 2);
                              % distractors
texT = Screen('MakeTexture', windowPtr, circle, 0, 0, 2);
                              % target
tex = texD;

if nargin && C_in_Os, tex = texT; end
Screen('FillRect', tex, p.ScreenBackground, [m/2 m/4 m m/4*3]);
       % C open to right
```

Display 6.7 (continued)

```
% Initialize a table to set up experimental conditions
p.recLabel = {'trialIndex' 'sizeIndex' 'targetPresent'…
       'respCorrect' 'respTime'};
rec = nan(nTrials, length(p.recLabel));
        % matrix rec initialized with NaN
rec(:, 1) = 1 : nTrials;
        % count the trial numbers from 1 to nTrials
sizeIndex = repmat(1 : nDispSize, [2 1 repeats]);
targetPresent = zeros(2, nDispSize, repeats);
        % first set all to 0
targetPresent(:, :, 1 : repeats / 2) = 1;
        % change first half to 1
[rec(:, 2) ind] = Shuffle(sizeIndex(:)); % shuffle size index
rec(:, 3) = targetPresent(ind);
        % shuffle present index in the same order

% Initialize Eyelink, return struct containing parameters
el = EyelinkInitDefaults(windowPtr);
EyelinkInit(1);                % dummy mode
Eyelink('Openfile', 'trackData.edf');
                               % open file to record data on eye PC
EyelinkDoTrackerSetup(el);     % calibrate the eye tracker

% Prioritize display to optimize display timing
Priority(MaxPriority(windowPtr));
str = ['Press the left arrow for target absent and the right '…
       'arrow for target present responses\n\n' …
       'Please respond as quickly and accurately as possible.\n\n'
       'Press SPACE to start.'];
DrawFormattedText(windowPtr, str, 'center', 'center', 1);
       % Draw Instruction text string
Screen('Flip', windowPtr);
       % flip the text image into the active buffer
Beeper;

WaitTill('space');      % wait till space bar is pressed
Secs = Screen('Flip', windowPtr);
p.start = datestr(now);          % record start time
Eyelink('StartRecording');       % start recording eye position
WaitSecs(0.1); % record for 100 ms
Eyelink('Message', 'SYNCTIME'); % mark zero-plot time in data file

% Run nTrials trials
for i = 1 : nTrials
    % parameters for this trial
```

Display 6.7 (continued)

```
    sz = p.dispSize(rec(i, 2));   % display size
    angles = theta + rand * 360;
    x = radius * cosd(angles) + p.ScreenRect(3) / 2;
    y = radius * sind(angles) + p.ScreenRect(4) / 2;
        % center of rec for item locations
    x = x + (rand(15, 1) - 0.5) * 2 * p.jitter; % apply jitter
    y = y + (rand(15, 1) - 0.5) * 2 * p.jitter;
    rects = CenterRectOnPoint([0 0 m m], x, y); % 15 rects
    nLoc = sz + sz / 4 - 1;
    tex(1 : nLoc) = texD;         % set up distractors
    tex([5 10 nLoc+1:15]) = texB; % 5, 10 and higher ones set as
                                  % background
    if rec(i, 3)                  % target present
        ind = randi(nLoc);
        if mod(ind, 5) == 0, ind = ind + 1; end
                                  % avoid blank location
        tex(ind) = texT;
    end
    WaitTill(Secs + p.ITI);
    Screen('DrawLines', windowPtr, fixXY, 3, 0); % fixation
    t0 = Screen('Flip', windowPtr, 0, 1);
    str = sprintf('trial=%g;angle=%g;displaySize=%g; …
          targetPresent=%g', i, angles(1), sz, rec(i, 3));
    Eyelink('Message', str); % mark trial onset

    disp(str)
    Screen('DrawTextures', windowPtr, tex, [], rects'); % C and O
    t0 = Screen('Flip', windowPtr, t0 + p.fixDuration);
    [key Secs] = WaitTill(keys); % wait till response
    Screen('Flip', windowPtr);    % remove stimulus
    if iscellstr(key), key = key{1}; end
        % take the first in case of multiple key presses
    if strcmp(key, 'esc'), break; end
    rec(i, 4) = strcmp(key, 'right') == rec(i, 3);
        % record correctness
    rec(i, 5) = Secs - t0;        % record respTime
    if rec(i, 4), Beeper; end     % beep if correct
end
p.finish = datestr(now);          % record finish time
save RT_Eyelink_rst.mat rec p;    % save the results
Eyelink('StopRecording');         % stop eye data recording
Eyelink('CloseFile');             % close data file
Eyelink('ReceiveFile'); % download data file to current folder
Eyelink('Shutdown');    % close the connection to eye PC
```

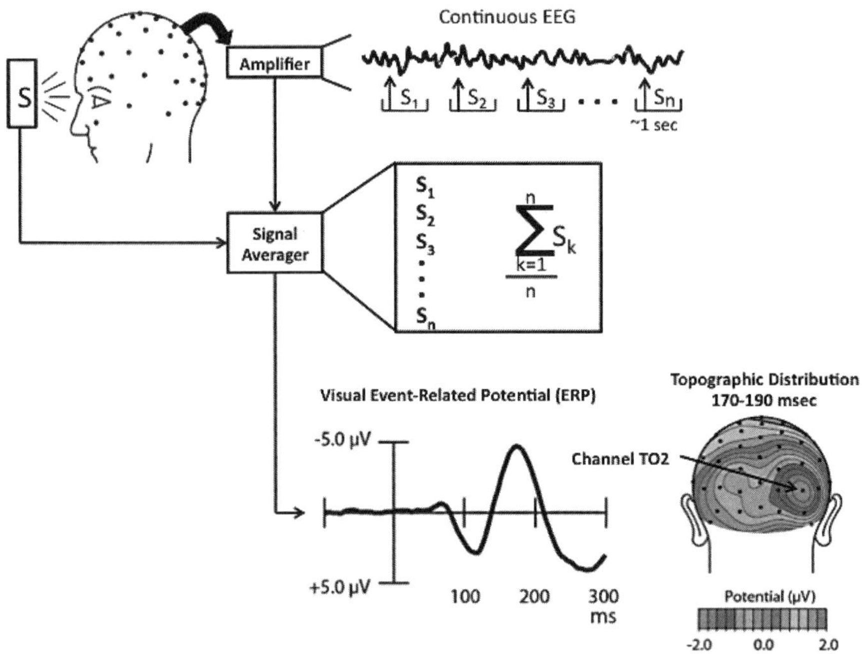

Figure 6.5
Continuous EEG traces time-locked to the onset of a visual stimulus are averaged to obtain the ERP. Topographic map of the ERP at selected latencies.

6.5.2 Magnetoencephalography

MEG[37,38] provides a converging method for measuring brain activity through the induced magnetic fields of underlying brain activity; it is a complementary measure of neuronal activities to the electrical traces of the EEG. Magnetic fields of the brain are very small compared with the ambient magnetic field induced by the earth's magnetic field. MEG devices require magnetically shielded facilities and arrays of superconducting quantum interference devices—SQUIDs—operated at low temperatures using liquid helium as a cooling agent. The experimental laboratory situations involve special considerations for removing electromagnetic activity of computers or display devices from the measurement room. For these reasons, MEG facilities are typically regionally shared devices that are operated at major university or industry research centers.

Magnetic signals that are detectable reflect currents that are perpendicular to cortical surfaces of the brain and likely involve 50,000 active neurons or more.[45] MEG responses can be measured at about 1 ms accuracy. The neurogenerators of MEG signals are the same as those of EEG signals, and source localization is also performed through a mathematical source localization model. MEG has better source localization than the EEG because magnetic fields are less susceptible to distortions from measurement through the skull. MEG is most accurate in localizing sources in brain regions near the scalp, such as primary auditory, somatosensory, and motor cortices, where the localization exceeds that of the EEG.[46] Figure 6.6 shows two sample topographic maps of MEG responses to auditory tones.[47]

Like EEG, synchronization is one important experimental issue. Here too, synchronization is typically managed through providing a time-mark input signal for the stimulus onset. The time mark often comes from a signal from the parallel port of the main computer that is managing the stimulus display and collecting response data into the computer that is controlling the imaging device and recording imaging data. The details are specific to the device and setup in the laboratory.

6.5.3 Functional Magnetic Resonance Imaging

Currently, one of the most widely used brain-imaging modalities, fMRI, infers activity in different parts of the brain based on the level of energy consumption from changes in blood oxygen.[39–41] This blood oxygenation level–dependent (BOLD) contrast is an indirect measure based on the assumption that regions of higher neural activity need more oxygen for energy and that the oxygenation level changes rapidly in response to the need. Blood containing more or less oxygen has different magnetic properties that can be measured by MRI. For many purposes, fMRI has become preferred to previous imaging technologies such as positron emission tomography (PET) that required the use of short-lived radioactive agents or other substances marking fuel consumption. It is often also preferred to X-rays for structural images. Radioactive and marking substances are more invasive, involving injection or intravenous delivery, and can be difficult to produce and have on hand. X-rays are associated with radiation exposure.

Relative BOLD activity is localized in 3D graphs or maps of brain activation. The fMRI images are often overlaid on structural MRI images that show the tissue and bone structures of the brain. The localization of activity to brain regions from MRI/fMRI can be quite precise, localizing to within a few millimeters. However, the time that it takes for blood oxygenation to change in response to neural activity is relatively sluggish, and typical dynamic brain signatures evoked by changes in the stimulus—or changes in what the brain is thinking or doing—occur in seconds, not milliseconds. So, fMRI is known for its excellent localization in brain geography. It is much more difficult to infer the timing of brain activity from fMRI, where either EEG or MEG are preferable measures.

An MRI machine has a powerful magnetic field that aligns the magnetization tissues in the body, and radio-frequency fields that alter this alignment. The timed sequences of altered magnetization are called *pulse sequences*. Different pulse sequences are selected to measure different properties or different kinds of tissue contrasts depending upon the sensitivity to MRI. Like the MEG, the MRI systems are specialized equipment requiring

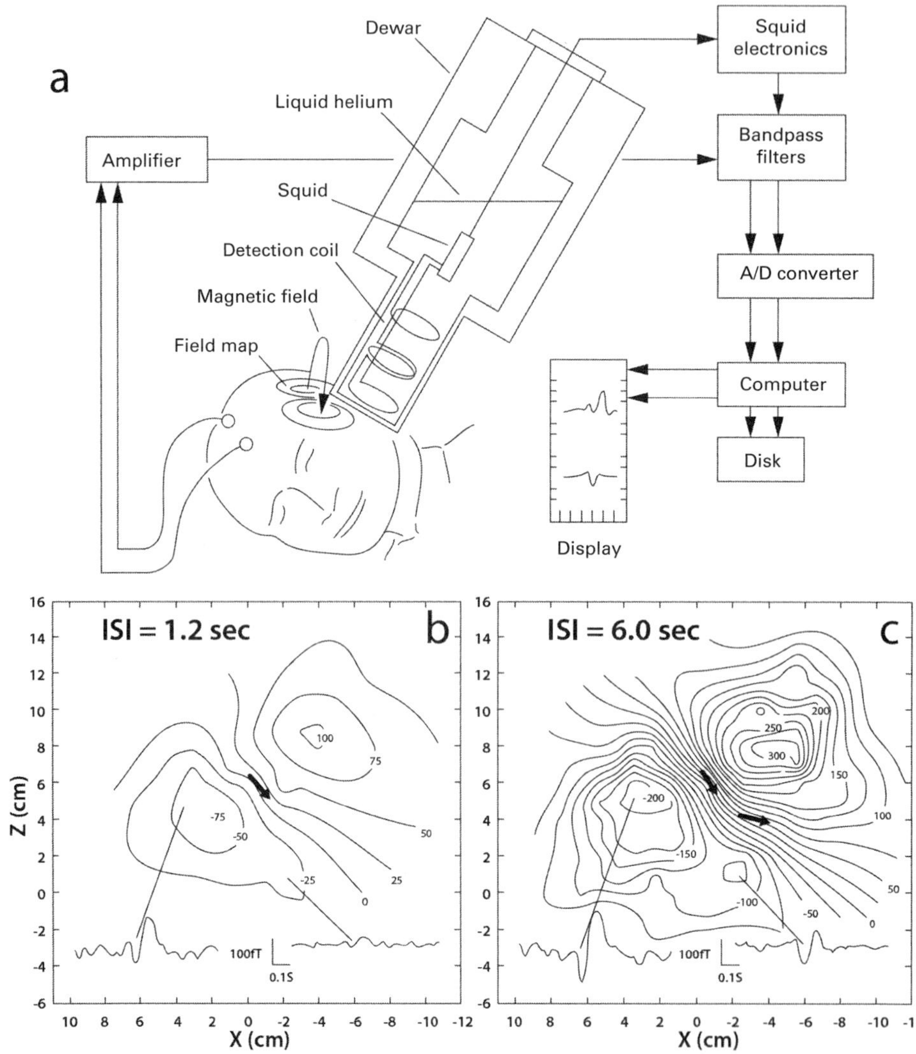

Figure 6.6
(a) Schematic of a single-channel neuromagnetometer measuring the brain's magnetic field, or MEG. (b) and (c) Isofield contours for subject ZL characterizing the measured field pattern over the left hemisphere 100 ms following the onset of a tone burst stimulus with interstimulus intervals of 1.2 and 6 s. Arrows denote the direction of current dipole sources that account for each pattern, with their bases placed at the dipole's surface location. In the right panels, the upper arrow is the N100m source and the lower arrow is the L100m source. Insets illustrate response waveforms obtained at the indicated positions. Both waveforms also exhibit a 200-ms component (after Williamson, Lu, Karren, and Kaufman[46] and Lu, Williamson, and Kaufman[47]).

lots of electricity, special cooling, magnetic shielding, and special protocols to protect subjects who may have metals within their bodies or special medical conditions that make them more vulnerable. Access to fMRI is typically provided in centralized research facilities. There are special control systems and computers to program the timing of changes in magnetic fields within the MRI machine and to record data and provide on-site checks of 3D reconstruction and signal verification.

These systems—if they are outfitted for psychophysical experimentation—will also have systems for the projection of images from a shielded source for the observer to see, noise isolation and sound presentation through suitable earphones, and response devices, all designed to work in the magnetic environment (figure 6.7).

The responsibility of the psychophysical experimenter using fMRI is to control the stimulus display in the correct temporal relationship to the sequence of MRI samples, to collect and time observer responses, and to store the sequences and timings of new stimulus response cycles. To preserve the integrity of the MRI system, synchronization is usually managed by having the MRI system send out time-mark signals to the experimental computer, and for the experimental computer to arrange to deliver the stimulus at the correct time and measure RT in relation to stimulus delivery.

Typically, MRI systems provide a TTL trigger at the beginning of each repetition time (TR), defined as the amount of time between successive pulse sequences applied to the

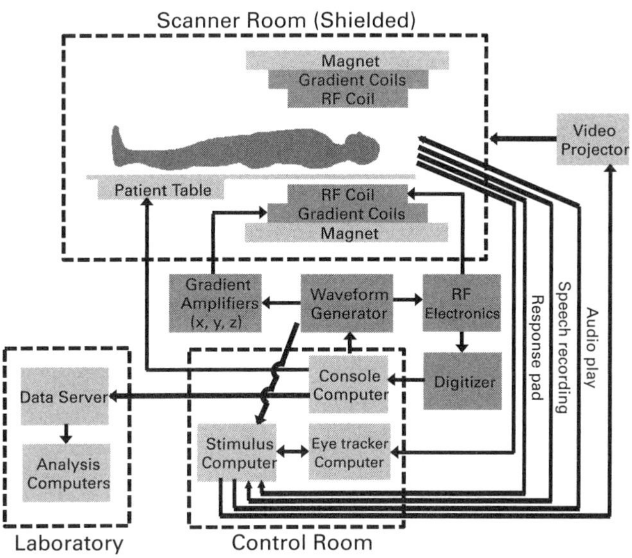

Figure 6.7
Schematics of the functional MRI setup in the Center for Cognitive and Behavioral Brain Imaging at The Ohio State University.

same brain volume. In fMRI brain imaging, the TR is typically 1–3 s. The TTL trigger input is recorded on the stimulus presentation computer. In some cases, the TTL trigger is connected via a keyboard, which has the same limitations on timing as the keyboard has for measuring RT. For better timing, the TTL trigger of a MRI system can be connected to an RTBox, in which case timing will be at millisecond accuracy.

6.6 Summary

This chapter provided a general introduction to the collection of observer-generated responses including simple key-press responses, specialized but very important behaviors such as eye movements, and physiological responses. It considered the issue of accurate timing of discrete responses and precise synchronization of displays and responses with physiological measurements. The treatment of different response collection devices and their characteristics could provide background for selecting hardware and software for a new or newly augmented experimental setup. We provide a sample program for experiments involving eye movements and gaze contingent displays that could serve as an example and template for such programs that interface with other kinds of measurement technologies. Finally, we introduced some of the major modalities of physiological measurement with special focus on the critical aspects of synchronization with visual and auditory displays. With these tools, a wide range of simple to complex responses of a human agent can be measured.

References

1. Jastrow J. *The time-relations of mental phenomena*. New York: Hodges; 1890.

2. Luce RD. *Response times: Their role in inferring elementary mental organization*. Oxford: Oxford University Press; 1991.

3. Reed AV. 1973. Speed-accuracy trade-off in recognition memory. *Science* 181(4099): 574–576.

4. Wickelgren WA. 1977. Speed-accuracy tradeoff and information processing dynamics. *Acta Psychol (Amst)* 41(1): 67–85.

5. Dosher BA. 1976. The retrieval of sentences from memory: A speed-accuracy study. *Cognit Psychol* 8(3): 291–310.

6. Li X, Liang Z, Kleiner M, Lu ZL. 2010. RTbox: A device for highly accurate response time measurements. *Behav Res Methods* 42(1): 212–225.

7. Li X, Lu ZL. 2012. Enabling high grayscale resolution displays and accurate response time measurements on conventional computers. *J Vis Exp* 60: e3312.

8. Empirisoft Corporation. Available at: www.empirisoft.com/directin.aspx.

9. Stewart N. 2006. A PC parallel port button box provides millisecond response time accuracy under Linux. *Behav Res Methods* 38(1): 170–173.

10. Psychtoolbox-3. [computer program] Available at: http://psychtoolbox.org/HomePage.

11. RTBox. Available at: http://lobes.osu.edu/rt-box.php.

12. Dosher BA, Han S, Lu ZL. 2004. Parallel processing in visual search asymmetry. *J Exp Psychol Hum Percept Perform* 30(1): 3–12.

13. Treisman A, Gormican S. 1988. Feature analysis in early vision: Evidence from search asymmetries. *Psychol Rev* 95(1): 15–48.

14. Wolfe JM, Friedman-Hill SR. 1992. Visual search for oriented lines: The role of angular relations between targets and distractors. *Spat Vis* 6(3): 199–207.

15. Wolfe JM. 2001. Asymmetries in visual search: An introduction. *Atten Percept Psychophys* 63(3): 381–389.

16. Nagy AL, Cone SM. 1996. Asymmetries in simple feature searches for color. *Vision Res* 36(18): 2837–2847.

17. Rosenholtz R. 2001. Search asymmetries? What search asymmetries? *Atten Percept Psychophys* 63(3): 476–489.

18. Williams D, Julesz B. 1992. Perceptual asymmetry in texture perception. *Proc Natl Acad Sci USA* 89(14): 6531–6534.

19. Rubenstein BS, Sagi D. 1990. Spatial variability as a limiting factor in texture-discrimination tasks: Implications for performance asymmetries. *JOSA A* 7(9): 1632–1643.

20. Kowler E. *Eye movements and their role in visual and cognitive processes*, Vol. 4. Amsterdam: Elsevier Science Ltd; 1990.

21. Hoffman JE, Subramaniam B. 1995. The role of visual attention in saccadic eye movements. *Atten Percept Psychophys* 57(6): 787–795.

22. Rayner K. 1998. Eye movements in reading and information processing: 20 years of research. *Psychol Bull* 124(3): 372–422.

23. Hayhoe M, Ballard D. 2005. Eye movements in natural behavior. *Trends Cogn Sci* 9(4): 188–194.

24. Allopenna PD, Magnuson JS, Tanenhaus MK. 1998. Tracking the time course of spoken word recognition using eye movements: Evidence for continuous mapping models. *J Mem Lang* 38(4): 419–439.

25. Kremenitzer JP, Vaughan HG, Jr, Kurtzberg D, Dowling K. 1979. Smooth-pursuit eye movements in the newborn infant. *Child Dev* 50(2): 442–448.

26. Young LR, Sheena D. 1975. Survey of eye movement recording methods. *Behav Res Methods* 7(5): 397–429.

27. Crane HD, Steele CM. 1985. Generation-V dual-Purkinje-image eyetracker. *Appl Opt* 24(4): 527–537.

28. Woestenburg J, Verbaten M, Slangen J. 1983. The removal of the eye-movement artifact from the EEG by regression analysis in the frequency domain. *Biol Psychol* 16(1): 127–147.

29. Guestrin ED, Eizenman M. 2006. General theory of remote gaze estimation using the pupil center and corneal reflections. *IEEE Trans Biomed Eng* 53(6): 1124–1133.

30. SR Research. Available at: http://www.sr-research.com/.

31. Jung R, Kornhuber H. *Neurophysiologie und Psychophysik des visuellen Systems*. [The visual system: Neurophysiology and psychophysics] Berlin: Springer-Verlag; 1961.

32. Jung R. 1961. Neuronal integration in the visual cortex and its significance for visual information. In: Rosenblith W, ed. *Sensory communication*. Cambridge, MA: MIT Press; 1961: 627–674.

33. Fechner G. Elemente der psychophysik. Leipzig: Breitkopf & Härtel; 1860.

34. Nunez PL, Srinivasan R. *Electric fields of the brain: The neurophysics of EEG*. New York: Oxford University Press; 2006.

35. Regan D. *Human brain electrophysiology: Evoked potentials and evoked magnetic fields in science and medicine*. New York: Elsevier; 1989.

36. Niedermeyer E, Da Silva FHL. *Electroencephalography: Basic principles, clinical applications, and related fields*. Philadelphia: Lippincott Williams & Wilkins; 2005.

37. Lu ZL, Kaufman L. *Magnetic source imaging of the human brain*. Hillsdale, NJ: Lawrence Erlbaum Associates; 2003.

38. Hansen PC, Kringelbach ML, Salmelin R. *MEG: An introduction to methods*. Oxford: Oxford University Press; 2010.

39. Huettel SA, Song AW, McCarthy G. *Functional magnetic resonance imaging*, Vol. 1. Sunderland, MA: Sinauer Associates; 2004.

40. Poldrack RA, Mumford J, Nichols T. *Handbook of functional MRI data analysis*. Cambridge, UK: Cambridge University Press; 2011.

41. Buxton RB. *Introduction to functional magnetic resonance imaging: Principles and techniques*. Cambridge, UK: Cambridge University Press; 2002.

42. Villringer A, Planck J, Hock C, Schleinkofer L, Dirnagl U. 1993. Near infrared spectroscopy (NIRS): A new tool to study hemodynamic changes during activation of brain function in human adults. *Neurosci Lett* 154(1): 101–104.

43. Kato T, Kamei A, Takashima S, Ozaki T. 1993. Human visual cortical function during photic stimulation monitoring by means of near-infrared spectroscopy. *J Cereb Blood Flow Metab* 13: 516–520.

44. Michel CM, Murray MM, Lantz G, Gonzalez S, Spinelli L, Grave de Peralta R. 2004. EEG source imaging. *Clin Neurophysiol* 115(10): 2195–2222.

45. Lu ZL, Williamson S. 1991. Spatial extent of coherent sensory-evoked cortical activity. *Exp Brain Res* 84(2): 411–416.

46. Williamson SJ, Lu ZL, Karron D, Kaufman L. 1991. Advantages and limitations of magnetic source imaging. *Brain Topogr* 4(2): 169–180.

47. Lu ZL, Williamson SJ, Kaufman L. 1992. Human auditory primary and association cortex have differing lifetimes for activation traces. *Brain Res* 572(1–2): 236–241.

III THEORETICAL FOUNDATIONS

7 Scaling

The goal of visual psychophysics is the quantification of perceptual experience—understanding the relationship between the external stimulus, the internal representation, and the overt responses. Scaling is one of the major tools in psychophysics to measure the perceived intensity or magnitude of different physical stimuli or to understand the relationships of different stimuli in perceptual space as a function of physical variation. Scaling allows us to infer the relationship between stimuli from their distances within this space. In this chapter, we illustrate and discuss several of the most powerful and commonly used scaling methods, such as direct scaling, indirect scaling, multidimensional scaling, and their theoretical underpinnings. We anticipate that applications of scaling to brain imaging will provide new methods of constraining theories from both approaches.

7.1 The Scaling Enterprise

Every visual stimulus generates an internal representation in the perceptual system of the observer. The goal of scaling in psychophysics is to quantify the relationship between perceptual intensities or qualities and properties of the physical stimulus.[1–5] This is accomplished through various empirical and theoretical methods of scaling including magnitude estimation, magnitude production, and similarity/dissimilarity judgments. The goal of the scaling enterprise is to derive an internal representation of the stimuli in either a single-dimensional or multiple-dimensional space. This internal representation is then used to predict performance and brain responses for other stimulus combinations.

Every psychophysical experiment involves the creation of an internal representation in the brain from the physical stimulus, and also the production of one or more responses using a decision structure or decision rule (figure 7.1). In visual psychophysics, we specify and control the physical stimulus and measure the response. To infer accurately the relationship between the physical stimulus and the internal representation, we must also understand the transformation from the internal representation to the response.[6]

The advancement of neurophysiology and brain imaging by electroencephalography (EEG), magnetoencephalography (MEG), and functional magnetic resonance imaging

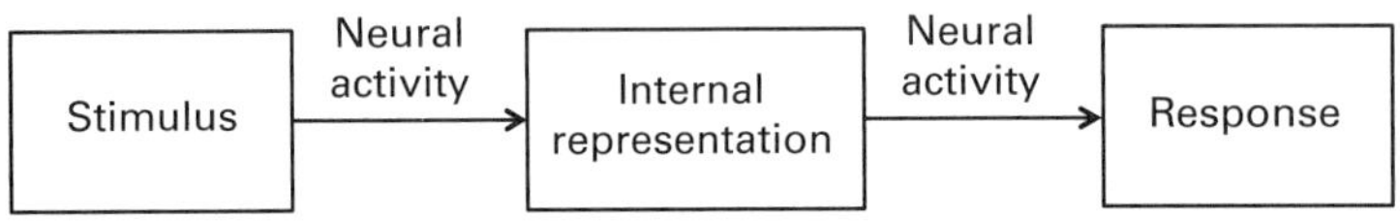

Figure 7.1
Mapping the physical stimulus onto an internal representation and then an overt response.

(fMRI) created new tools that may provide insights into both the internal representations of stimuli and the decision rules underlying responses. A combination of the information obtained from psychophysical scaling and neurophysiology, or *neuropsychophysics*, will prove useful in resolving ambiguities in both domains.[1,7,8]

Some of the most important quantitative "laws" or relationships between simple stimulus variations and perceptual qualities that model human performance have been discovered through scaling. In cases where the perceptual representations are more complicated, such as for complex objects, characterizing stimulus variations in multidimensional psychological spaces can yield important metrics for the stimulus variations themselves. Understanding these complicated representational spaces may improve the interpretation of the neural codes that represent the stimuli. This in turn may be critical to the development of devices that augment human perception and cognition.

7.2 Single-Dimensional Scaling

7.2.1 Direct Estimation

Many psychophysical experiments vary stimuli along a single dimension, such as intensity, contrast, physical length, and so on. Single-dimensional scaling is used to measure the subjective quantity induced by the changes in the physical stimulus along a corresponding psychological dimension.

One classical method for doing unidimensional scaling is magnitude estimation.[9,10] An example of this for luminance intensity was illustrated in section 2.1 of chapter 2, where we showed hypothetical magnitude rating data and the resulting functional relationship between the physical variation of intensity and the rated intensity on a numerical scale. Often, a sample test stimulus and intended corresponding magnitude (i.e., an intensity corresponding to a scale value of 10) is shown to anchor the responses. Observers are required to provide numerical values in reference to the anchor. Magnitude estimation has been widely used to estimate the functional relationships between the intensity of the physical stimulus and the magnitude rating.

Another scaling method is magnitude production.[11–15] In magnitude production, an observer produces a stimulus value that meets some criterion, such as the perceived midpoint between two presented stimuli of different intensities. This approach uses successive interpolations to estimate the function relating the magnitude rating responses to manipulations of stimulus intensity.

Magnitude estimation and magnitude production experiments have been widely studied and have generated well-known functional relationships between the stimulus variation and the magnitude response. The most famous of these is the *power-law* relationship known as Steven's law. The power function is

$$g(I) = \alpha I^{\beta}, \tag{7.1}$$

where I is the physical intensity or magnitude of the stimulus, β is the power exponent, and α is a scaling factor. There are three kinds of relationships (figure 7.2): linear when β is 1, compressive when β is less than 1, and expansive when β is greater than 1. In a compressive scale, larger and larger stimulus differences are needed to yield a given difference on the internal scale as the stimulus intensity increases. In an expansive scale, the opposite is true. The measured exponents are typically compressive for dimensions

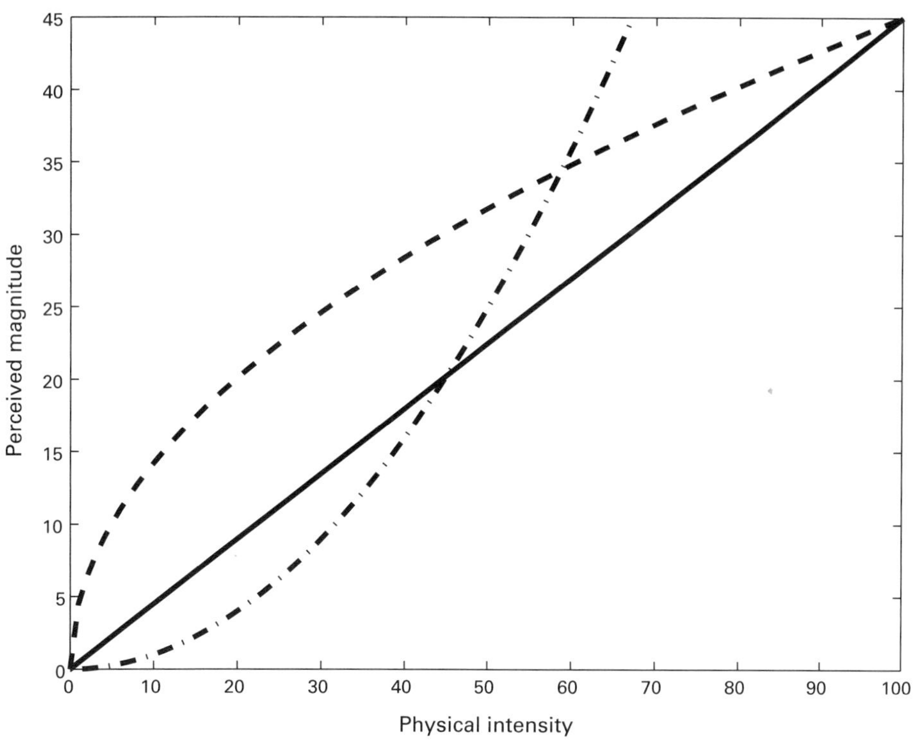

Figure 7.2
Illustrations of compressive, linear, and expansive relationships between stimulus values and perceptual strength.

such as loudness and brightness, have been reported occasionally to be near 1 for line length, and are often expansive when dealing with some aspects of taste or noxious stimuli.[10,16]

Direct scaling seeks to understand how changes in the physical stimulus are reflected in the internal representation that underlies the psychological percept. However, the direct scaling methods actually estimate the relationship between the magnitude response and the stimulus. Inferring the internal representation from this stimulus–response relationship requires that we know how the internal representation is mapped to the overt response. Usually, it is implicitly assumed that the magnitude score directly accesses the internal representation—that is, the function that relates the internal representation to response is linear. Discovery of the accurate relationship between the external stimulus and the internal representation depends upon the validity of this assumption.[6] This assumption is rarely tested, although there is evidence that the function is nonlinear in at least some circumstances.[17]

One way to check the validity of a linear linking function between the internal representation and the reported magnitude is to relate the results of one power-law to another by measuring the relationship through either within-modality matching or cross-modality matching.[17] For example, if loudness and brightness are each described by their own power-law functions with different exponents β_A and β_V and the linking function is linear in both cases, then this implies specific relationships between pairs of auditory and visual stimuli. So, if you start with one loudness and one brightness that are perceived to be matched on intensity, and if you double the perceived loudness by using the auditory power function and do the same for brightness, then the new loudness and the new brightness should also match. Consistency in such tests provides evidence that the two linking functions are the same and are more likely to be linear. Empirically, such tests sometimes fail to provide consistent results.[2]

Recent advancements in neurophysiology and brain imaging may bring tools that will contribute to understanding both the representation and the linking functions or decision processes that take the internal representation(s) to responses. This is a point we return to later in the chapter.

7.2.2 Weber–Fechner Scaling

The idea of Weber–Fechner scaling is to use measurements of the *discriminability* of stimuli to estimate the magnitude of the internal response of stimuli of different intensities.[1,18] Discriminability is the ability to perceive that two stimuli are different. The basic idea is to measure the discriminability between two near stimuli, I and $I+\Delta I$, that differ by increment ΔI, at several levels of intensity I. The stimuli generate internal responses $g(I)$ and $g(I+\Delta I)$. By measuring the probability that the more intense stimulus is judged to be more intense, $[g(I+\Delta I)>g(I)]$, it is possible to infer properties of the internal representation.[19]

Weber originally measured the increment ΔI in the physical stimulus intensity needed for the observer to perceive the larger stimulus to be greater 75% of the time, defined to be the "just noticeable difference" (JND). He found that the increment ΔI was proportional to the magnitude of the baseline stimulus, or $\Delta I/I = k$ over a wide range of intensities I. This relationship has been called "Weber's law."[18]

One example experiment measures the JND for line length. On each trial, line segments with length L_i and $L_i + \Delta L_i$ are presented, and the subject judges which is longer. A staircase procedure (see section 11.2 in chapter 11) or a method of limits (see section 8.2 of chapter 8) procedure is used to find ΔL_i such that $L_i + \Delta L_i$ is perceived as longer 75% of the time. The experiment measures this for eight different baseline lengths L_i over a wide range. The experimenter then graphs ΔL_i as a function of L_i. This functional relationship is approximately linear and can be fit by a straight line to estimate the Weber fraction, k (figure 7.3).

Weber's law is an empirical statement about the constancy of the ratio of the JND to stimulus intensity. However, the law by itself does not specify the internal scale,

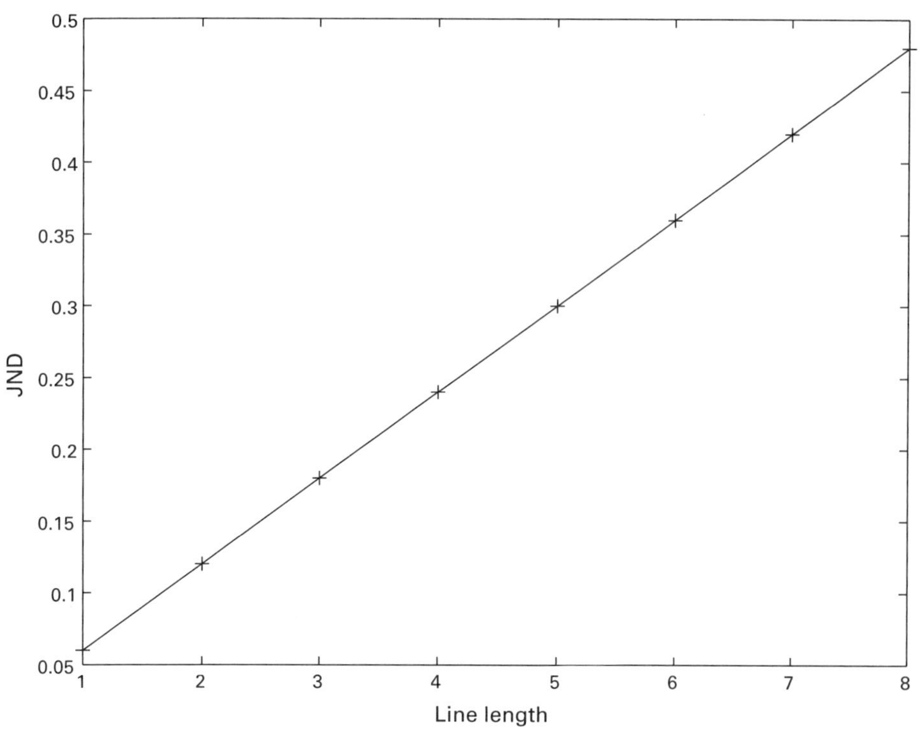

Figure 7.3
Just noticeable difference (JND) as a function of baseline length.

$g(I_i)$.Fechner reformulated Weber's law to create a psychological scale from the measured JNDs. He showed that if Weber's law holds, then the function of intensity is logarithmic: $g = A\ \log(I) + C$. This formulation is called Fechner's law.[1] Fechner's reformulation of Weber's law makes a critical assumption that a function F maps the internal representations of a pair of stimuli to their discriminability in a way that depends only on the difference between the internal scale values: $[g(I+\Delta I) - g(I)]$, and that this is a constant for a particular level of discriminability. In this formulation, $P(g(I+\Delta) > g(I)) = F[g(I+\Delta I) - g(I)] = k_1$ and $[g(I+\Delta I) - g(I)] = F^{-1}(k_1) = k_2$, or

$$\frac{\delta g}{\delta I} = \frac{k_2}{\Delta I} = \frac{k_2}{k}\frac{1}{I},$$

corresponding with a logarithmic scale.

This critical assumption of Fechner's law still must be tested empirically. Some experiments suggest that the function F may also depend in a subtle way on the base intensity I and other factors.[20–23] There are cases in which Weber's law does not hold. In auditory perception, the Weber fraction k is not exactly a constant. It is slightly lower for more intense auditory stimuli, a phenomenon called the "near miss to Weber's law."[13]

7.2.3 Thurstone Scaling

Thurstone introduced the law of comparative judgment for psychophysical scaling.[24] This approach uses pairwise comparative judgments for close stimuli to estimate the distance between pairs of distant stimuli in the internal representation. Thurstone was one of the first researchers to explicitly acknowledge that internal representations have variability or are noisy.

Thurstone assumed that stimuli S_i and S_j have mean internal representation values of $g(S_i)$ and $g(S_j)$ with variances σ_i^2 and σ_j^2. The noise associated with stimuli S_i and S_j may in the most general case be correlated with correlation coefficient r_{ij}. Thurstone defined the distance between the mean internal representations of the two stimuli as[24]

$$[g(S_i) - g(S_j)] = d_{ij}\sqrt{\sigma_i^2 + \sigma_j^2 - 2r_{ij}\sigma_i\sigma_j}\ , \tag{7.2}$$

where d_{ij} is the z-score of the probability that S_i is judged greater than S_j.

In Thurstone scaling, the experimenter measures the probability that one stimulus is judged greater than another for many different stimulus combinations. These data are used to estimate scale values $g(S_i)$ for all tested stimuli. Most researchers simplify the estimation of scale values by assuming that the variabilities of the internal representations are independent and identically distributed—Thurstone's case V[25–27]:

$$[g(S_i) - g(S_j)] = d_{ij}\sqrt{2\sigma^2}\ . \tag{7.3}$$

This means that all the noise variances are equal, and all the noise correlations are 0. Other cases considered by Thurstone (cases I to IV) relax these simplifying assumptions in various ways.

Thurstone's approach could be used in visual psychophysics to measure the positions of stimuli along an arbitrary single internal dimension. For example, Thurstone scaling could estimate the distance on an internal representation between sine waves of different spatial frequencies. The experimenter selects sine waves with a number of different spatial frequencies between the lowest and the highest spatial frequency to be tested and performs pairwise comparative judgments of which has higher frequency for each pair. These data are used to estimate the distances between near pairs, and the distances between highly separated pairs are estimated by cumulating shorter distances.

In figure 7.4, we present data from a hypothetical spatial frequency scaling experiment. In this experiment, seven different spatial frequencies (indexed by numbers 1 to 7, from

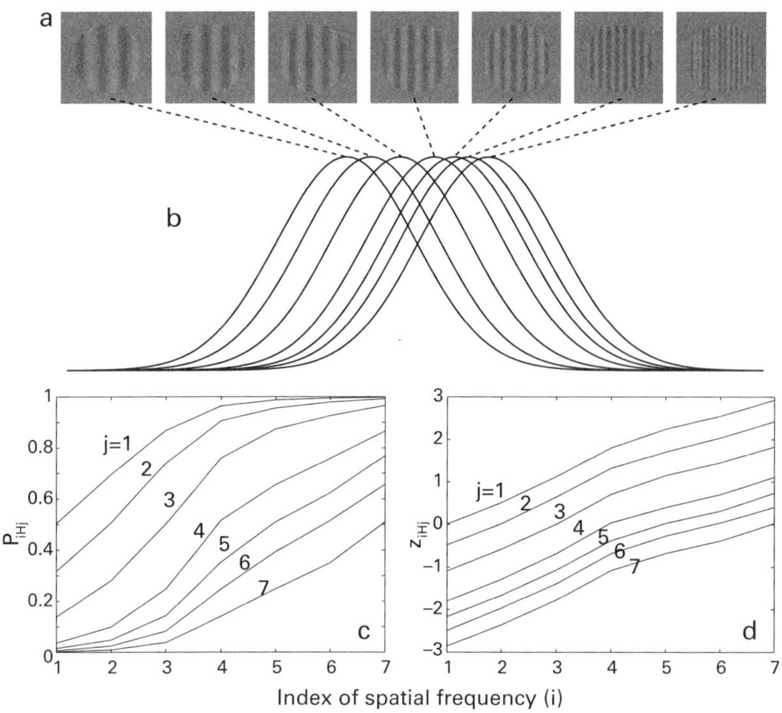

Figure 7.4
An example of Thurstone scaling of different spatial frequency stimuli. (a) Sine-wave gratings of seven different spatial frequencies. (b) Internal representation of the seven sine-wave gratings on a single dimension assuming independence and equal variance. (c) and (d) Hypothetical probability and *z*-score data in which the spatial frequency indexed by *i* is judged to be higher than the spatial frequency indexed by *j*.

low to high) are used. In each trial, the observer is presented with a pair of spatial frequencies and asked to judge which one is higher. This is repeated many times. At the end of the experiment, we can summarize the data in terms of P_{iHj}, the observed proportion that spatial frequency i is judged to be higher than spatial frequency j (figure 7.4a). The proportions are also converted into corresponding z-scores, z_{iHj} (figure 7.4b). If we set the first spatial frequency (i = 1) at the origin and estimate the distance of all the other spatial frequencies (i = 2,..., 7) to the first spatial frequency as d_i, we can express z_{iHj} as

$$z_{iHj} = \sqrt{2}(d_j - d_i). \tag{7.4}$$

Displays 7.1 and 7.2 present MATLAB programs for recovering the internal psychological distances from the data shown in figure 7.4. The recovered distances are 0.347, 0.776, 1.268, 1.558, 1.764, and 2.050. For Thurstone case V, these distances are in units of the common standard deviation. The program enters (hypothetical) P_{iHj} data and provides initial guesses for the locations of stimuli on the internal dimension. A fitting routine, `fminsearch`, is used to find the internal locations that optimize the fit to the data. This kind of model estimation is treated in detail in chapter 10.

Display 7.1

```
%%% Program ThurstoneScaling.m
P_iHj = …
        [0.5181  0.6937  0.8740  0.9665  0.9863  0.9940  0.9983
         0.3256  0.5192  0.7284  0.9113  0.9598  0.9785  0.9927
         0.1372  0.2813  0.5183  0.7702  0.8745  0.9240  0.9642
         0.0394  0.1050  0.2527  0.5151  0.6690  0.7641  0.8713
         0.0142  0.0480  0.1360  0.3497  0.5009  0.6198  0.7707
         0.0068  0.0236  0.0881  0.2425  0.3905  0.5076  0.6694
         0.0021  0.0084  0.0379  0.1406  0.2522  0.3577  0.5151];

initial_guess = [0.7 1.0 1.5 2.1 2.8 3.5];
oldopts=optimset('fminsearch');
newopts=optimset('Tolx', 1.e-10, 'Tolfun', 1.e-10, 'MaxIter', …
        100000,  'TolFun', 0.00000000001, 'MaxFunEvals', …
        100000, 'Diagnostics', 'on');
options = optimset(oldopts, newopts);
Distances = fminsearch('costfunction', initial_guess, …
        options, P_iHj)
```

Display 7.2

```
%%% Program costfunction.m
function L = costfunction(initial_guess, P_iHj)

d = [0 initial_guess];       % distance of the first spatial
                             % frequency is defined to be 0.
zData = norminv(P_iHj);      % z_iHj is obtained from P_iHj
through z-transformation

L = 0;
for i = 1:7
    for j = 1:7
        z(i, j) = sqrt(2) * (d(j) - d(i));
        L = L + (zData(i, j) - z(i,j))^2;
    end
end

r2 = 1 - L/sum(sum((z(i,j) - mean2(zData)).^2));
```

Thurstone scaling relies on pairwise measurements and only close pairs yield meaningful probability data that are not saturated to near 0 or near 1. Long-range scales that are constructed from these local measures usually assume that the scale is unidimensional and on a straight line embedded in a Euclidean space. This corresponds to assuming that you can add many small distances to estimate a long distance—but these long distances cannot be directly tested in the data in a meaningful way.

In general Thurstone scaling, not just the mean scale values but also the variances and the correlations are free parameters in the solution. Most meaningful comparisons—those where one stimulus does not completely dominate another—are limited to near neighbors. For this reason, the simplifying assumptions of Thurstone case V of equal variances and zero correlations may not be easy to test. The estimated distances of highly separated stimuli may be altered, for example, by variances that change systematically with the mean, so as to compress or expand the distance scale at one end. Yet these may not be empirically detectable. Also, the assumption of unidimensionality may be violated in some instances. Researchers have developed other techniques to place stimuli in a multidimensional space (see section 7.3).

7.2.4 Measurement Theory

In all of these approaches to scaling, we hope to understand the internal or psychological scale that corresponds with variations in the physical stimulus. Stimulus manipulations often are changes along physical scales that embody interval or ratio information. A related but different approach to inferring the scales of the internal representation is the area of measurement theory. Measurement theory asks whether the internal or psychological scale preserves the strong properties of a physical scale, such as length which has a physical ratio scale, or instead has weaker properties with respect to various mathematical operations.[28–30] In other words, to quote Ashby[31](p. 103): "When are we justified in representing phenomena with properties X by numerical structure Y?" or "How much are we allowed to read into the numbers that result?"

The field of measurement theory often tests relationships between stimuli, or between pairs of stimuli, that probe necessary mathematical relations implied by ratio, interval, or ordinal scale representations.[32] Good primary references to the extensive mathematical development in measurement theory can be found in the three volumes of *Foundations of Measurement* and in others.[13,33–36] Several examples are considered next.

In direct estimation, subjects are asked to assign numbers to stimuli. However, numbers are not just words, but carry with them mathematical operations. Numbers can be used to indicate names or categories (nominal scale), in which case they provide no information about relationships between corresponding stimuli. Or, numbers may indicate order or ranking information between examples (ordinal scale). Numbers may also preserve sums and differences (interval or ratio scales). Understanding the relational and other properties of the assigned numbers provides information about the internal representations. If the assigned numbers of direct estimation obey interval scale properties, then we infer that the internal representation must also preserve those same interval properties. In contrast, if the internal representations preserve only ordinal properties, then there would be no basis for assigned numerical values to show interval scale consistency.

Conjoint measurement theory asks the question whether it is possible to measure the corresponding psychological scales for objects with distinct attributes A and X through stimulus combinations and judgments.[37–39] The theory assumes that stimulus attributes A and X combine (represented by the symbol $\oplus$) non-interactively to determine the value of a perceived variable P. The psychological scales for A and X are unknown. The theory tests whether P's resulting from different combinations of different values of A and X are consistent in the data. For example, if $a_i \oplus x_k$ is judged greater than $a_j \oplus x_k$, then $a_i \oplus x_l$ should be judged greater than $a_j \oplus x_l$. Elaborate versions of these and other similar tests—the so-called cancellation theorems—are sometimes used to infer interval properties of the scales of A and X as well as to validate the assumption of non-interactive combination. If all these properties hold, then we conclude that the internal representations of A and X are on an interval scale.

Conjoint measurement has been widely used in marketing and other applications where people are asked to judge objects with multiple attributes.[40–42] For example, fuel efficiency (mpg*)* and engine power (rpm) of cars both contribute to people's judgment of their value or desirability. Conjoint measurements have also been used in psychophysical applications. In classic examples from audition, for example, the two attributes were the loudness of a stimulus in the two ears.[43,44] Ideas from conjoint measurement have recently been applied to vision[45] to examine the perceptual combination of different levels of gloss (shiny appearance) and surface texture.

7.3 Multidimensional Scaling

The earlier sections of this chapter focused on understanding the internal representations of stimuli (such as a series of increasing magnitude; i.e., line lengths) that differ along a single dimension. However, internal perceptual representations are often multidimensional. Multidimensional scaling (MDS) is a set of statistical procedures or algorithms that position stimuli within a multidimensional geometric representation that expresses the psychological distances between the stimuli based on dissimilarity judgments or other measures of proximity or distance. The MDS algorithms explain many pairwise measures of similarity/dissimilarity by estimating the locations of the stimuli within a multidimensional space. An MDS solution provides a compact interpretation of the pairwise data and allows predictions about all combinations from a small set of estimated parameters.

Metric MDS assumes a particular distance metric, so that distances between items in the internal representation are computed as Euclidean, city-block, or some other Minkowski distance. The Minkowski distance between two points $(x_1, x_2, ..., x_n)$ and $(y_1, y_2, ..., y_n)$ is

$$\left(\sum_{i=1}^{n} |x_i - y_i|^p\right)^{1/p}.$$

Setting $p = 2$ corresponds to Euclidean or straight-line distance; $p = 1$ corresponds to the "city-block" distance, or the distance between two objects along edges of a grid. As $p \to \infty$, the computed distance approaches the single largest difference; as $p \to -\infty$, the computed distance approaches the single smallest difference. The most commonly used metric is the Euclidean distance metric, followed by the city-block metric. In non-metric MDS, ordinal data are predicted; only the rank order of distances is considered in the predictions.[3–5,46–48]

One famous example of MDS by Shepard[49] recovered the positions of 14 colored lights of different wavelengths in a two-dimensional (2D) internal representation (the third dimension, luminance, is not considered here). The wavelengths ranged from violet (434 nm) through blue, green, yellow, to red (674 nm). The judged similarities between all 91 possible color pairs were taken from data by Ekman (table 7.1).[50] Shepard's MDS solution for the positions of the colors in the 2D space is qualitatively very similar to the standard color wheel (figure 7.5a).

The MDS solution also provided an estimate of the function relating the judged similarity to the Euclidean distances in the internal representation. This so-called *transfer function* or linking function between estimated internal Euclidean distances and the judged similarity is not linear in Shepard's solution (see Figure 7.5b).

Table 7.1
Judged similarities between 14 spectral colors from Ekman[50]

	Judged Similarity													
434	1.00	0.86	0.42	0.42	0.18	0.06	0.07	0.04	0.02	0.07	0.09	0.12	0.13	0.16
445	0.86	1.00	0.50	0.44	0.22	0.09	0.07	0.07	0.02	0.04	0.07	0.11	0.13	0.14
465	0.42	0.50	1.00	0.81	0.47	0.17	0.10	0.08	0.02	0.01	0.02	0.01	0.05	0.03
472	0.42	0.44	0.81	1.00	0.54	0.25	0.10	0.09	0.02	0.01	0.00	0.01	0.02	0.04
490	0.18	0.22	0.47	0.54	1.00	0.61	0.31	0.26	0.07	0.02	0.02	0.01	0.02	0.00
504	0.06	0.09	0.17	0.25	0.61	1.00	0.62	0.45	0.14	0.08	0.02	0.02	0.02	0.01
537	0.07	0.07	0.10	0.10	0.31	0.62	1.00	0.73	0.22	0.14	0.05	0.02	0.02	0.00
555	0.04	0.07	0.08	0.09	0.26	0.45	0.73	1.00	0.33	0.19	0.04	0.03	0.02	0.02
584	0.02	0.02	0.02	0.02	0.07	0.14	0.22	0.33	1.00	0.58	0.37	0.27	0.20	0.23
600	0.07	0.04	0.01	0.01	0.02	0.08	0.14	0.19	0.58	1.00	0.74	0.50	0.41	0.28
610	0.09	0.07	0.02	0.00	0.02	0.02	0.05	0.04	0.37	0.74	1.00	0.76	0.62	0.55
628	0.12	0.11	0.01	0.01	0.01	0.02	0.02	0.03	0.27	0.50	0.76	1.00	0.85	0.68
651	0.13	0.13	0.05	0.02	0.02	0.02	0.02	0.02	0.20	0.41	0.62	0.85	1.00	0.76
674	0.16	0.14	0.03	0.04	0.00	0.01	0.00	0.02	0.23	0.28	0.55	0.68	0.76	1.00
	434	**445**	**465**	**472**	**490**	**504**	**537**	**555**	**584**	**600**	**610**	**628**	**651**	**674**

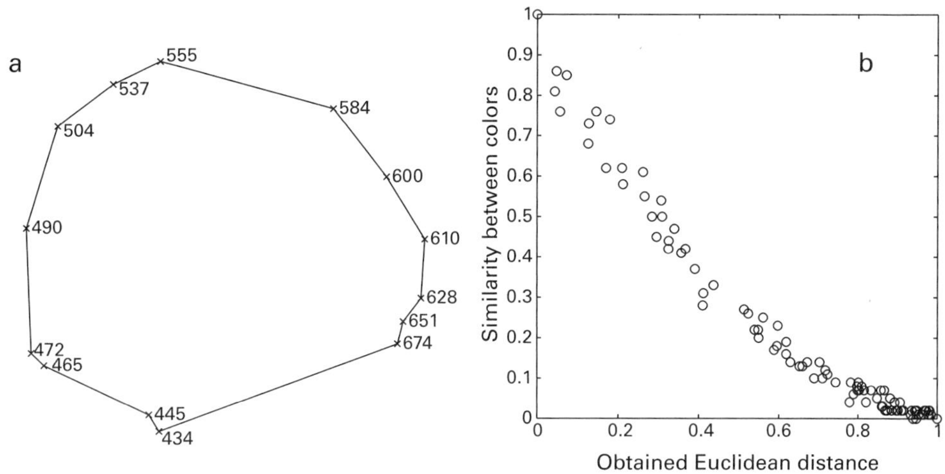

Figure 7.5
MDS solution for the similarity judgment data in Ekman[50] for colored lights. (a) Two-dimensional configuration obtained by MDS. (b) A graph of the relation between judged similarities and corresponding Euclidean distances between points in the MDS solution.

Display 7.3 enters the similarity data used by Shepard and uses a simple MATLAB function called `mdscale`, specifying a 2D solution, to recover the representation of the 14 colored lights. The solution is expressed as 14 (x_i, y_i) locations in a two-dimensional coordinate system. In display 7.3, we show a simple MATLAB program that calls `mds`, the multidimensional scaling routine, to recover the color space using Ekman's data. We closely recapitulate Shepard's analysis,[49] with essentially identical results. What was a demanding project in 1980 is now routine due to the development of faster computer hardware and algorithms.

Another good example of the application of MDS is the shape-variation study of Cutzu and Edelman[51] (figure 7.6). They studied the perception of several sets of artificial animal-like shapes created from generalized cylinders in different configurations (figure 7.6a).

The different artificial animal stimulus sets were created with distinct kinds of spacing within the 2D variations of shape along physical dimensions. All these animal-like shapes had the same parts, modeled by generalized cylinders, and were controlled by 70 parameters that determined the geometry of the parts and their mutual arrangement, via nonlinear functions. MDS did an excellent job of recovering the relationships in all these different sets using empirical similarity scores collected in several different ways. The empirical methods included pairwise ratings, comparisons of one pair to another pair, long-term memory confusions, and delayed match to sample tests. These tests served as a validation of the MDS method.

Display 7.3

```
%%% Program shepard.m
Judged_similarities = …
[1.0 .86 .42 .42 .18 .06 .07 .04 .02 .07 .09 .12 .13 .16
 .86 1.0 .50 .44 .22 .09 .07 .07 .02 .04 .07 .11 .13 .14
 .42 .50 1.0 .81 .47 .17 .10 .08 .02 .01 .02 .01 .05 .03
 .42 .44 .81 1.0 .54 .25 .10 .09 .02 .01 .00 .01 .02 .04
 .18 .22 .47 .54 1.0 .61 .31 .26 .07 .02 .02 .01 .02 .00
 .06 .09 .17 .25 .61 1.0 .62 .45 .14 .08 .02 .02 .02 .01
 .07 .07 .10 .10 .31 .62 1.0 .73 .22 .14 .05 .02 .02 .00
 .04 .07 .08 .09 .26 .45 .73 1.0 .33 .19 .04 .03 .02 .02
 .02 .02 .02 .02 .07 .14 .22 .33 1.0 .58 .37 .27 .20 .23
 .07 .04 .01 .01 .02 .08 .14 .19 .58 1.0 .74 .50 .41 .28
 .09 .07 .02 .00 .02 .02 .05 .04 .37 .74 1.0 .76 .62 .55
 .12 .11 .01 .01 .01 .02 .02 .03 .27 .50 .76 1.0 .85 .68
 .13 .13 .05 .02 .02 .02 .02 .02 .20 .41 .62 .85 1.0 .76
 .16 .14 .03 .04 .00 .01 .00 .02 .23 .28 .55 .68 .76 1.0];

Wavelength = {'434' '445' '465' '472' '490' '504' '537'…
     '555' '584' '600' '610' '628' '651' '674'};
Representation = mdscale(Judged_similarities, 2);

figure;
for i=1:14
     plot(Representation(i, 1), -Representation(i, 2),'kx');
     text(Representation(i,1),   -Representation(i,2), …
          Wavelength(i));
     hold on;
end
plot([Representation(:, 1); Representation(1,1)], …
     -[Representation(:,2); Representation(1, 2)], 'k-');
axis('square');
axis('off');

sdRelation = [];
for i = 1:14
     for j = 1:14
       Distance = sqrt(…
            (Representation(i, 1) - Representation(j, 1)).^2 …
          + (Representation(i, 2) - Representation(j, 2)).^2);
       sdRelation = [sdRelation; Distance …
            Judged_similarities(i, j)];
     end
end
```

Display 7.3 (continued)

```
figure;
plot(sdRelation(:, 1), sdRelation(:, 2), 'ko');
axis('square');
xlabel('Obtained Euclidean distance','Fontsize', 24, ...
      'Fontname', 'Arial');
ylabel('Similarity between colors','Fontsize', 24, ...
     'Fontname', 'Arial');
```

MDS algorithms recover two things. The first and primary thing is the positions of all the tested stimuli within the multidimensional space (figure 7.6) for the two artificial animal sets with different patterns of similarity between objects. The second important aspect is the functional relationship, or the transfer function, between the internal distances and the behavioral measurement such as rated similarity or rate of confusion errors. The quality of the model in relation to the data is summarized by an index of stress, or a fidelity criterion that weights the errors between predicted and observed distances.

Often, the dimensionality of the internal representation is unknown when applying MDS to a new experimental example. An MDS solution with a smaller dimensionality is generally preferred as it provides an account with fewer estimated parameters in the representation and also is more easily visualized and interpreted.[52] For example, n objects located in 2D space require $2n$ parameters and are visualized by locations in a 2D plane, whereas n objects in a three-dimensional (3D) space require $3n$ parameters and are visualized in 3D projections; and so on for higher-dimensional spaces. Dimensions are added or subtracted from the MDS solution until the best solution with the smallest number of dimensions is identified. Because MDS depends upon distances only, there are many equivalent solutions that differ in the orientations in the dimensional space. The spatial configuration can sometimes be rotated or realigned to find some set of axes in which the stimulus variations make the dimensions easier to interpret.

The number of dimensions in an MDS solution can be known only approximately. Indeed, MDS is viewed by some researchers primarily as a simple form of data reduction—a solution that takes a large set of pairwise data and displays it in terms of a small number of locations in a small number of dimensions. In practice, researchers find the minimum dimensionality that usefully describes the data and can be interpreted. This recovered dimensionality may only approximate the relevant psychological dimensions in a particular experimental situation.

MDS is more complex than unidimensional scaling because it requires the choice of a distance metric, such as Euclidean or city-block, and the determination of the dimensionality of the recovered similarity space. The MDS method in common metrics such as Euclidean or city-block implicitly makes assumptions of independent and equal variance

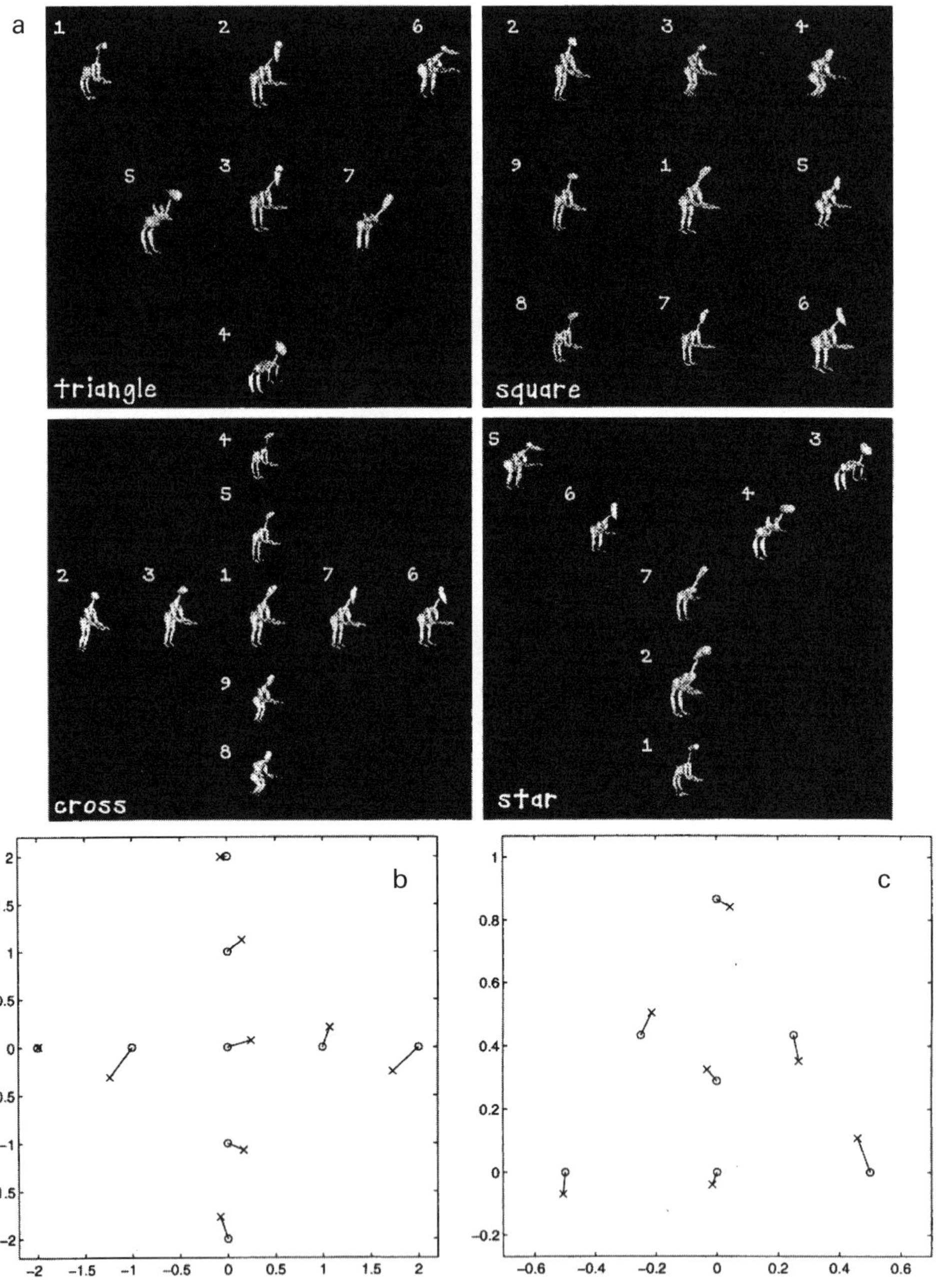

Figure 7.6
MDS analysis of similarity data on sets of artificial animal-like shapes. (a) The four 2D parameter-space configurations illustrating the similarity patterns built into the experimental stimuli used by Cutzu and Edelman. (b) The 2D MDS solution for all subjects in the pairwise comparison experiment. (c) The 2D MDS solution for all subjects in the long-term memory experiment (after Cutzu and Edelman,[51] figures 1 and 2).

in internal representations, and the distances in the solutions are de facto scaled as signal-to-noise ratios.

This brief introduction to MDS only touches on a vast set of technical developments. Classical MDS led to several more complicated implementations designed to incorporate, for example, the variations between the ratings of different observers[46] or use of only ordinal properties of the data.[48] The original MDS algorithms were deterministic and ignored variability in the data and the model. Probabilistic MDS[53,54] incorporated a more serious treatment of error variation. Some of these methods were recently extended using Bayesian estimation algorithms.[55–58] The probabilistic and Bayesian approaches may relax the assumptions of independence and equal variance. However, data from typical paradigms may not provide sufficient constraints to model internal noises effectively. For some kinds of stimuli, such as word meanings or knowledge spaces, non-geometric representations such as graphs or tree structures may provide an alternative and better representation of the relationships between objects embodied in the data.[59–63]

7.4 Scaling in Neuropsychophysics

Neurophysiology and brain imaging provide a new window into the relationship between the physical stimulus and the internal representations. They may also identify the brain activity associated with making a decision. The field is just beginning to see an interactive development of brain-imaging data and methods of scaling.

For example, consider the internal response to stimuli varying in intensity—the *contrast response function*. Single-unit recordings measure the contrast response function of a single neuron.[64–68] Multi-electrode arrays measure the simultaneous responses of hundreds of neurons.[69–72] fMRI measures the contrast response function of a cortical area.[73–76] EEG and MEG measure population responses.[77–79] All these measurements provide evidence of internal neural representations of the stimulus intensity within certain stages of neural processing. Exploring the relationship between these neural functions and the psychological scales is a new direction in neuropsychophysics.

Another important direction will be to understand the relationship between the internal representation and the overt response, the linking or transfer function. The transfer function is the relationship between the internal representation and the magnitude estimate in direct scaling, or the relationship between rated similarity and internal distances in MDS. Recent neuropsychological and imaging studies are beginning to address these issues.[74,80–83]

Multivariate pattern analysis in fMRI and emerging methods in analyzing data from multi-array neural recordings offer neural measures of similarity of internal representations for different stimuli and brain regions.[69,70,84] Exploring the relationship between neural similarity measures and MDS and other related scaling technologies represents a major direction in cognitive neuroscience. For example, Haushofer, Livingstone, and Kanwisher[85] used fMRI to measure the physical, behavioral, and neural similarity between pairs of novel shapes. They obtained perceptual similarity measures for each pair of shapes in a psychophysical same–different task, physical similarity measures from stimulus parameters, and neural similarity measures from multivoxel pattern analysis methods applied to the anterior and posterior lateral occipital complex (LOC). Pattern analysis of the similarity between the responses to two stimuli in the same locations or brain regions found that the pattern of pairwise shape similarities in the posterior LOC is highly correlated with physical shape similarities, whereas shape similarities in the anterior LOC most closely matched perceptual shape similarities. Results from such experiments could be used to specify stimulus manipulations and critical validation for neuropsychophysical investigations and identify the brain regions that underlie the psychological scales. Also, results from neuropsychophysical investigations could provide an important basis for further investigation of multidimensional representations of visual stimuli.

7.5 Summary

Scaling methods were among the earliest of the empirical methods applied in psychophysics, dating back to the end of the nineteenth century. The goal of the scaling enterprise is to quantify perceptual experience. In this chapter, we reviewed early advances in the development of perceptual scales and the extension of methods of scaling over large stimulus variations. We treated the scaling of multidimensional stimuli along with some key examples. We also pointed to alternative developments in measurement theory. Finally, we considered the amazing possiblities for future research in neuropsychophysics that combines new technologies in brain imaging with the power of quantitative scaling methods. Together, these approaches have the potential to identify and specify the translation from stimuli to internal representations and from internal representations to observer responses. Understanding these relationships may provide some of the keys that unlock our further understanding of the brain mechanisms in information processing.

References

1 Fechner, G. *Elemente der psychophysik*. Leipzig: Breitkopf & Härtel; 1860.

2. Gescheider GA. 1988. Psychophysical scaling. *Annu Rev Psychol* 39: 169–200.

3. Carroll JD, Arabie P. 1980. Multidimensional scaling. *Annu Rev Psychol* 31: 607–649.

4. Cliff N. 1973. Scaling. *Annu Rev Psychol* 24: 473–506.

5. Young FW. 1984. Scaling. *Annu Rev Psychol* 35: 55–81.

6. Shepard RN. 1981. Psychological relations and psychophysical scales: On the status of. *J Math Psychol* 24: 21–57.

7. Ehrenstein WH, Ehrenstein A. 1999. Psychophysical methods. *Modern Techniques in Neuroscience Research* 3: 1211–1241.

8. Scheerer E. Fechner's inner psychophysics: Its historical fate and present status. In: Geissler HG, Link SW, Townsend JT, eds. *Cognition, Information Processing and Psychophysics*. Hillsdale, NJ: Lawrence Erlbaum; 1992: pp. 3–22.

9. Stevens SS. 1946. On the theory of scales of measurement. *Science* 103: 677–680.

10. Stevens SS. 1957. On the psychophysical law. *Psychol Rev* 64: 153–181.

11. Pfanzagl J. *Theory of measurement*. New York: John Wiley & Sons; 1968.

12. Narens L. 1996. A theory of ratio magnitude estimation. *J Math Psychol* 40: 109–129.

13. Falmagne JC. *Elements of psychophysical theory*. Oxford: Oxford University Press; 2002.

14. Green DM, Luce RD, Duncan JE. 1977. Variability and sequential effects in magnitude production and estimation of auditory intensity. *Atten Percept Psychophys* 22: 450–456.

15. Krantz DH. 1972. A theory of magnitude estimation and cross-modality matching. *J Math Psychol* 9: 168–199.

16. Stevens SS. *Psychophysics: Introduction to its perceptual, neural, and social prospects*. Piscataway, NJ: Transaction Publishers; 1975.

17. Krantz DH. 1972. A theory of magnitude estimation and cross-modality matching. *J Math Psychol* 9: 168–199.

18. Weber EH. *De Pulsu, resorptione, auditu et tactu: Annotationes anatomicae et physiologicae*. Leipzig: CF Koehler; 1834.

19. Luce RD, Galanter E. Discrimination. In: Luce RD, Bush RR, Galanter E, eds. *Handbook of mathematical psychology*, Vol. 1. New York: Wiley; 1963: pp. 191–243.

20. Foley JM, Legge GE. 1981. Contrast detection and near-threshold discrimination in human vision. *Vision Res* 21: 1041–1053.

21. Wilson HR, Humanski R. 1993. Spatial frequency adaptation and contrast gain control. *Vision Res* 33: 1133–1149.

22. Lu ZL, Sperling G. 1996. Contrast gain control in first-and second-order motion perception. *JOSA A* 13: 2305–2318.

23. Watson AB, Solomon JA. 1997. Model of visual contrast gain control and pattern masking. *J Opt Soc Am* 14: 2379–2391.

24. Thurstone LL. 1927. A law of comparative judgment. *Psychol Rev* 34: 273–286.

25. Reeves A, Sperling G. 1986. Attention gating in short-term visual memory. *Psychol Rev* 93: 180–206.

26. Galanter E, Messick S. 1961. The relation between category and magnitude scales of loudness. *Psychol Rev* 68: 363–372.

27. Woods RL, Satgunam PN, Bronstad PM, Peli E. 2010. Statistical analysis of subjective preferences for video enhancement. *Human Vision and Electronic Imaging* XV, vol. 7527, article 14.

28. Hand D. *Measurement theory and practice: The world through quantification*. London: Arnold Publishers; 2004.

29. Michell J. *Measurement in psychology: Critical history of a methodological concept*. Cambridge, UK: Cambridge University Press; 1999.

30. Narens L. *Theories of meaningfulness*. Hillsdale, NJ: Lawrence Erlbaum; 2002.

31. Ashby F. 1991. Book reviews, *Foundations of Measurement*, Volume II and III. *Appl Psychol Meas* 15: 103–108.

32. Díez JA. 1997. A hundred years of numbers. An historical introduction to measurement theory 1887–1990: Part I: The formation period. Two lines of research: Axiomatics and real morphisms, scales and invariance. *Studies in History and Philosophy of Science Part A* 28: 167–185.

33. Krantz DH, Suppes P, Luce RD. *Foundations of measurement, Vol. 1*. New York: Academic Press; 1971.

34. Suppes P, Krantz DM, Luce RD, Tversky A. *Foundations of measurement, Vol. 2: Geometrical, threshold, and probabilistic representations*. New York: Academic Press; 1989.

35. Luce RD, Krantz DH, Suppes P, Tversky A. *Foundations of measurement, Vol. 3: Representation, axiomatization, and invariance*. New York: Academic Press; 1990.

36. Narens L. *Abstract measurement theory*. Cambridge, MA: The MIT Press; 1985.

37. Luce RD, Tukey JW. 1964. Simultaneous conjoint measurement: A new type of fundamental measurement. *J Math Psychol* 1: 1–27.

38. Gustafsson A, Herrmann A, Huber F. *Conjoint measurement: Methods and applications*. Berlin: Springer-Verlag; 2007.

39. Krantz DH, Tversky A. 1971. Conjoint-measurement analysis of composition rules in psychology. *Psychol Rev* 78: 151–169.

40. Steenkamp JBEM. 1987. Conjoint measurement in ham quality evaluation. *J Agric Econ* 38: 473–480.

41. Reid, GB, Shingledecker, CA, Eggemeier, FT. 1981. Application of conjoint measurement to workload scale development. *Proceedings of the 1981 Human Factors Society Annual Meeting,* 522–526.

42. Green PE, Rao VR. 1971. Conjoint measurement for quantifying judgmental data. *J Mark Res* 8(3): 355–363.

43. Falmagne JC, Iverson G, Marcovici S. 1979. Binaural" loudness" summation: Probabilistic theory and data. *Psychol Rev* 86: 25–43.

44. Falmagne JC, Iverson G. 1979. Conjoint Weber laws and additivity. *J Math Psychol* 20: 164–183.

45. Ho YX, Landy MS, Maloney LT. 2008. Conjoint measurement of gloss and surface texture. *Psychol Sci* 19: 196–204.

46. Carroll JD, Chang JJ. 1970. Analysis of individual differences in multidimensional scaling via an N-way generalization of "Eckart-Young" decomposition. *Psychometrika* 35: 283–319.

47. Borg I, Lingoes JC. *Multidimensional similarity structure analysis*. Berlin: Springer-Verlag; 1987.

48. Borg I, Groenen PJF. *Modern multidimensional scaling: Theory and applications*. Berlin: Springer-Verlag; 2005.

49. Shepard RN. 1980. Multidimensional scaling, tree-fitting, and clustering. *Science* 210: 390–398.

50. Ekman G. 1954. Dimensions of color vision. *J Psychol* 1054(38): 467–474.

51. Cutzu F, Edelman S. 1996. Faithful representation of similarities among three-dimensional shapes in human vision. *Proc Natl Acad Sci USA* 93: 12046–12050.

52. Lee MD. 2001. Determining the dimensionality of multidimensional scaling representations for cognitive modeling. *J Math Psychol* 45: 149–166.

53. MacKay DB. 1989. Probabilistic multidimensional scaling: An anisotropic model for distance judgments. *J Math Psychol* 33: 187–205.

54. MacKay DB, Zinnes JL. 1986. A probabilistic model for the multidimensional scaling of proximity and preference data. *Mark Sci* 5(4): 325–344.

55. Oh MS, Raftery AE. 2001. Bayesian multidimensional scaling and choice of dimension. *J Am Stat Assoc* 96: 1031–1044.

56. Okada K, Shigemasu K. 2010. Bayesian multidimensional scaling for the estimation of a Minkowski exponent. *Behav Res Methods* 42: 899–905.

57. Lee MD. 2008. Three case studies in the Bayesian analysis of cognitive models. *Psychon Bull Rev* 15: 1–15.

58. Okada K, Shigemasu K. 2009. BMDS: A collection of R functions for Bayesian multidimensional scaling. *Appl Psychol Meas* 33: 570–571.

59 Buneman, OP. The recovery of trees from measures of dissimilarity. In: Hodson FR, Kendall DG, Tautu PT, eds. *Mathematics in the archaeological and historical sciences*. Edinburgh: Edinburgh University Press; 1971, pp. 387–395.

60. Pruzansky S, Tversky A, Carroll JD. 1982. Spatial versus tree representations of proximity data. *Psychometrika* 47: 3–24.

61. Carroll JD, Pruzansky S. 1975. Fitting of hierarchical tree structure (HTS) models, mixtures of HTS models and hybrid models, via mathematical programming and alternating least squares. Proceedings of the U.S.-Japan Seminar on Multidimensional Scaling, 9–19.

62. Corter JE. *Tree models of similarity and association*. Thousand Oaks, CA: Sage Publications; 1996.

63. Falmagne JC, Koppen M, Villano M, Doignon JP, Johannesen L. 1990. Introduction to knowledge spaces: How to build, test, and search them. *Psychol Rev* 97: 201–224.

64. Tolhurst DJ, Movshon JA, Thompson ID. 1981. The dependence of response amplitude and variance of cat visual cortical-neurons on stimulus contrast. *Exp Brain Res* 41: 414–419.

65. Albrecht DG, Hamilton DB. 1982. Striate cortex of monkey and cat: Contrast response function. *J Neurophysiol* 48(1): 217–237.

66. Albrecht DG, Geisler WS. 1991. Motion selectivity and the contrast-response function of simple cells in the visual cortex. *Vis Neurosci* 7: 531–546.

67. Sclar G, Maunsell JHR, Lennie P. 1990. Coding of image contrast in central visual pathways of the macaque monkey. *Vision Res* 30: 1–10.

68. Tolhurst D, Movshon J, Thompson I. 1981. The dependence of response amplitude and variance of cat visual cortical neurones on stimulus contrast. *Exp Brain Res* 41: 414–419.

69. Cohen MR, Kohn A. 2011. Measuring and interpreting neuronal correlations. *Nat Neurosci* 14: 811–819.

70. Averbeck BB, Latham PE, Pouget A. 2006. Neural correlations, population coding and computation. *Nat Rev Neurosci* 7: 358–366.

71. Smith MA, Kohn A. 2008. Spatial and temporal scales of neuronal correlation in primary visual cortex. *J Neurosci* 28: 12591–12603.

72. Montani F, Kohn A, Smith MA, Schultz SR. 2007. The role of correlations in direction and contrast coding in the primary visual cortex. *J Neurosci* 27: 2338–2348.

73. Li X, Lu ZL, Tjan BS, Dosher BA, Chu W. 2008. Blood oxygenation level-dependent contrast response functions identify mechanisms of covert attention in early visual areas. *Proc Natl Acad Sci USA* 105: 6202–6207.

74. Boynton GM, Demb JB, Glover GH, Heeger DJ. 1999. Neuronal basis of contrast discrimination. *Vision Res* 39: 257–269.

75. Buracas GT, Boynton GM. 2007. The effect of spatial attention on contrast response functions in human visual cortex. *J Neurosci* 27: 93–97.

76. Gardner JL, Sun P, Waggoner RA, Ueno K, Tanaka K, Cheng K. 2005. Contrast adaptation and representation in human early visual cortex. *Neuron* 47: 607–620.

77. Nunez PL, Srinivasan R. *Electric fields of the brain: The neurophysics of EEG*. Oxford: Oxford University Press; 2006.

78. Lu ZL, Kaufman L. *Magnetic source imaging of the human brain*. New York: Psychology Press; 2003.

79. Hansen PC, Kringelbach ML, Salmelin R. *MEG: An introduction to methods*. Oxford: Oxford University Press; 2010.

80. Ress D, Backus BT, Heeger DJ. 2000. Activity in primary visual cortex predicts performance in a visual detection task. *Nat Neurosci* 3: 940–945.

81. Ress D, Heeger DJ. 2003. Neuronal correlates of perception in early visual cortex. *Nat Neurosci* 6: 414–420.

82. Gold JI, Shadlen MN. 2007. The neural basis of decision making. *Neuroscience* 30: 535–574.

83. Shadlen MN, Newsome WT. 1998. The variable discharge of cortical neurons: Implications for connectivity, computation, and information coding. *J Neurosci* 18: 3870–3896.

84. Kriegeskorte N, Mur M, Bandettini P. 2008. Representational similarity analysis—connecting the branches of systems neuroscience. *Front Systems Neurosci* 2: 4.

85. Haushofer J, Livingstone MS, Kanwisher N. 2008. Multivariate patterns in object-selective cortex dissociate perceptual and physical shape similarity. *PLoS Biol* 6: e187.

8 Sensitivity and Signal-Detection Theory

This chapter treats the measurement of visual sensitivity, or how we detect threshold or near-threshold stimuli. It begins with a brief review of the classical methods of measuring thresholds. It then presents the signal-detection theory framework and an analysis of several classical and modern procedures. The chapter ends with the signal-detection approach to more complex situations involving inputs from multiple detectors and tasks that involve stimuli in multidimensional space.

8.1 Sensitivity and Threshold

Characterizing the sensitivity of the visual system to environmental stimulation provides not only an important way to understand the limitations of the perceptual system but also fundamental building blocks for models of functional vision. Sensitivity is sometimes defined as the inverse of threshold (1/threshold). By threshold we mean the minimum physical stimulus level or minimum difference between physical stimuli that can be detected at some target accuracy level, such as 75% correct. Measurements of sensitivity to stimuli that vary systematically, such as the ability to detect sine waves of different spatial frequencies, provide one basic characterization of limiting visual processes.[1–4] Systematic measurements of sensitivity provide a basis for predicting the sensitivity of the visual system to many other simple and more complex visual patterns.[5,6]

Consider the measurement of sensitivity for a set of sine waves of different spatial frequencies, or the contrast sensitivity function, CSF (see chapters 1 and 2). We can estimate the sensitivity for many other spatial frequencies by interpolation. By decomposing visual stimuli into spatial frequency components or parts, we may be able to use the CSF to predict the responses to more complex stimuli from their component parts.[7] Of course, the actual resulting perception may be complicated somewhat by interactions and nonlinearities, but a simple non-interactive principle provides important baseline predictions. Indeed, the CSF has been used as the "front end" of visibility in many models of functional vision.[5,6]

This chapter treats many of the most common methods for measuring the sensitivity in a single condition. In later chapters, we consider how multiple measurements in several conditions are combined to estimate sensitivity functions or sensitivity surfaces over several dimensions of stimulus variation. Signal-detection theory[8–12] provides the theoretical framework that separates decision factors (bias, etc.) from sensitivity and is used in this chapter to analyze and understand the different paradigms for the measurement of sensitivity and threshold.

8.2 Classical Psychophysical Procedures

Three of the most common methods first introduced to measure threshold are the method of limits, the method of adjustment, and the method of constant stimuli.[13,14] These methods date to the turn of the past century. In this section, we discuss each of these in turn.

8.2.1 Method of Limits

In the method of limits, a stimulus value such as contrast or luminance is varied from trial to trial until the observer can barely detect the stimulus. The method changes the stimulus value in two different directions. The ascending method of limits starts with stimuli at very low, invisible values and gradually increases the stimulus value until the observer first reports that the stimulus is barely visible. The descending method of limits starts with stimuli at high, visible values and gradually decreases the stimulus value until the observer reports that the stimulus is barely visible. For both procedures, the final stimulus value is the estimate of the threshold. Often, both ascending and descending methods are used in an experiment, and the average of their end values is used to estimate the threshold.

An example of the method of limits is shown in the program `MethodofLimits.m` (display 8.1). A square-wave grating is shown first at very low or very high contrast. The observer can adjust the contrast of the square up (ascending) or down (descending) in increments to measure his or her threshold contrast.

Although the method of limits is easy to use, it may suffer from two kinds of errors that were discussed in the early literature: (1) the observer may continue to report that a stimulus is visible beyond the threshold in the descending method or invisible above the threshold in the ascending method (the error of habituation), and (2) the observer may anticipate by reporting invisibility above the threshold in a descending method or by reporting visibility below the threshold in the ascending method (the anticipation error) (figure 8.1). Also, each stimulus sequence involves a small number of trials, and typically the number of repetitions that are averaged to estimate threshold is relatively small. These methods of assessing threshold are relatively rapid due to the small number of trials, but may be limited in accuracy for the same reason.

The method can also be used to measure difference thresholds. Starting with a very small or no difference between two visible stimuli, one can gradually increase the difference until the difference is visible. Or starting with a large difference between two stimuli, one can gradually reduce the difference until the difference is invisible.

8.2.2 Method of Adjustment

The method of adjustment requires the observer to change the level of the stimulus until it is just barely visible to measure absolute threshold, or to change the value of one stimulus until it is perceived to be just different from another stimulus to measure difference threshold. The procedure is often repeated multiple times, and the mean end value of the repeats is used to estimate the threshold.

The program `MethodofAdjustment.m` (display 8.2) presents a sine wave of one frequency and a target contrast on the left and a test sine wave of a different frequency on the right. The contrast of the test sine wave is increased or decreased by pressing the "u" or "d" keys until the perceived contrasts match (figure 8.2).

Usually, the adjustment would be made several times with different starting values, and the matching points would be averaged to improve the quality of the estimated values.

8.2.3 Method of Constant Stimuli

The method of constant stimuli measures the performance of an observer for a fixed set of stimuli over a range of stimulus values. It estimates the threshold from the function that relates performance accuracy to stimulus variation, the so-called psychometric function (see section 2.2 of chapter 2 for a detailed example). The method randomly samples from the set of stimuli at preselected values and presents them to the observer in different trials. The observer responds in each trial (e.g., is the stimulus visible?). The probability of a response category (e.g., "yes") is tabulated for each stimulus value. A plot of the probability of detection versus stimulus value is the psychometric function. The threshold is that stimulus value that would generate a particular performance level (e.g., 75% "yes"). Estimation of the threshold almost always requires interpolation from the data for the tested contrasts.

In display 8.3, we present an example (`MethodofConstantStimuli.m`) for an experiment where the observer reports whether a Gabor is visible, and the stimuli are selected from a set of Gabor contrasts that span the psychometric function. The method of constant stimuli requires more test trials—here we selected seven contrasts with 100 test trials each, which produces a reasonable psychometric function. Methods for interpolation of the threshold value by fitting a function to the psychometric function data and for the interpretation of the estimate are developed in section 10.4.2 of chapter 10.

Display 8.1

```
%%% Program MethodofLimits.m
function MethodOfLimits(descending)

%% Experimental Module
p.stimSize = [6 6];       % horizontal and vertical stimulus
                          % size in visual angle
p.stimDuraion = 0.1;      % stimulus duration in seconds
p.sf = 2 ;                % spatial frequency in cycles/degree
p.ITI = 0.5;              % seconds between trials
if nargin < 1, descending = true; end
if descending
    respKey = 'down';
    p.startContrast = 0.02; % starting visible grating contrast
    inc = -0.001;         % contrast increment
else
    respKey = 'up';
    p.startContrast = 0; % starting invisible grating contrast
    inc = 0.001;
end
keys = {respKey 'esc'};   % allowed response keys

% Compute stimulus parameters
ppd = pi / 180 * p.ScreenDistance / p.ScreenHeight * ...
      p.ScreenRect(4);    % pixels per degree
ppc = round(ppd / p.sf); % pixels per cycle
m(1) = round(p.stimSize(1) * ppd / ppc / 2) * ppc * 2;
                          % horizontal size in pixels
m(2) = round(p.stimSize(2) * ppd); % vertical size in pixels
fixRect = CenterRect([0 0 1 1] * 8, p.ScreenRect);
                          % 8 x 8 fixation square
p.randSeed = ClockRandSeed; % use clock to set the seed of the
                            % random number generator

img = ones(1, m(1) / ppc) * 0.5;
img(1 : 2 : end) = -0.5;
img = Expand(img, ppc, m(2));

% Prioritize display to optimize display timing
Priority(MaxPriority(windowPtr));

% Start experiment with instructions
str = ['Use ' respKey ' arrow to change contrast.\n\n' ...
       'ESC to exit when the stimulus becomes barely ' ...
       'visible.\n\n' 'Press SPACE to start.'];
DrawFormattedText(windowPtr, str, 'center', 'center', 1);
```

Display 8.1 (continued)

```
        % Draw Instruction text string centered in window
Screen('Flip', windowPtr);
WaitTill('space');              % wait till space is pressed
Screen('FillOval', windowPtr, 0, fixRect);
        % create a black fixation box
Secs = Screen('Flip', windowPtr);
        % flip the fixation image into active buffer
p.start = datestr(now);     % record start time

% Run trials until user finds threshold
i = 1;                          % for record
thre = p.startContrast;
while 1
    tex = Screen('MakeTexture', windowPtr, img * thre + 0.5, ...
          0, 0, 2);
    Screen('DrawTexture', windowPtr, tex);
    Screen('FillOval', windowPtr, 0, fixRect); % black fixation
    t0 = Screen('Flip', windowPtr, Secs + p.ITI); % stim on
    Screen('FillOval', windowPtr, 0, fixRect);
    Screen('Flip', windowPtr, t0 + p.stimDuraion);% stim off
    Screen('Close', tex);
    [key Secs] = WaitTill(keys);      % wait till response
    rec(i, :) = [i thre Secs-t0];     % trial #, contrast,  RT
    i = i + 1;
    if strcmp(key, 'esc'), break; end
    thre = thre + inc;         % increase or decrease contrast
end
p.finish = datestr(now);       % record finish time
save MethodOfLimits_rst.mat rec p;   % save results
```

8.3 Signal-Detection Theory in Psychophysics

8.3.1 Subjective Criteria in Performing Psychophysical Tasks

The classical methods in psychophysics, including the method of limits, the method of adjustment, and the method of constant stimuli, involve responses of the observer to stimuli.[13,14] In each case, a stimulus is presented, and the observer makes a decision about how to respond. Selecting the response depends not only on the internal sensory response to the stimulus but on a decision process as well. In the method of limits, the observer decides how much evidence is required to say that the stimulus is visible. In the method of adjustment, the observer again decides whether the stimulus is visible or whether the adjusted stimulus is sufficiently close to matching a test stimulus. Similarly, decisions are made in the method of constant stimuli in each trial about whether an individual

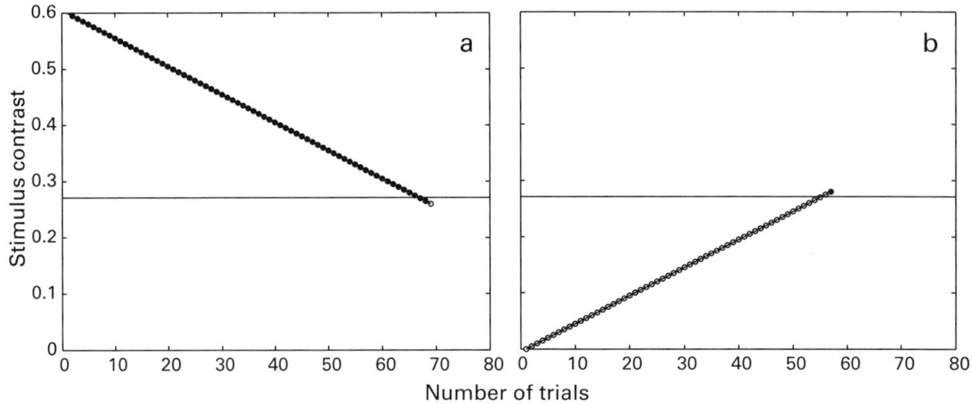

Figure 8.1
The method of limits. (a) A descending sequence. (b) An ascending sequence. Filled circles indicate "visible" responses; open circles indicate "invisible" responses. The horizontal lines indicate the true threshold.

stimulus is visible. In each case, the sensory stimulation is one ingredient in performance, but the decision about how much evidence is required—the criterion for decision—is another.

The method of limits and the method of adjustment are both useful in getting an approximate threshold measurement relatively quickly, but these estimates can be systematically affected by how the observer chooses to make decisions. The individual could require that the stimulus is clearly visible before he or she indicates a "visible" response. A high criterion shifts the estimate of the threshold higher. Or, the individual could give a "visible" response whenever there is a hint of visibility. Such a low criterion shifts the estimate of threshold lower.

Psychologists have developed a formal theory—signal-detection theory—to analyze and separate the sensory and the decision factors in responses.[8–12] The application of these methods can improve the estimation of sensory threshold separate from other properties of the observer.

8.3.2 Introduction to Signal-Detection Theory

Signal-detection theory (SDT) is easiest to introduce for a "Yes/No" task. Our example measures the Yes/No response to stimuli such as Gabor patches of a specific contrast. The observer performs a series of trials in which the stimulus is presented on some, usually one-half of all trials. On each trial, the observer responds with a decision that "yes" the stimulus was present or "no" the stimulus was not present.

The observer's responses fall into four different categories (table 8.1). If the stimulus was presented and the observer responds "yes," this is a *hit*. If the stimulus was presented

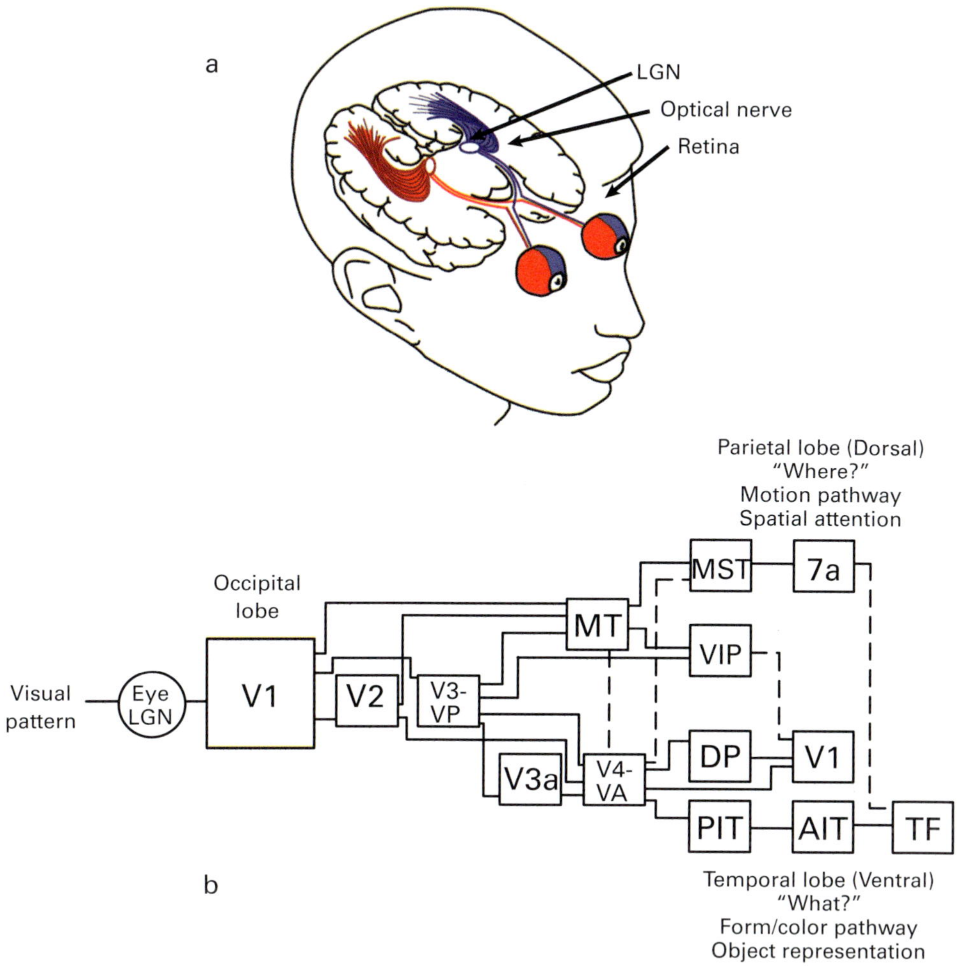

Plate 1 (figure 1.1)

(a) A picture of the visual system starting from the eyes and going through to various regions in visual cortex. (b) A two-pathway architecture for visual processing (adapted from Movshon[21]). Each box in the diagram refers to a functional region of the visual processing pathway. V1, V2, V3-VP, V3a, V4-VA, and 7a all refer to areas of visual cortex, each with distinct sensitivities. Higher processing areas involved in further processing, decision, and memory include the medial superior temporal cortex (MST), the ventral intraparietal cortex (VIP), the posterior inferior temporal cortex (PIT), the dorsal prelunate gyrus (DP), the anterior inferior temporal cortex (AIT), and area TF of parahippocampal cortex. The dashed lines refer to connections across the dorsal and ventral processing pathways thought to subserve analyses of "where" and "what," respectively.

Plate 2 (figure 1.2)

An image of the coffee shop where we are writing, with several processed versions emphasizing luminance, edges, and color blocks. The upper left panel shows the original digital photograph; the upper right is the same image rendered in grayscale (black and white); the lower left is a version of the grayscale image that shows edges (thresholded); the lower right shows the large regions of color in the image without detailed texture (filtered).

Plate 3 (figure 3.1)

A checkerboard pattern shown with three different color lookup tables.

Plate 4 (figure 3.8)
A true color image defined with RGB values.

Plate 5 (figure 3.12)
A digital color photograph of a church.

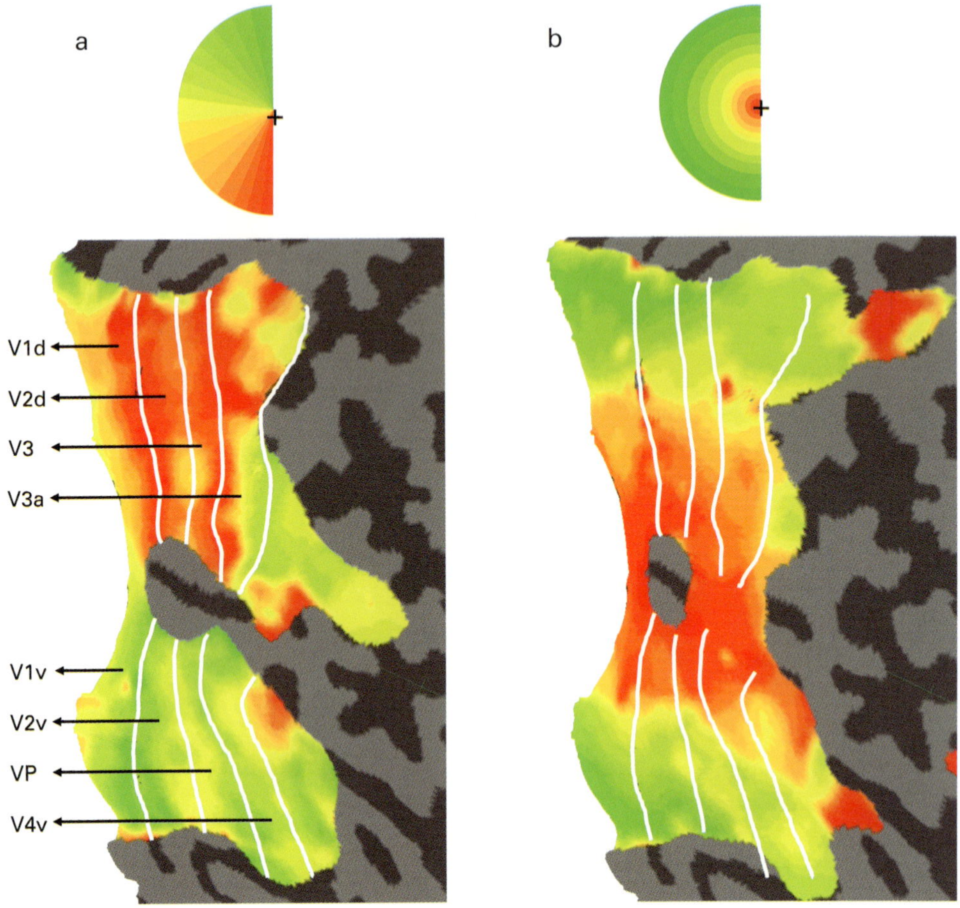

Plate 6 (figure 4.9)

Flattened representations of the visual cortex labeled with retinotopic regions corresponding to different visual areas from an fMRI retinotopy experiment. (a) Retinotopy from the flickering wedge displays. (b) Retinotopy from the flickering ring displays. The activity in different cortical regions is color coded to the positions in the wedges and rings respectively. Retinotopy data from Li, Lu, Tjan, Dosher, and Chu (2008).[14]

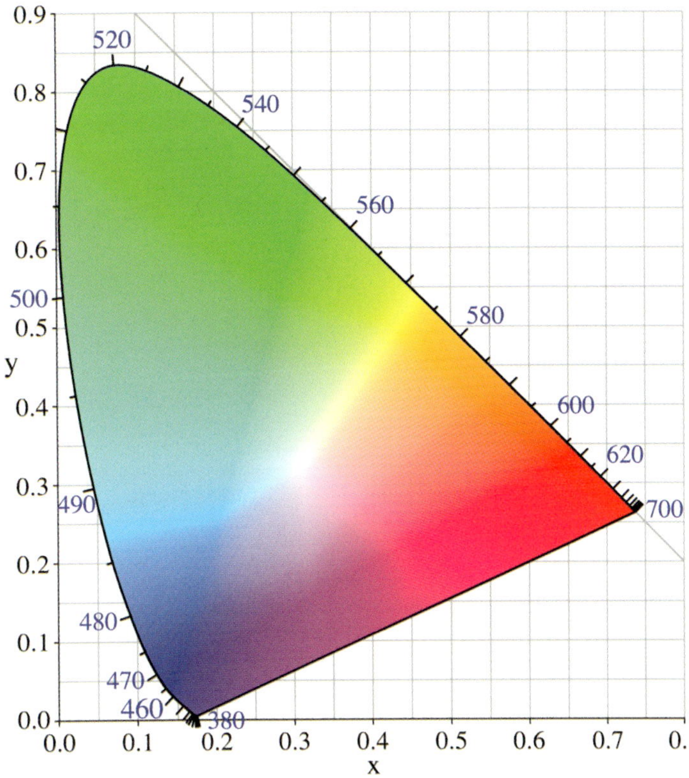

Plate 7 (figure 5.10)

CIE 1931 xy chromaticity diagram. The diagram shows the positions of different chromaticities in the CIE space, with positions varying in hue and in saturation or depth of color. The positions of pure wavelength lights are indicated around the edges of the color manifold, and a white point is in the middle.

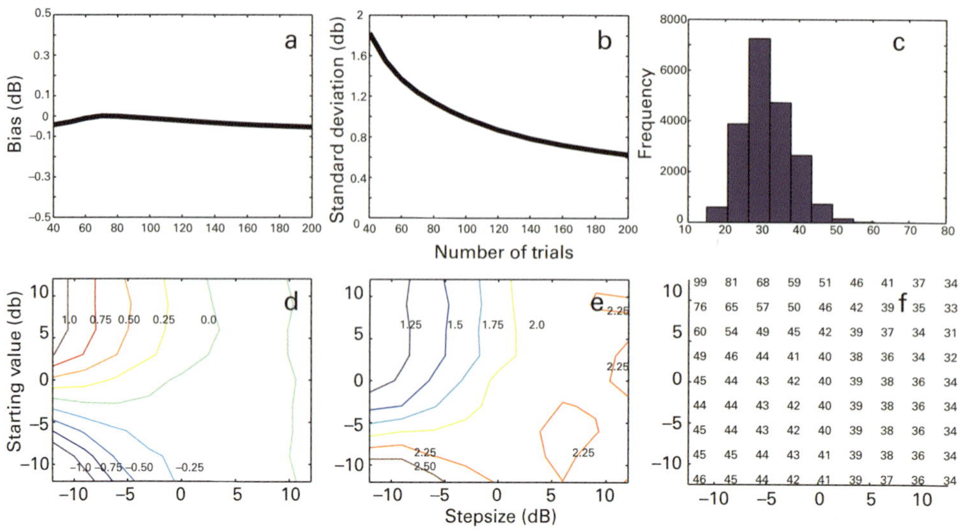

Plate 8 (figure 11.3)

Bias, precision, and sampling demands of a 2/1 staircase. (a) Bias of the estimated threshold as a function of number of trials. (b) Precision of the estimated threshold from these staircases as a function of number of trials. (c) Histogram of the number of trials needed to reach 10 reversals in the staircase. (d) Contour plot of the biases of a 2-down, 1-up staircase as a function of the starting value and step-size. (e) Contour plot of the precision of the 2-down, 1-up staircase as a function of the starting value and step-size. (f) Number of trials needed to obtain 10 reversals in the staircase with 90% probability. Bias dB is referred to 0 at the actual threshold; step-size dB is referred to 0 at the recommended step-size, or 0.5/slope of the psychometric function near threshold.

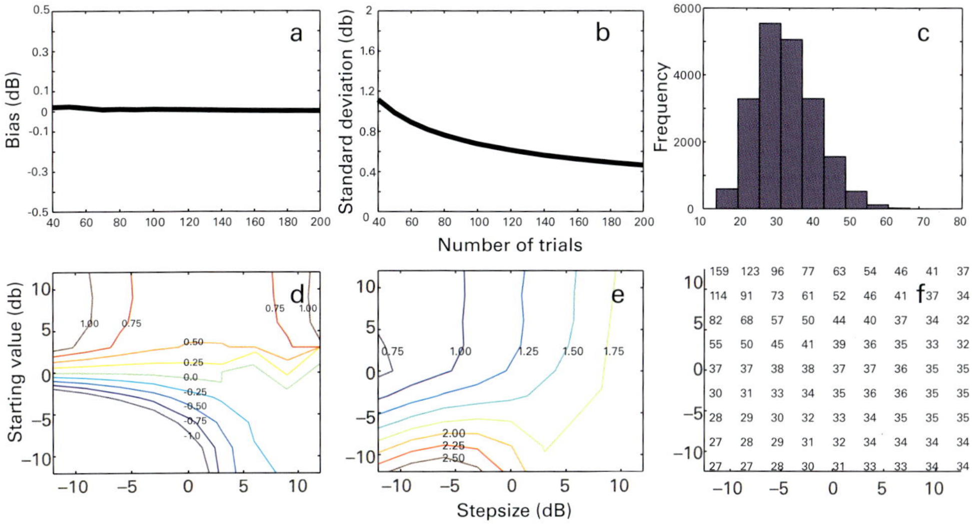

Plate 9 (figure 11.5)

Bias, precision, and trial requirements of the accelerated stochastic approximation (ASA) staircase tracking 75%, and its sensitivity to starting value and step-size. (a) Bias of the estimated threshold from an ASA staircase as a function of the number of trials. (b) Precision of the estimated threshold from the same staircase as a function of number of trials. (c) Histogram of the number of trials needed to reach 10 reversals in the staircase. (d) Contour plot of the bias of an ASA staircase as a function of the starting value and step-size. (e) Contour plot of the prevision of the staircase as a function of the starting value and step-size. (f) Number of trials needed to obtain 10 reversals in the staircase with 90% probability.

Plate 10 (figure 13.5)
A picture of the Airbus A380 cockpit (from http://www.flickr.com/photos/83823904@N00/64156219).

Display 8.2

```
%%% Program MethodOfAdjustment.m
function MethodOfAdjustment

%% Experimental Module

% Specify the stimulus
p.stimSize = [6 6];           % horizontal and vertical
                              % stimulus size in visual angle
p.sf = 1 ; % spatial frequency in cycles/degree
p.contrast = [1 2] * 0.3;     % target and test contrasts
p.dist = 2;                   % distance between edges of the
                              % two stim in degrees
keys = {'up' 'down' 'esc'};   % keys to respond and break

% Compute stimulus parameters
 ppd = pi / 180 * p.ScreenDistance / p.ScreenHeight * …
       p.ScreenRect(4);      % pixels per degree
ppc = round(ppd / p.sf);      % pixels per cycle
m(1) = round(p.stimSize(1) * ppd); % horizontal size in pixels
m(2) = round(p.stimSize(2) * ppd / ppc / 2) * ppc * 2;
                              % vertical size in pixels
sf = p.sf / ppd;              % cycles per pixel
rect = CenterRect([0 0 m], p.ScreenRect);
xOffset = p.dist * ppd / 2 + m(1) / 2;
rect1 = OffsetRect(rect, -xOffset, 0); % target stim rect
rect2 = OffsetRect(rect, xOffset, 0);  % tesst stim rect
params1 = [0 sf p.contrast(1) 0];
        % parameters for target stim: phase, sf, contrast
params2 = [0 sf p.contrast(2) 0];  % parameters for test stim
p.randSeed = ClockRandSeed;   % use clock to set random number
                              % generator

% Procedural sinewavegrating allows us to change its
% parameters very quickly
 tex = CreateProceduralSineGrating(windowPtr, m(1), m(2), …
       [1 1 1 0] * 0.5, [], 0.5);

% Prioritize display to optimize display timing
Priority(MaxPriority(windowPtr));

% Start experiment with instructions
str=['Use Up/Down arrow keys to increase/decrease the '…
     'contrast of the grating on right\n\n' 'Press ESC to '…
     'exit when both gratings have the same contrast.\n\n' …
     'Press SPACE to start.'];
```

Display 8.2 (continued)

```
DrawFormattedText(windowPtr, str, 'center', 'center', 1);
        % Draw Instruction text string centered in window
Screen('Flip', windowPtr);
WaitTill('space');                  % wait till space is pressed
Screen('Flip', windowPtr);  % turn off instruction
p.start = datestr(now);     % record start time

% Run trials until user finds threshold
i = 1;                      % initial record #
while 1
    Screen('DrawTexture', windowPtr, tex, [], rect1, 90, [],...
           [], [], [], [], params1);
    Screen('DrawTexture', windowPtr, tex, [], rect2, 90, [], ...
           [], [], [], [], params2);
    t0 = Screen('Flip', windowPtr); % update contrast
    KbReleaseWait;              % avoid continous change by single
                                % key press
    [key Secs] = WaitTill(keys);        % wait till response
    rec(i, :) = [i params2(3) Secs-t0];
    i = i + 1;
    if strcmp(key, 'up')
        params2(3) = params2(3) * 1.1; % increase contrast
    elseif strcmp(key, 'down')
        params2(3) = params2(3) / 1.1; % decrease contrast
    else
        break;
    end
end
p.finish = datestr(now);                % record finish time
save MethodOfAdjustment_rst.mat rec p; % save results
```

and the observer responds "no," this is a *miss*. If the stimulus was not presented but the observer responds "yes," this is a *false alarm*. If the stimulus was not presented and the observer responds "no," this is a *correct rejection*. The proportion of hits and the proportion of false alarms summarize the data. The proportion of misses and correct rejections are not independent quantities, but are directly related to hits and false alarms. The proportion of misses = 1 – proportion of hits, and the proportion of correct rejections = 1 – proportion of false alarms.

The primary assumption of SDT is that the strength of the internal representation of a stimulus is not identical on each trial, but is variable. That is, the observer's internal response to the stimulus is not exactly repeatable, but is noisy, presumably due to randomness in neural responses.[8–12] The distribution of internal responses on trials where the

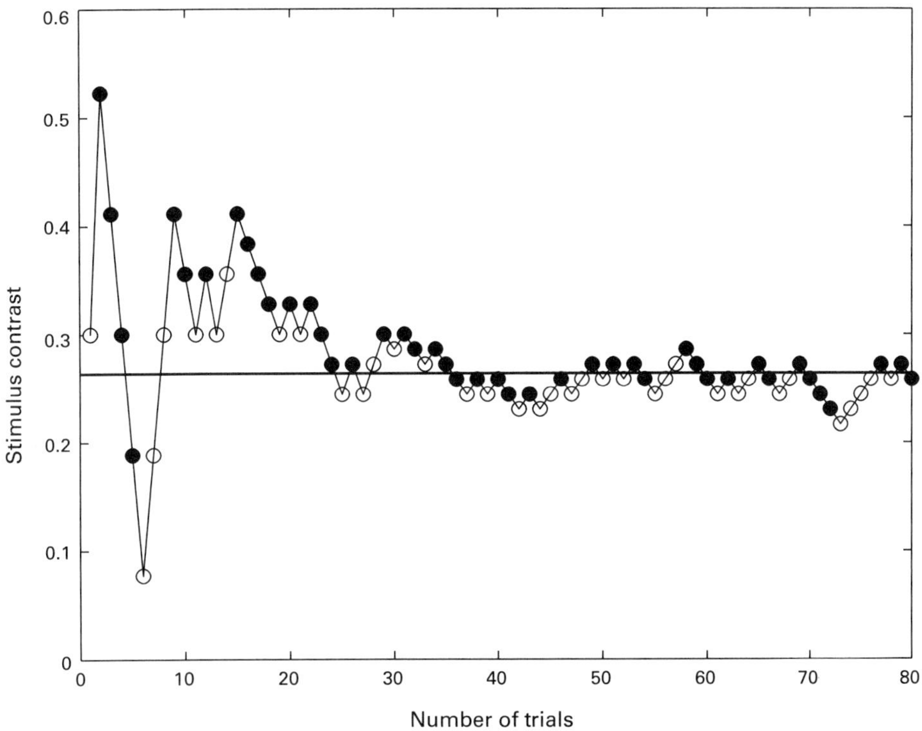

Figure 8.2
The method of adjustment. Filled circles indicate "target higher" responses; open circles indicate "target lower" responses. The horizontal line indicates the true matching contrast.

stimulus is not presented is the "noise" distribution, while the distribution of internal responses on trials where the stimulus is presented is the "signal + noise" distribution. The signal + noise distribution has a higher mean. (In this labeling, the "noise" refers to noise or variability in the internal response, not to external masking noise that is part of the physical stimulus.)

In most cases, the noise and the signal + noise distributions overlap. So, the particular internal response value or strength sensed by the observer on a given trial could arise either from a signal-present or a signal-absent stimulus presentation. Signal-present trials are more likely to produce larger values in the internal representation.

Hit and false alarm rates provide key data for testing and applying SDT. A simple detection task measures performance for a single contrast. On each trial of an example experiment, the observer first sees a fixation frame followed after 150 ms by a very brief presentation of the low-contrast Gabor ("signal") or a blank frame ("noise" or "blank"). Observers participate in 200 trials with the signal Gabor appearing randomly in half of

Display 8.3

```
%%% Program MethodofConstantStimuli.m
function MethodofConstantStimuli

%% Experimental Module

% Specify general experimental parameters
nContrast = 7;      % number of contrast levels in experiment
repeats = 100;      % number of trials to repeat for each contrast
nTrials = repeats * nContrast;
                    % compute total number of trials
p.randSeed = ClockRandSeed;
                    % use clock to set random number generator
keys = {'1' '2' 'esc'};    % specify response keys for yes,
                           % no and to break

% Specify the stimulus
p.stimSize = 6;     % image diameter in visual degree
p.sf = 1;           % spatial frequency in cycle/degree
conRange = [0.0025 0.04]; % lowest and highest tested
                          % contrasts
p.stimDuration = 0.1;     % stimulus duration in seconds
p.ISI = 0.5;        % seconds between trials

% Compute stimulus parameters
ppd = pi/180 * p.ScreenDistance / p.ScreenHeight * …
      p.ScreenRect(4);     % compute pixels per degree from
                    % p.ScreenDistance and p.ScreenHeight
m = round(p.stimSize * ppd /2) * 2;
                    % stimulus size in pixels
fixRect = CenterRect([0 0 1 1] * 8, p.ScreenRect);
                    % define an 8 x 8 pixel fixation
% create another fixation stimulus of horizontal and vertical
% cross hairs outside the area of the sine wave pattern
fixLen = 32;        % length of fixation in pixels
fixXY = [[-1 0 0 1]*fixLen + [-1 -1 1 1 ] * m/2 + …
       p.ScreenRect(3)/2 [1 1 1 1] * p.ScreenRect(3)/2;…
       [1 1 1 1]*p.ScreenRect(4)/2 [-1 0 0 1]*fixLen + …
       [-1 -1 1 1 ]*m/2 + p.ScreenRect(4) / 2];

p.contrasts = logspace(log10(conRange(1)),
log10(conRange(2)), nContrast);
       % use logspace function to choose nContrast contrasts
       % (7) at log intervals between the minimum and
       % maximum specified contrasts
```

Display 8.3 (continued)

```
sf = p.sf / ppd;   % compute spatial frequency in cycles
                   % per pixel
tex = CreateProceduralSineGrating(windowPtr, m, m, …
      [[1 1 1]*p.ScreenBackground 0], m/2, 0.5);
      % "CreateProceduralSineGrating" is a psychtoolbox
      % function to create a procedural texture for drawing
      % sine grating stimulus patches on the GPU

% Initialize a table to set up experimental conditions
p.recLabel = {'trialIndex' 'contrastIndex' 'anserYes' …
      'respTime'}; % labels of the columns of the data
                   %  recording array rec
rec = nan(nTrials, length(p.recLabel));
        % matrix rec is initialized as an nTrials x 5 of NaN
rec(:, 1) = 1 : nTrials;
        % initialize trial numbers from 1 to nTrials
contrastIndex = repmat(1 : nContrast, repeats, 1);
        % use repmat to cycle thru contrast #s
rec(:, 2) = Shuffle(contrastIndex(:));
        % shuffle contrast indexes to randomize

% Prioritize display to optimize display timing
Priority(MaxPriority(windowPtr));

% Start experiment with instructions
str = sprintf('Press 1 for stimulus present, 2 for absent.' …
      '\n\nPress SPACE to start.');
DrawFormattedText(windowPtr, str, 'center', 'center', 1);
      % Draw Instruction text string centered in window
Screen('Flip', windowPtr);
      % flip the text image into active buffer
WaitTill('space');
      % wait till space bar is pressed
Screen('FillOval', windowPtr, 0, fixRect);
      % create a black fixation dot
Secs = Screen('Flip', windowPtr);
      % flip the fixation image into active buffer
p.start = datestr(now);      % record start time

% Run nTrials trials
for i = 1 : nTrials
    con = p.contrasts(rec(i, 2));  % use contrast index from
                   % rec to set contrast for this trial

    Screen('DrawTexture', windowPtr, tex, [], [], 0, [], …
```

Display 8.3 (continued)

```
            [], [], [], [], [180 sf con 0]);
          % draw the sine grating with phase 180, spatial
          % frequency, and contrast
     Screen('DrawLines', windowPtr, fixXY, 3, 0.3);
          % add the fixation crosshairs
     t0 = Screen('Flip', windowPtr, Secs + p.ISI);
          % show the stimulus and return the time
     Screen('Flip', windowPtr, t0 + p.stimDuration);
          % turn off the stimulus by flipping to background
        % image after p.stimDuration secs

     Screen('FillOval', windowPtr, 0, fixRect);
        % draw the smaller centered fixation
     Screen('Flip', windowPtr, t0 + 0.25 + p.stimDuration);
        % show small fixation briefly to cue the observer to
        % respond with the interval

     [key Secs] = WaitTill(keys);
        % wait for response, return key and response time
     if iscellstr(key), key = key{1}; end
        % take the first response in case of multiple
        % key presses
     if strcmp(key, 'esc'), break; end
        % check if response is <escape> to stop experiment
     rec(i, 3 : 4) = [strcmp(key, '1') Secs-t0];
        % record answer and respTime
end

p.finish = datestr(now); % record finish time
save MethodofConstantStimuli_rst.mat rec p;
        % save the results
```

Table 8.1
Definitions of hit, miss, false alarm and correct rejection

	Stimulus Present	Stimulus Absent
"Yes"	Hit	False alarm
"No"	Miss	Correct rejection

the trials. On each trial, the observer says "signal present" or "yes" if he or she believes the Gabor was present in the display and "signal absent" or "no" otherwise, and we tabulate the number of hits, misses, false alarms, and correct rejections. Display 8.4 shows a simple program that carries out this experiment and tabulates the responses. The data from one observer are shown in table 8.2. Further SDT treatment of the data is considered in section 8.3.3, along with the resulting response table.

8.3.3 Application of SDT to Yes/No Tasks

SDT specifies the decision process for the "Yes/No" paradigm as the choice of a criterion value c. Some researchers describe SDT using an alternative derived likelihood measure as the decision axis, in which case the criterion value is set on the likelihood axis. This alternative leads to a related and largely parallel development of SDT.[8,9] We choose the more direct formulation here.[15] In this and the next several sections, we describe the major equations of SDT, discuss how they are applied, and provide some examples.

In SDT, the input stimulus on a given trial generates an internal response x, which usually is assumed to be drawn from a Gaussian distribution, or normally distributed. The internal responses to a signal-absent stimulus have probability density $g(x, \mu_N, \sigma_N)$ with mean $\mu_N = 0$ and standard deviation σ_N. Setting μ_N to 0 anchors the scale. The internal responses to a signal-present stimulus have probability density $g(x, \mu_{S+N}, \sigma_{S+N})$ with mean μ_{S+N} and standard deviation σ_{S+N}. On a given test trial, if the internal response exceeds the criterion, $x > c$, the observer responds "Yes," otherwise, he or she responds "No."

If the variances of the two distributions are identical, $\sigma_{S+N} = \sigma_N = \sigma$, then the discriminability d' is defined as:

$$d' = \frac{d}{\sigma} = \frac{\mu_{S+N} - \mu_N}{\sigma}. \tag{8.1}$$

This discriminability of the observer is the signal-to-noise ratio for the internal response. The mean shift d between the noise and the signal + noise internal response distributions is scaled by their standard deviation. The discriminability d' improves as the distance d between the two distributions increases or as the standard deviation σ (noisiness) of the internal representations decreases.

Figure 8.3 shows the internal response distributions for stimulus-present and stimulus-absent conditions, with the shaded areas corresponding to the four types of response. The equations for correct rejections, misses, false alarms, and hits are written here in a general form that allows the standard deviations of target-present and target-absent distributions to differ.

The proportion of correct rejections corresponds with the cumulative distribution function of the noise condition below the criterion c, or

$$P_{CR}(c) = \int_{-\infty}^{c} g(x, \mu_N, \sigma_N)\,dx = G\left(\frac{c}{\sigma_N}, 0, 1\right), \tag{8.2}$$

Display 8.4

```
%%% Program YesNoTask.m
function YesNoTask

%% Experimental Module

% Specify the stimulus
p.stimSize = 6;     % image diameter in visual degree
p.contrast = 0.01; % gabor contrast
p.sf = 0.9; % c/d
p.sigma = 1.1;      % degree
p.stimDuration = 0.1;      % stim duration in seconds
p.fixDuration = 0.15;      % fixation duration in seconds
p.ITI = 0.5;        % seconds between trials
nTrials = 200; % total number of trials
keys = {'1' '2' 'esc'};    % response keys

% Compute stimulus parameters
ppd = pi / 180 * p.ScreenDistance / p.ScreenHeight *
p.ScreenRect(4);    % pixels per degree
m = round(p.stimSize * ppd);      % stimulus size in pixels
sf = p.sf / ppd;   % cycles per pixel
sc = round(p.sigma * ppd); % sigma in pixels
params = [180 sf sc 0 1 0 0 0]; % gabor parameters: phase
                                % sf sigma contrast
fixLen = 32;        % length of fixation in pixels
[xc yc] = RectCenter(p.ScreenRect);
fixXY = [[-1 0 0 1]*fixLen + [-1 -1 1 1 ]*m/2 + xc …
       [1 1 1 1]*xc; [1 1 1 1]*yc [-1 0 0 1]*fixLen + …
       [-1 -1 1 1 ]*m/2 + yc];

p.randSeed = ClockRandSeed;       % use clock to set seed
                      % for the random number generator

% procedural gabor allows us to change its parameters
% very quickly
tex = CreateProceduralGabor(windowPtr, m, m, 0, …
      [1 1 1 0] * 0.5, 1, 0.5);

% Initialize a table to set up experimental conditions
p.recLabel = {'trialIndex' 'haveGabor' 'respYes' …
      'respTime'};
rec = nan(nTrials, length(p.recLabel));
        % matrix rec is made of nTrials x 4 of NaN
rec(:, 1) = 1 : nTrials;
        % count trial numbers from 1 to nTrials
haveGabor = repmat([0 1], 1, ceil(nTrials / 2));
rec(:, 2) = Shuffle(haveGabor(1 : nTrials));
        % shuffle 0s and 1s
```

Display 8.4 (continued)

```
% Prioritize display to optimize display timing
Priority(MaxPriority(windowPtr));

% Start experiment with instructions
str = ['Press 1 for Gabor absent and 2 for Gabor present ' …
       'responses.\n\n' 'Press SPACE to start.'];
DrawFormattedText(windowPtr, str, 'center', 'center', 1);
Screen('Flip', windowPtr);
        % flip the text image into active buffer
WaitTill('space');       % wait till space bar is pressed
Secs = Screen('Flip', windowPtr);    % turn off instruction
p.start = datestr(now); % record start time

% Run nTrials trials
for i = 1 : nTrials
    Screen('DrawLines', windowPtr, fixXY, 3, 0.3);
        % fixation crosshairs
    t0 = Screen('Flip', windowPtr, Secs + p.ITI, 1);
    if rec(i, 2), params(4) = p.contrast;
    else params(4) = 0;
    end
    Screen('DrawTexture', windowPtr, tex, [], [], 0, …
          [], [], [], [], 2, params);
    t0 = Screen('Flip', windowPtr, t0 + p.fixDuration);
    Screen('Flip', windowPtr, t0 + p.stimDuration);
    [key Secs] = WaitTill(keys);        % wait till response
    if iscellstr(key), key = key{1}; end
        % take the first key in case of multiple key presses
    if strcmp(key, 'esc'), break; end % to stop
    rec(i, 3 : 4) = [strcmp(key, '2') Secs-t0];
        % record yes/no and respTime
end

p.finish = datestr(now);      % record finish time
save YesNoTask_rst rec p;     % save the results

% Tabulate the result
rec(i : end, :) = [];         % remove unfinished trials from
                              % trial record in case of early
                              % exit
nPresent = sum(rec(:, 2) == 1);
nAbsent = sum(rec(:, 2) == 0);
nHit = sum(rec(rec(:, 2) == 1, 3));
nFA = sum(rec(rec(:, 2) == 0, 3));
fprintf('\n         Stimulus Present    Stimulus Absent\n');
fprintf('"Yes"   %3g                 %3g\n', nHit, nFA);
fprintf('"No"    %3g                 %3g\n', nPresent - nHit,…
       nAbsent - nFA);
```

Table 8.2
Frequency and percentage (in parentheses) of hits, misses, false alarms, and correct rejections from an observer

	Stimulus Present	Stimulus Absent
"Yes"	80 (80%)	30 (30%)
"No"	20 (20%)	70 (70%)
Total	100 (100%)	100 (100%)

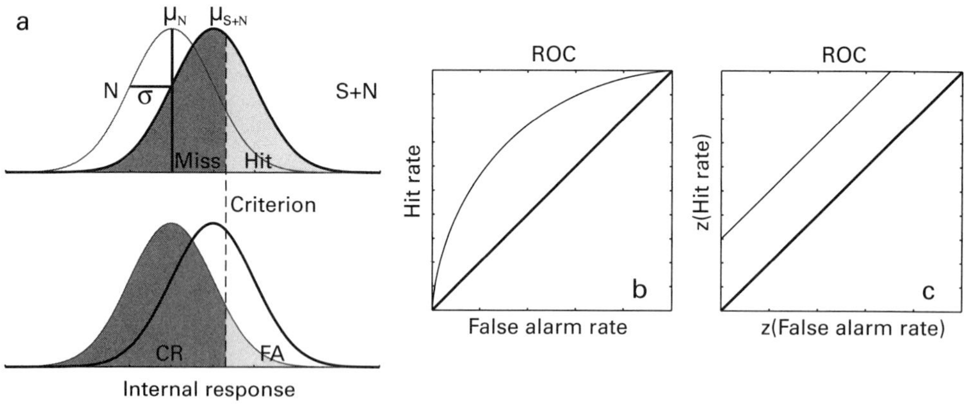

Figure 8.3
SDT in a "Yes/No" task. (a) The two distribution curves in each panel represent internal response distributions for signal present (S + N) and for signal absent (N) trials. A subjective criterion determines whether the observer reports that the signal is present ("Yes") or not ("No"). Internal responses greater than the criterion lead the observer to decide that the signal is present. The probabilities of hit, false alarm, miss, and correct rejection responses, as indicated by the shaded areas, are jointly determined by the internal response distributions and the criterion. (b) The ROC plots the hit rate as a function of the false alarm rate as the observer varies his or her criterion. (c) ROC shown transformed in a *z*–*z* plot.

where $g(x,\mu,\sigma)$ is the Gaussian probability density function and $G(x,\mu,\sigma)$ is the cumulative Gaussian distribution (proportion of shaded area under curves) (see figure 8.3a). The probability density function is the relative probability of observing a particular value. A cumulative distribution function is the proportion of the distribution below a particular value.

The proportion of misses corresponds to the cumulative distribution function of the signal plus noise distribution below the criterion value c:

$$P_{\text{Miss}}(c) = \int_{-\infty}^{c} g(x,\mu_{\text{S+N}},\sigma_{\text{S+N}})\,dx = G\left(\frac{c}{\sigma_{\text{S+N}}},\frac{\mu_{\text{S+N}}}{\sigma_{\text{S+N}}},1\right). \tag{8.3}$$

The proportions of false alarms and hits are the complement of correct rejections and misses, respectively:

$$P_{\text{FA}}(c) = 1 - P_{\text{CR}}(c) = 1 - G\left(\frac{c}{\sigma_{\text{N}}},0,1\right) \tag{8.4}$$

$$P_{\text{Hit}}(c) = 1 - P_{\text{Miss}}(c) = 1 - G\left(\frac{c}{\sigma_{\text{S+N}}}, \frac{\mu_{\text{S+N}}}{\sigma_{\text{S+N}}}, 1\right). \tag{8.5}$$

Assuming $\sigma = \sigma_{\text{S+N}} = \sigma_{\text{N}}$, the SDT parameters d' and c can be estimated from the observed hits and false alarms:

$$d' = \frac{\mu_{\text{S+N}}}{\sigma} = z(P_{\text{Hit}}) - z(P_{\text{FA}}) \tag{8.6}$$

$$c = z(P_{\text{FA}})\sigma . \tag{8.7}$$

Here, the mean of the stimulus-absent or noise distribution μ_{N} is set to 0. The z refers to the z-score of the corresponding probability; it converts a cumulative probability for the normal distribution into units of standard deviation.

In SDT, the measure of discrimination d' is described as a "pure" index of discrimination sensitivity, while criterion c estimates the decision bias. Discrimination d' in SDT provides a true measure of the sensitivity of the observer independent of criterion. (Note, this use of the word *sensitivity* is related to but not the same as the sensitivity that is estimated as the inverse of threshold in other contexts in psychophysics.)

To see how SDT is implemented in the example, consider again the response tabulation in table 8.2 for the simple low-contrast Gabor detection experiment in section 8.3.2. We can transform the observed proportions of hits and false alarms (or correspondingly of misses and correct rejections) into z-scores with the MATLAB function `norminv`. These z-scores are used to compute the observed discriminability d' and the criterion c. For this example, with hit rate of 80% and false alarm rate of 30%, corresponding to an overall percent correct of 75%, the estimated discriminability d' and criterion c are 1.366 and –0.524, respectively. These calculations are shown in display 8.5. Alternatively, MATLAB has a function `dprime` that performs these computations (`[d,c]=dprime(pHit,pFA)`).

To provide an intuition about SDT with different parameters, we use MATLAB functions for the probability density function of the normal distribution, `normpdf`, in a program that plots the signal and noise distribution for the d' and c estimated for this example data, and we use the function for the cumulative distribution function, `normcdf`, to show the corresponding proportions for hits and false alarms (display 8.6). This program `plotsdt.m` can be used with different values of d' and c to visualize and better understand the functional relations between these model parameters and the observable response categories.

Display 8.5

```
%%% Program dPrime1.m
function [dPrime, c] = dPrime1(pHit, pFA)
dPrime=norminv(pHit) - norminv(pFA);
c = norminv(pFA);
```

Display 8.6

```
%%% Program PlotSDT.m
function [pHit, pFA] = plotSDT(dprime, c)
x = (-300:(300+dprime*100))/100;
N = normpdf(x, 0, 1);
SN = normpdf(x, dprime, 1);
xc = [c c];
yc = [0 0.5];
plot (x, N, 'k-', x, SN, 'k-', xc, yc, 'k-');
pHit = 1 - normcdf(c, dprime, 1);
pFA = 1 - normcdf(c, 0, 1);
```

All of these examples assume that the variances of the signal and noise distributions are the same. We consider unequal variance examples in the following section on receiver operating characteristics.

8.3.4 Receiver Operating Characteristic Curves

Receiver operating characteristic (ROC) curves show the systematic relationship between hits and false alarms over a range of different criteria.[8,9] To measure an ROC, the observer is asked to use a number of different criteria, and hits and false alarms are measured for each criterion condition. One experimental approach asks the observer to provide a confidence rating on each trial that gives a higher confidence response for a higher criterion. Another experimental approach uses different instructions on different trials, for example by instructing observers to use a "lax," "strict," or "average" criterion. Or, in other experiments, ROCs are measured by manipulating the probability of the signal-present trials, which in turn should alter the criterion that the observer chooses to use in decision.[9]

The function describing the relationship between the hit and false alarm rates for different criteria, the ROC curve (figure 8.3b), provides an important test of SDT. In particular, it can be used to test equal variance assumptions and to provide enhanced estimates of discriminability when the variances of the internal response distributions are unequal.

The functional relationship between hits and false alarms (derived from equations 8.4 and 8.5) is defined by this general equation:

$$z(P_{\mathrm{Hit}}) = \frac{\sigma_{\mathrm{N}}}{\sigma_{\mathrm{S+N}}} z(P_{\mathrm{FA}}) - \frac{\mu_{\mathrm{S+N}}}{\sigma_{\mathrm{S+N}}}. \tag{8.8}$$

The relationship depends upon the distance between the noise and signal + noise internal response distributions and the ratio of their variances. The distance between the two distributions is $\mu_{\mathrm{S+N}}$ because μ_{N} is defined as 0.

The discrimination sensitivity of the observer can be empirically observed by measuring the ROC curve. The graph of hits and false alarms can be displayed on probability–

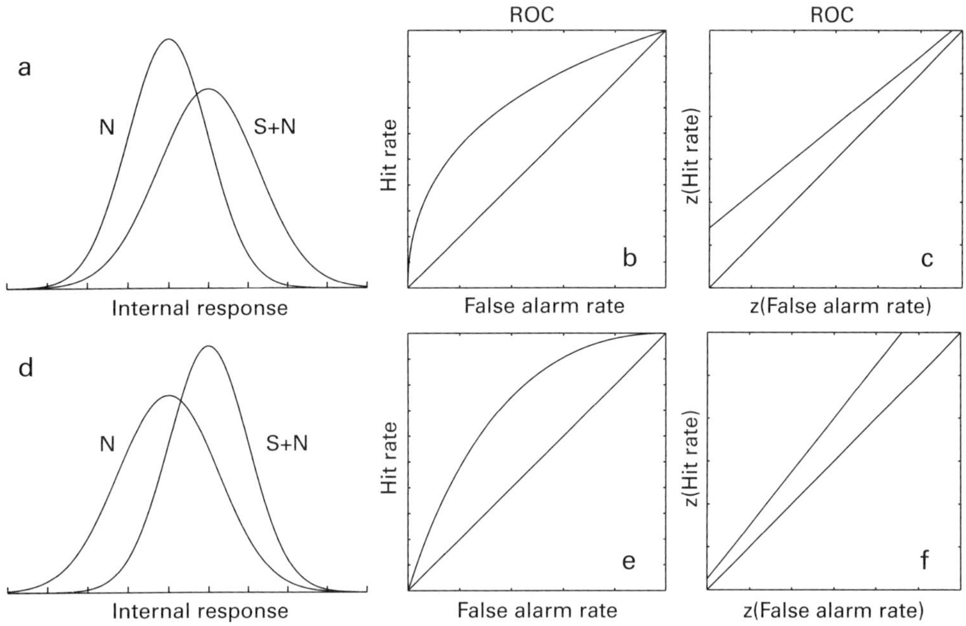

Figure 8.4
SDT in a "Yes/No" task for signal and noise distributions with different variances. (a) S + N has a higher variance than N. (b, c) ROC curves for the distributions in (a). (d) N has a higher variance than S + N. (e, f) ROC curves for the distributions in (d).

probability axes, in which case the ROC is a concave curve. The ROC can also be graphed on $z(P_{\text{Hit}})$ versus $z(P_{\text{FA}})$ axes (figure 8.3c). This is convenient because the ROC corresponds to a straight line in z–z space for normally distributed internal responses.

The example shown in figure 8.3c assumes a simple case in which the variance of the internal response on signal-present and signal-absent trials is equal—the *equal-variance SDT*. In this case, the slope of the ROC curve on z–z axes is 1.

If the variance of the internal responses on signal-present and signal-absent trials is not equal—the *unequal-variance SDT*—then the slope of the ROC on z–z axes will deviate from 1. Figure 8.4 shows two cases. If the variance of the internal responses on signal-present trials is larger, then the slope will be less than 1 (figure 8.4a–c); if the variance of the internal responses on signal-present trials is smaller, then the slope will be greater than 1 (figure 8.4d–f). The slope of the ROC has been considered a signature feature for certain perceptual processes[16–18] and so has been extensively studied.

Whether or not the variances are equal, larger discrimination d' values have ROC lines that are further from the negative diagonal. ROC functions for higher d's move toward the upper left of the ROC graph, which corresponds with more hits and fewer false alarms.

The definition of d' is altered slightly for the unequal variance case to scale the difference between the means of signal + noise and noise distributions by the square root of the *average* of the two variances:

$$d'_{\mathrm{a}} = \frac{\mu_{\mathrm{S+N}} - \mu_{\mathrm{N}}}{\sqrt{\dfrac{\sigma^2_{\mathrm{S+N}} + \sigma^2_{\mathrm{N}}}{2}}} = z(P_{\mathrm{Hit}}) - \frac{\sigma_{\mathrm{N}}}{\sigma_{\mathrm{S+N}}} z(P_{\mathrm{FA}}), \tag{8.9}$$

where $\sigma_{\mathrm{N}}/\sigma_{\mathrm{S+N}}$ is the slope of the z–z ROC.

Our ROC experimental example, `RatingExperiment.m` (display 8.7), uses a rating scale variant of the simple Gabor detection experiment from display 8.4. Instead of simple "yes" and "no" responses, we collect 6-point rating scale responses, corresponding to high-, middle-, and low-confidence "target-present" and high-, middle-, and low-confidence "target-absent" responses. We increase the number of trials to 300 to allow better estimates of the probabilities of the different response categories. Hypothetical data for this experiment and the corresponding proportions are shown in table 8.3. Display 8.8a–c provides an ROC analysis program that estimates the standard deviation of the signal + noise distribution, which is equal to the slope of the z–z ROC curve, and the corresponding d' and the five c_i criteria that separate the different rating responses. The results are shown in figure 8.5. Estimating the slope of the ROC involves issues about how to treat deviations between predicted and observed proportions. We simply present this estimation in displays 8.8a–c; the issues behind the computation will be considered again in chapter 10 on data analysis and model fitting.

An alternative index of discriminability based on the measured ROC is A', the area under the ROC curve, or

$$A' = 1 - \sum_{k=1}^{n} (X_k - X_{k-1})(Y_k + Y_{k-1})/2, \tag{8.10}$$

where n reflects the number of points on the ROC curve from (0, 0) to (1, 1), and Y_k and X_k are the hits and false alarm rates (figure 8.5a).

A' is an alternative measure to d'. It estimates the proportion of times that a randomly chosen response from the internal signal + noise distribution exceeds a randomly chosen response from the noise distribution. Researchers may prefer the A' measure in some situations because it is independent of the exact distribution assumed or whether the variances of the two distributions are equal. The value of A' for the data shown in table 8.3 is 0.7376.

8.4 Unidimensional Signal-Detection Tasks in SDT

The previous section introduced SDT using the Yes/No paradigm. The theoretical framework of SDT is designed to separate discriminability from shifts in criterion. However, the Yes/No paradigm itself is unusually susceptible to criterion shifts, as exposed by an SDT analysis. It may be desirable in some circumstances to reduce or eliminate the impact of criterion on performance. A number of other paradigms, such as forced-choice paradigms, have been developed to reduce the impact of criterion shifts on performance. These other paradigms also can be analyzed within the SDT framework. This section considers a number of representative paradigms.

8.4.1 Two-Interval Forced-Choice/Two-Alternative Forced-Choice

In a two-interval forced-choice/two-alternative forced-choice (2IFC/2AFC) task, an observer is presented with two input stimuli, one from each of two stimulus categories. Two-interval forced-choice (2IFC) refers to cases where stimuli from the two stimulus categories appear in different temporal intervals, first one and then the other, and the observer responds by choosing an interval containing a stimulus from the target category. In the case of visual 2AFC, the two alternatives are often presented simultaneously, perhaps on two sides of fixation. The two stimulus categories may be "signal present" and "signal absent," but might also be, for example, stimuli of lower or higher luminance.

SDT for 2IFC/2AFC assumes that on each trial, there are two internal responses, x_1 and x_2, corresponding to interval 1 and interval 2 or location 1 and location 2. The probability density function of the internal response x generated by an A stimulus is $g(x,\mu_A,\sigma_A)$ and by a B stimulus is $g(x,\mu_B,\sigma_B)$, where we choose labels A and B so that $\mu_B > \mu_A$ (figure 8.6). To decide whether x_1 is generated by an A stimulus (and therefore x_2 is generated by a B stimulus) or x_1 is generated by a B stimulus (and therefore x_2 is generated by an A stimulus), an observer compares x_1 and x_2.

The decision in the 2IFC/2AFC paradigm can be computed in two different ways that are equivalent in many situations. The first decision rule assumes that a comparison is carried out by computing the difference between the two samples and comparing to a criterion at 0. This is usually labeled the *difference* rule. So, if $x_1 - x_2 > 0$, the observer concludes that x_1 is generated by a stimulus in category B; otherwise, he or she concludes that x_1 is generated by a stimulus in category A.

For the difference rule decision, the probability correct P_c is the probability that $x_1 - x_2 > 0$, given that x_1 is generated by a stimulus in category B and x_2 is generated by a stimulus in category A:

$$
\begin{aligned}
P_c &= P(x_1 - x_2 > 0 \mid x_1 \in C_B \;\&\; x_2 \in C_A) \\
&= 1 - \int_{-\infty}^{0} g\left(x, \mu_B - \mu_A, \sqrt{\sigma_A^2 + \sigma_B^2}\right) dx \qquad (8.11) \\
&= 1 - G\left(0, \frac{\mu_B - \mu_A}{\sqrt{\sigma_A^2 + \sigma_B^2}}, 1\right).
\end{aligned}
$$

Display 8.7

```
%%% Program RatingExperiment.m
function RatingExperiment

%% Experimental Module

% Specify the stimulus
p.stimSize = 6;       % image diameter size in visual degree
p.contrast = 0.01;   % Gabor contrast
p.sf = 0.9; % c/d
p.sigma = 1.1; % degree
p.stimDuration = 0.1;   % stim duration in seconds
p.fixDuration = 0.15;    % fixation duration in seconds
p.ITI = 0.5;            % seconds between trials
nTrials = 300;          % total number of trials
keys = {'1' '2' '3' '4' '5' '6' 'esc'}; % response keys

% Compute stimulus parameters
ppd = pi / 180 * p.ScreenDistance / p.ScreenHeight * ...
      p.ScreenRect(4);            % pixels per degree
m = round(p.stimSize * ppd);     % stimulus size in pixels
sf = p.sf / ppd;                 % cycles per pixel
sc = round(p.sigma * ppd);       % sigma in pixels
params = [180 sf sc 0 1 0 0 0]; % gabor parameters: phase sf
                                 % sigma contrast
fixLen = 32;              % length of fixation in pixels
[xc yc] = RectCenter(p.ScreenRect);
fixXY = [[-1 0 0 1]*fixLen + [-1 -1 1 1 ]*m/2 + xc ...
       [1 1 1 1]*xc; [1 1 1 1]*yc [-1 0 0 1]*fixLen + ...
       [-1 -1 1 1 ] * m / 2 + yc];

p.randSeed = ClockRandSeed;      % use clock to set seed for
                                 % the random number generator

% procedural gabor allows us to change its parameters
% very quickly
tex = CreateProceduralGabor(windowPtr, m, m, 0, ...
      [1 1 1 0] * 0.5, 1, 0.5);

% Initialize a table to set up experimental conditions
p.recLabel = {'trialIndex' 'haveGabor' 'rating' 'respTime'};
rec = nan(nTrials, length(p.recLabel));
        % matrix rec is made of nTrials x 4 of NaN
rec(:, 1) = 1 : nTrials;
        % count the trial numbers from 1 to nTrials
haveGabor = repmat([0 1], 1, ceil(nTrials / 2));
rec(:, 2) = Shuffle(haveGabor(1 : nTrials));
```

Display 8.7 (continued)

```
        % shuffle 0s and 1s

% Prioritize display to optimize display timing
Priority(MaxPriority(windowPtr));

% Start experiment with instructions
str = ['Press 1, 2, 3 as high, middle, and low confidence '...
      'Gabor absent responses\n\n' 'Press 4, 5, 6 as low, ' ...
      'middle and high confidence Gabor present responses ' ...
      '\n\n' 'Press SPACE to start.'];
DrawFormattedText(windowPtr, str, 'center', 'center', 1);
        % Draw Instruction text string centered in window
Screen('Flip', windowPtr);
        % flip the text image into active buffer
WaitTill('space');      % wait till space bar is pressed
Secs = Screen('Flip', windowPtr);    % turn off instruction
p.start = datestr(now); % record start time

% Run nTrials trials
for i = 1 : nTrials
    Screen('DrawLines', windowPtr, fixXY, 3, 0.3);
        % fixation crosshairs
    t0 = Screen('Flip', windowPtr, Secs + p.ITI, 1);

    if rec(i, 2), params(4) = p.contrast;
    else params(4) = 0;
    end
    Screen('DrawTexture', windowPtr, tex, [], [], 0, [], ...
          [], [], [], 2, params);
    t0 = Screen('Flip', windowPtr, t0 + p.fixDuration);
    Screen('Flip', windowPtr, t0 + p.stimDuration);

    [key Secs] = WaitTill(keys);       % wait till response
    if iscellstr(key), key = key{1}; end
        % take the first in case of multiple keys
    if strcmp(key, 'esc'), break; end % to stop
    rec(i, 3 : 4) = [str2double(key)  Secs-t0];
        % record rating and respTime
end

p.finish = datestr(now);              % record finish time
save RatingExperiment_rst rec p;      % save the results

% Tabulate the result
nPresent = zeros(1, 6);
nAbsent  = zeros(1, 6);
```

Display 8.7 (continued)

```
rec(i : end, :) = [];      % remove unfinished trials in
                           % case of early exit
for i = 1 : 6
    nPresent(i)  = sum(rec(rec(:, 2) == 1, 3) == i);
    nAbsent(i)   = sum(rec(rec(:, 2) == 0, 3) == i);
end
fmtStr = repmat('%5.0f', 1, 6);
fprintf('\n');
fprintf(['Rating ' fmtStr '    total\n'], 1 : 6);
fmtStr = repmat('%5.0f', 1, 7);
fprintf(['Present' fmtStr '\n'], nPresent, sum(nPresent));
fprintf(['Absent ' fmtStr '\n'], nAbsent,  sum(nAbsent));
```

The expression $g(x, \mu_B - \mu_A, \sqrt{\sigma_A^2 + \sigma_B^2})$ represents the probability density function of the difference between two Gaussian random variables (figure 8.6b) with respective probability density functions $g(x, \mu_A, \sigma_A)$ and $g(x, \mu_B, \sigma_B)$ (figure 8.6a).

The second decision rule assumes that the observer compares the two internal responses directly and chooses the interval or the location with the larger value as coming from category B. Because the decision rule takes the maximum value, it is sometimes called the *max* rule. The probability density of obtaining an internal response x from a B stimulus is $g(x, \mu_B, \sigma_B)$. The probability that the internal response x is greater than any random sample from the distribution for a stimulus sampled from category A is $G(x, \mu_A, \sigma_A)$. The probability that all possible internal responses from stimuli in the B category are greater than those from stimuli in the A category is the product of the two probability functions, integrated over all the possible values of x, or the probability of correct response, P_c:

$$P_c = \int_{-\infty}^{+\infty} g(x, \mu_B, \sigma_B) G(x, \mu_A, \sigma_A) dx . \tag{8.12}$$

If $\sigma_A = \sigma_B = \sigma$, we can define $d' = (\mu_B - \mu_A) / \sigma$, and simplify equation 8.12:

$$P_c = \int_{-\infty}^{\infty} g(x - d', 0, 1) G(x, 0, 1) dx . \tag{8.13}$$

Equations 8.11 and 8.12 are mathematically equivalent for 2IFC/2AFC. Although equation 8.11 is more intuitive, equation 8.12 is more readily extended to situations with more intervals or samples (see section 8.4.2 for n-alternative forced-choice paradigms) or to situations with decision uncertainty (see section 8.5.2 for a full discussion of uncertainty).

8.4.2 *n*-Alternative Forced-Choice/*n*-Interval Forced-Choice

The 2AFC/2IFC procedures are sometimes generalized to n-alternative forced-choice (n-AFC) or n-interval forced-choice (n-IFC). In n-AFC, a trial displays one stimulus

Display 8.8a

```
%%% Program ROCanalysis.m
function ROC = ROCanalysis(Srating, Nrating);
freqHits(1) = Srating(6);
freqFA(1) = Nrating(6);
for i = 2:6
    freqHits(i) = freqHits(i-1) + Srating(7-i);
    freqFA(i) = freqFA(i-1) + Nrating(7-i);
end

pHits = freqHits(1:5)/freqHits(6);
pFA = freqFA(1:5)/freqFA(6);

figure;
plot(pFA, pHits, 'ks', pFA, pHits, 'k-');
line([0 1], [0 1]);
axis('square');
zHits = norminv(pHits);
zFA = norminv(pFA);

figure;
plot(zFA, zHits, 'ks', zFA, zHits, 'k-');
line([-3 3], [-3 3]);
axis([-3 3 -3 3]);
axis('square');

data = [zHits; zFA];
guess = [1 1.5 -2 -1 0 1 2];
options = optimset('fminsearch');
ROC = fminsearch('costfunc', guess, options, data);
d = ROC(1);  s1=1;
s2  = ROC(2);
c = ROC(3:7);

x = (-500:750)/100;
y1 = normpdf(x, 0, s1);
y2 = normpdf(x, d, s2);
figure;     % plot of the S+N and N distributions and
            % criteria
plot(x, y1, 'k-', x, y2, 'k-'); hold on;
for i = 1:5
  x = [c(i) c(i)];
y = [0 0.2];
line(x, y);
end

% Cmpute the area under the ROC vurve
Y = [0 pHits 1];
X = [0 pFA 1];
A = 0;
for i = 2:7
    A = A + (X(i) - X(i-1))*(Y(i) + Y(i-1))/2;
end

ROC = [ROC A];
```

Display 8.8b

```
%%% Program costfunc.m
function L = costfunc(guess, data)

zHits_observed =data(1,:);
zFA_observed = data(2, :);
d = guess(1);
sigma = guess(2);
criteria = guess(3:7);

zHits_predicted = norminv(1-normcdf(criteria, d, sigma));
zFA_predicted   = norminv(1- normcdf(criteria, 0, 1));
L = sum((zHits_observed - zHits_predicted).^2 + ...
    (zFA_observed - zFA_predicted).^2);
```

Display 8.8c

```
>> Srating = [ 7    15    27    35    20    46 ];
>> Nrating = [ 5    36    68    38     4     1 ];
>> ROC = ROCanalysis(Srating, Nrating)
ROC = 1.494 1.994 2.484 1.831 0.579 -0.6112 -1.842 0.7376
```

Table 8.3
Data from a hypothetical rating procedure

	Absent3	Absent2	Absent1	Present1	Present2	Present3	Total
Signal present	9	15	24	31	30	41	150
Signal absent	32	43	43	7	24	1	150

from the target stimulus category, such as "stimulus present," and $n-1$ stimuli from the other category, such as "stimulus absent." An advantage of adding intervals or stimuli is that random guessing corresponds with lower accuracy, and so a larger n provides a larger range over which to measure performance accuracy. For example, pure guessing corresponds to 50% correct in 2IFC, but 25% in 4IFC, while high levels of discrimination approach 100% correct in both. A disadvantage is that the increased number of alternatives or intervals may pose additional processing or memory load on the observer.

The n-AFC/n-IFC paradigms that take one sample from category B and $n-1$ samples from category A are simplest to think about within the context of the *max rule* formulation introduced for 2AFC. In this case, the correct response occurs when the internal response to the singular category B stimulus exceeds the internal response to the $n-1$ samples of category A:

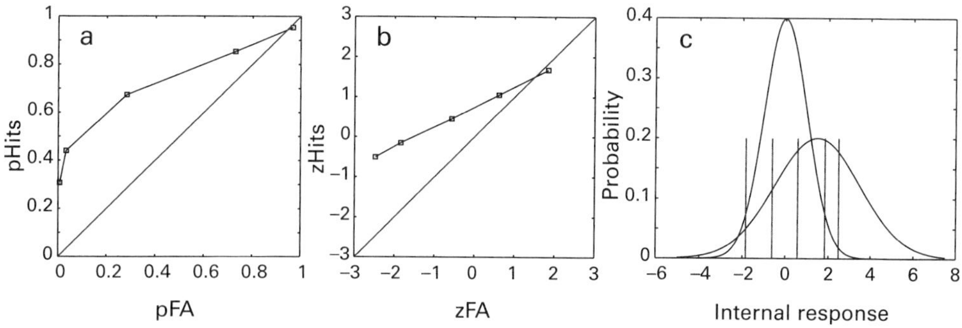

Figure 8.5
An example of a hypothetical ROC curve shown as (a) a probability plot and (b) a z–z plot and (c) graphs of the signal and noise distributions and the location of the five criteria recovered from the sample data. The mean and variance of the noise distribution are set to $\mu_N = 0$ and $\sigma_N = 1$, which scales the x-axis. In this example, $\sigma_{S+N} > \sigma_N$, corresponding with a z–z ROC slope of less than 1.

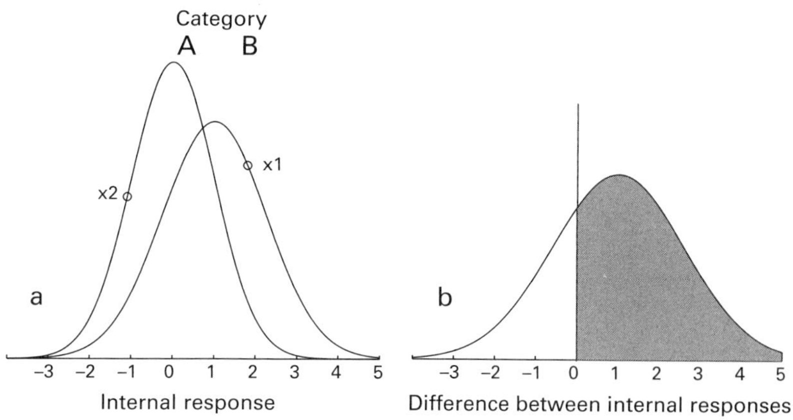

Figure 8.6
SDT in a 2AFC task. A 2AFC task presents two stimuli to an observer in each trial, one from each of two categories, and forces the observer to decide the correspondence between the stimuli and the categories. (a) The two bell curves represent the internal response distributions for the two stimulus categories with different means and standard deviations. The circles represent hypothetical values of x_1 and x_2 for a given trial. (b) The probability density distribution of the difference between the internal responses of stimuli in the two categories. The vertical line indicates the decision boundary.

$$P_c = \int_{-\infty}^{\infty} g(x, \mu_B, \sigma_B) G(x, \mu_A, \sigma_A)^{n-1} \, dx \, . \tag{8.14}$$

The probability correct for a n-AFC/n-IRC paradigm can be computed from the underlying theoretical discriminability d' that reflects the difference between the means of the internal response distributions from the two categories divided by the standard deviation. The equal variance case is usually assumed. The underlying discriminability is the key theoretical measure of sensitivity. This sometimes leads to confusion. The same percent correct observed in a 2AFC and an n-AFC experiment corresponds to a higher level of discrimination with larger n. Conversely, the identical underlying d' discriminability corresponds to different, lower, observed percent correct performance in designs with higher n.

This formulation assumes no observer biases toward particular locations or intervals. However, in principle it is possible for observers to show response biases or differential criteria for the different locations of simultaneously presented stimuli in n-AFC experiments or for particular interval(s) of the n-IFC experiment (see ref. 19). The equation must be generalized to include criterion terms if biases are present. Forced-choice experiments are generally believed to exhibit fewer issues of bias than the Yes/No experiments, and, in fact, in many cases where it can be checked, observed biases tend to be small while the number of alternatives or intervals is small. The data from an experiment can be examined by tabulating responses for each location or interval over all stimuli to look for major departures from unbiased performance. Other types of decision rules (than the max rule) have also been formulated for the n-AFC/n-IFC cases; these may lead to slightly different estimates.

8.4.3 Same–Different

The same–different paradigm compares two stimuli jointly and asks observers to judge whether they are the same. The simple same–different experiment consists of two categories of stimuli, A and B. Two stimuli are presented on each trial, and the observer responds either "same" or "different." On any given trial of the experiment, stimuli from AA, AB, BA, or BB trial types are presented. The correct response to AA or BB trial types is "same" and to AB or BA trial types is "different." The two stimuli in each trial are chosen independently and randomly across trials. Alternatively, the numbers of trials of the four types are set to be equal for a perfectly balanced design, and the order of the trials is randomized. As an example, four stimulus pairs manipulating the angle of Gabors are shown in figure 8.7.

The same–different paradigm is more complicated than some of the other tasks because it involves the use of multiple criteria or multiple decisions that are combined together. No single criterion (or linear boundary in two dimensions) can solve the problem.

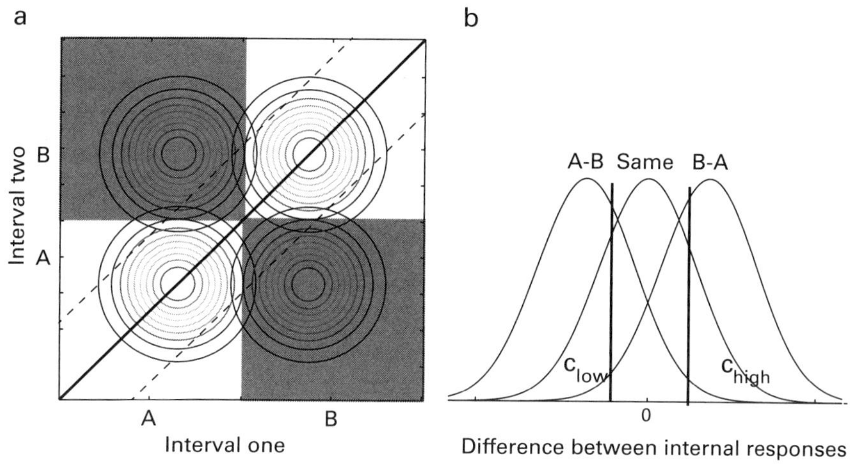

Figure 8.7
The same–different task. (a) In each interval, a stimulus from either category A or category B is shown, resulting in four possible trial types: AA, AB, BA, and BB. Observers must judge if the stimuli in the two intervals are the same or different. (b) Difference distributions.

There are several strategies—or ways for an observer to go about making a decision—in a same–different task. The most commonly cited decision rule (see ref. 9) takes the difference between the internal response to the stimuli appearing either successively in two intervals or simultaneously in two locations. Researchers usually assume that the variances σ^2 are equal and that the internal responses are uncorrelated. For trials where stimuli are the same, AA or BB, the distribution of the difference scores should be centered about zero and the variance is $2\sigma^2$. For trials where the stimuli are different, the distribution of difference scores is centered either around $+d$ or around $-d$ depending upon whether the pair is AB or BA; the variance is $2\sigma^2$. In this case, the subject should say "same" whenever the difference is sufficiently close to 0 and say "different" otherwise. This requires two criterion values, c_{low} and c_{high}. If the two criteria are symmetric, only one parameter $\pm c$ is estimated. One way the criteria may be displayed on a graph is as a pair of lines diagonal to the axes (figure 8.7).

Another different—and more optimal—way to make the same–different decision is to combine independent classifications of the two stimuli. The first stimulus is classified as an A or B, the second stimulus is classified as an A or B, and then these two classifications are combined to determine the "same" and "different" response. In the two-dimensional representation of the internal responses in figure 8.7, this corresponds to partitioning the four quadrants of the two-dimensional space. If the criteria applied to both dimensions are unbiased, then the multiple classification strategy is optimal.

The probability correct for the optimal independent decisions strategy is related to d' by:

$$P_c = \left[G(d'/2)^2 + G(-d'/2)^2\right]. \tag{8.15}$$

An approximation to d' from the standard equation is often used, $d' = z_{\text{Hits}} - z_{\text{FA}}$. For the same–different paradigms, hits are defined as responding "different" to AB and BA trials, or correctly detecting a difference. False alarms are defined as responding "different" to AA and BB trials, or incorrectly detecting a difference.

For the independent decisions rule,

$$d' = 2z\left(\frac{1}{2}\left[1 + (2P_c - 1)^{1/2}\right]\right).$$

For a given d' between the distribution of A's and B's, the observed probability correct is slightly higher for the (optimal) independent decisions rule than for the differencing rule. Both decision rules in same–different test designs will lead to percent correct performance that is noticeably less than the percent correct for a Yes/No design with the same two stimuli. For example, for a d' of 1, Yes/No performance yields 69% correct, independent decisions same–different yields 57% correct, and differencing same–different yields 55% correct.

A final approach to the same–different decision is to choose a response based on the likelihood ratio. Here, the subject would say "same" whenever the observed internal responses are more likely to have arisen from same than from different pairs. The likelihood ratio formulation is another way to define optimal performance. It has been argued that the likelihood ratio requires particularly complex processing on the part of the observer—it assumes that the observer has full knowledge of the form of the distributions of internal responses.

See ref. 9 for a more detailed treatment of the same–different tasks and issues of response bias. Several researchers have provided more detailed analyses of these different approaches to the same–different task and their optimality.[20–22] For a recent development of ways to test for various strategies, including those with asymmetric criteria, see ref. 23.

8.4.4 Identification

In an identification experiment, a single stimulus is presented on any trial, and the stimulus is drawn from one of two or more stimulus types or conditions. The subject must classify—identify—the stimulus. If the stimuli vary along a single dimension, such as contrast, the same single detector might process all the stimuli. In this case, different conditions are assumed to produce distributions shifted along the internal response axis.

Unidimensional identification tasks default to the SDT formulation of a Yes/No task. If there are only two stimulus conditions and the distribution of internal responses is higher for one and lower for the other, then identification is completely analogous to simple Yes/No. Simply respond "B" if the internal response sampled on that trial exceeds criterion c. If there are n stimulus categories corresponding to n shifted distributions on a single internal response dimension, then the subject will need n - 1 criteria to yield n response categories, and this is theoretically related to an n-point rating scale. Whether an identification task is intrinsically unidimensional depends upon the stimuli and how they are processed in the visual (or auditory or tactile) system. In some cases, what may seem like a variation of stimuli along a single dimension, such as orientation, may in fact be processed in the brain by a set of competing detectors tuned to different ranges of orientations. We take up the extensions of SDT to multiple dimensions in the next section.

8.5 General Recognition Theory

The standard signal detection experiments and analysis involve stimuli that differ in a single dimension—or whose differences are projected onto a single relevant decision dimension (section 8.4.4). However, stimuli often differ along more than one dimension of variation. For example, Gabor patches varying in contrast have other attributes such as spatial size and spatial frequency. In some circumstances, the experimenter—or the world—presents stimuli that vary in several dimensions, in the same experiment or setting, such as varying color and shape, or spatial frequency and orientation, or hue and saturation.

SDT provides a theoretical analysis for many detection and discrimination paradigms for single-dimensional situations. The general recognition theory (GRT) extends the principles of SDT to multidimensional stimuli for several different kinds of decisions.[24] Some cases of GRT are simple extensions of the original unidimensional SDT. Other cases can be quite complicated. This section describes some of the simpler ones.

Consider an experiment with four distinct stimuli created with two values or levels on each of two stimulus dimensions. For example, the stimuli might be a set of four lines made from one of two lengths and one of two orientations. As in SDT, the external stimuli are encoded in internal representations that vary from trial to trial due to internal variability or internal noise. Presenting a given stimulus leads to an internal value on each dimension, corresponding to a point in a two-dimensional (2D) space representing length and orientation. Over trials, the variability in the internal representations of length and orientation leads to four distributions in the 2D space, corresponding to the four stimuli that can be presented in different trials: $(\text{length}_1, \text{orientation}_1)$, $(\text{length}_1, \text{orientation}_2)$, $(\text{length}_2, \text{orientation}_1)$, and $(\text{length}_2, \text{orientation}_2)$.

Figure 8.8 illustrates the simple case where the internal response distributions for each of four stimuli (listed above) are drawn from bivariate normal distributions that are equivalent. The means and standard deviations for stimuli varying in length (dimension 1) are not dependent on orientation (dimension 2) and vice versa.

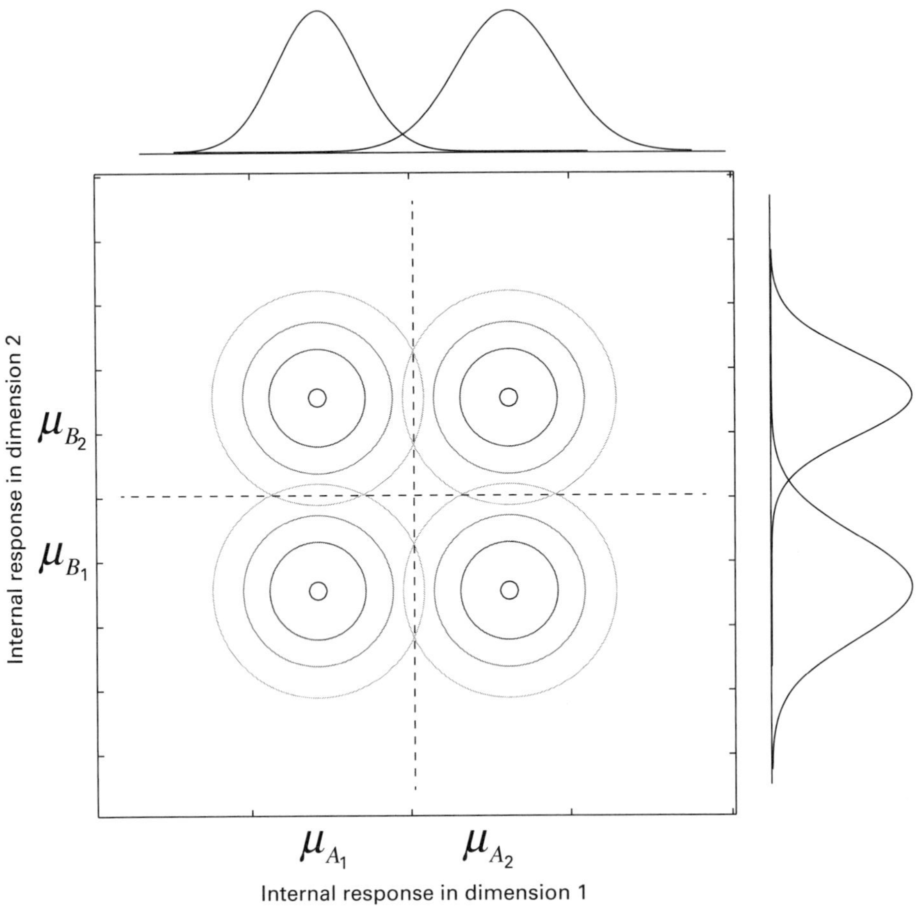

Figure 8.8
GRT representations of four stimulus types in two dimensions that exhibit perceptual separability and perceptual independence. Joint probability density distributions for the four stimuli in two dimensions are shown as contour circles along with the marginal probability density distributions shown above or to the right of the relevant dimensions. The dotted lines indicate the optimal decision boundaries that maximize proportion correct for identification of stimuli in the four categories.

When the values of the 2D internal representation are projected onto the length axis, the marginal distributions are identically and normally distributed for the two levels of orientation. Because they are identical, we see only a single marginal distribution curve. (Projection onto the length axis means that we look at the distribution of values on that dimension regardless of the value on the other one.) Similarly, when the values of the 2D internal representation are projected onto the orientation axis, the marginal distributions are identically and normally distributed for the two levels of length. If the aspect ratios of the two axes have been scaled to visually equate the standard deviations in the two dimensions, and the two dimensions are perceptually independent, then the equal-probability contour lines that represent the bivariate normal probability density distributions will appear circular in the (scaled) 2D space as in figure 8.8. The contour lines connect the points in the distribution that have the same value.

In standard SDT, a single decision criterion demarks the regions of the one-dimensional (1D) line associated with each response. By analogy, in the simple case of GRT depicted here, a single criterion for perceived length corresponds with a line in the 2D space that is perpendicular to the length axis—and does not depend upon the sensed orientation. The same is true for a decision about the stimulus orientation. Jointly, these two lines separate four regions in space that would correspond with an identification response.

In the most general case of the GRT, the internal representations of different stimulus conditions are described completely by their means and variances on the different dimensions and by the covariance matrix, all of which are free to vary. The GRT was developed to handle not just the simple special cases but also other more complicated cases. There are many ways in which four stimuli in two dimensions might depart from the simple case. The means and/or standard deviations on one dimension may depend upon the level of its own dimension and/or on the level of the other variable. Even if the marginal distributions are the same, the variability in the two dimensions may be correlated, exhibiting distributions with contours on the 2D graph that are oblique to the axes.

Discussions of the GRT have focused on three simplifying properties of the multidimensional description, perceptual independence, perceptual separability, and decision separability.[25,26] Two dimensions are *perceptually independent* if the bivariate distribution of the internal representation is the product of the marginal distributions. This only occurs if the internal values of the two dimensions are statistically independent—chosen independently and at random.[24,27]

Two dimensions are *perceptually separable* if the marginal distributions are independent of the level of the other variable. Dimensions that are perceptually separable are not necessarily perceptually independent, and vice versa.[28,29]

Decision separability corresponds to the simple cases where the decision boundaries for classifying each of the single dimensions are straight lines parallel to the coordinate axes and perpendicular to the axis of the decision. Decision boundaries are lines or curves in the 2D decision space that demark regions for different responses.[28,29]

Figure 8.8 showed a particularly simple case in which perceptual separability, perceptual independence, and decision separability all hold.

Figure 8.9 shows a special case where variances of the internal representations on a dimension depend upon the level of that dimension, but the representations still exhibit perceptual independence and perceptual separability. The perceptual independence corresponds to equal probability ellipses of the bivariate normal distributions that are not tilted relative to the axes in the 2D space. The variance in the internal representations of each individual dimension is larger for the larger level, yet does not depend upon the level of the other variable, meeting perceptual separability. Each individual decision (i.e., for longer or shorter or for more or less oriented) shows decision separability, and the two decision bounds (dashed lines) taken together provide decision separable bounds for identifying stimuli in the four categories.

Figure 8.10 shows a case in which perceptual separability holds, yet perceptual independence and decision separability fail. Indeed, the means and variances of the distributions are the same regardless of the level of either dimension. Yet, the variability in the internal representation is correlated; higher values on one dimension tend to go with higher values on another. This correlation suggests optimal decision boundaries that do not show decision separability—they are not parallel to the dimensional axes.

Empirically, observers may use simplified but nonoptimal decision strategies that exhibit decision separability in complicated perceptual situations in which there are failures of perceptual independence or perceptual separability. If the distributions in perceived lengths, for example, differ substantially for lines of different orientations, an observer might exhibit separable decision boundaries even though more complicated decision boundaries—no longer parallel to the coordinate axes or even straight lines—may be required to optimize accurate classification. The observer may default to a strategy of decision separability simply because he or she does not know how to set up the more complicated decision boundaries.

Developments in GRT have created tests for perceptual independence, perceptual separability, and decision separability based on response categorization data. Other tests have been developed based on response time signatures, in some cases by assuming that response times are inversely related to distance from a decision boundary,[26,30–33] or adding dynamic noise to the GRT to derive response time predictions.[34] Many of these developments are quite technical. Interested readers should consult key publications in this area, such as Refs. 24, 25, and 35.

8.6 Uncertainty in SDT

8.6.1 Identification with Channel Labels

The perceptual system has many channels, or types of detectors, that process different feature dimensions. In a typical identification experiment, one stimulus is presented and

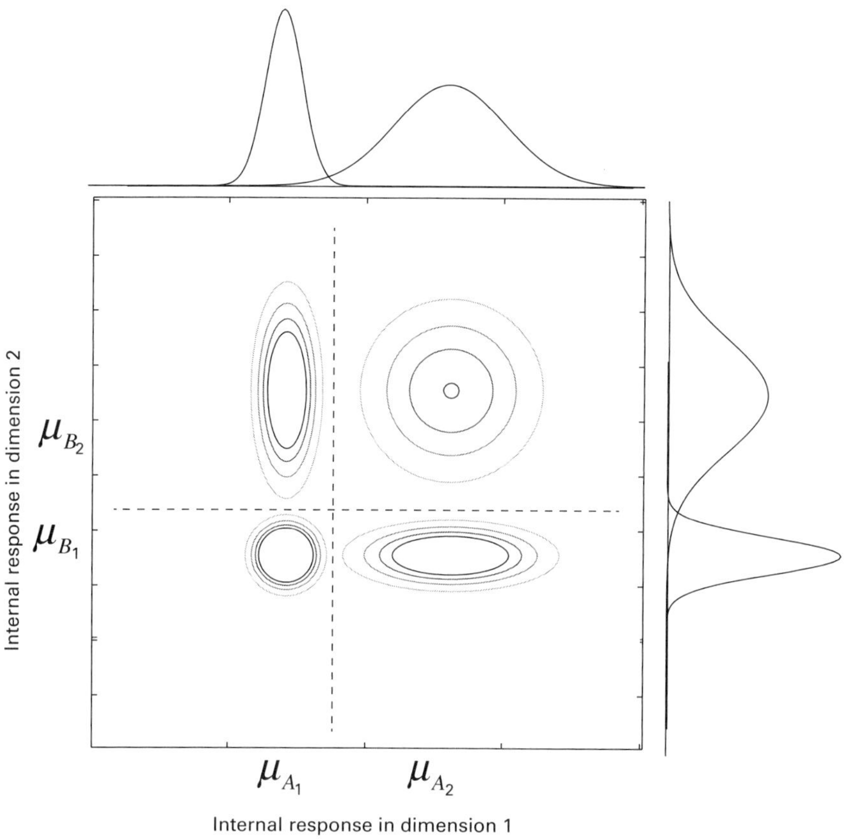

Figure 8.9
GRT representations of four stimuli in two dimensions that exhibit perceptual and decision separability and perceptual independence in the presence of variance differences. Joint probability density distributions for two stimulus values in each of two dimensions are shown as contour ellipses along with the marginal probability density distributions shown as curves to the right or above the relevant dimension. The dotted lines indicate the optimal decision boundaries that maximize proportion correct for identification of stimuli in the four categories.

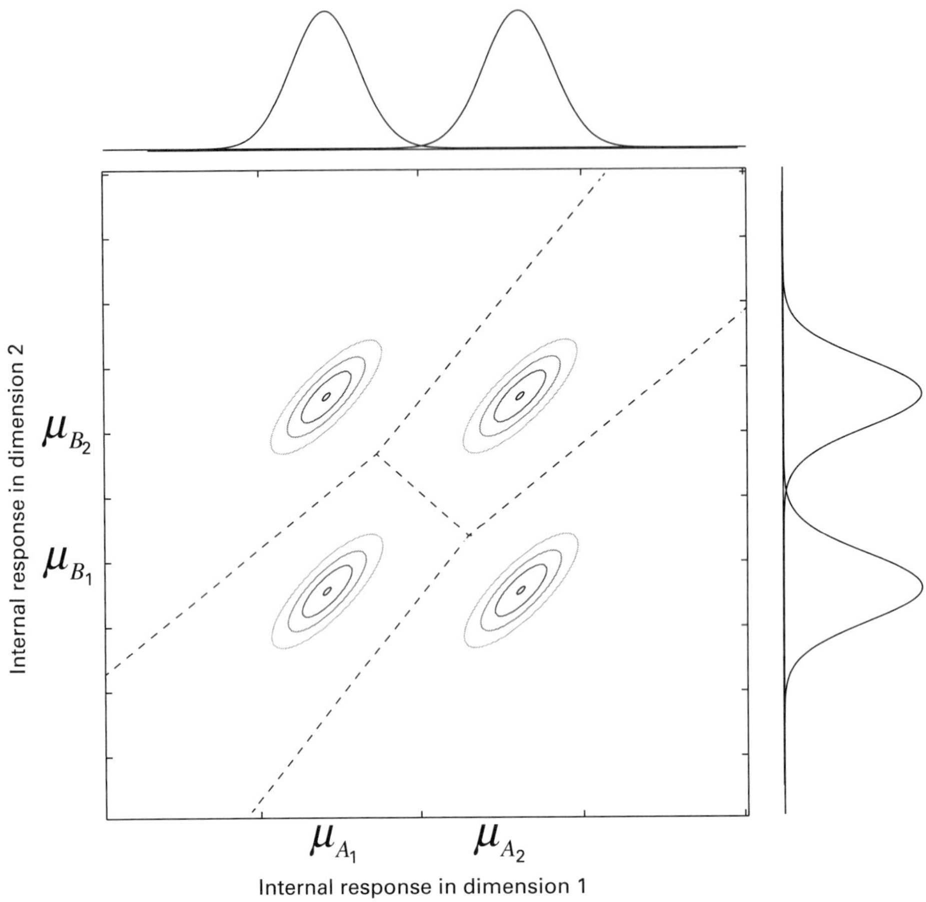

Figure 8.10
GRT representations of four stimuli in two dimensions that exhibit perceptual separability but neither perceptual independence nor decision separability. Joint probability density distributions for two stimulus values in each of two dimensions are shown along with the marginal probability density distributions. The dotted lines indicate the optimal decision boundaries that maximize proportion correct for recognition of stimuli in the four categories.

can be potentially processed by many different detectors or channels. Identifying the stimulus means responding with the label for the detector or channel that is activated most by the stimulus. So far, we have considered SDT for single-dimensional representations and GRT for stimuli with two or more features. Here we consider cases where a single stimulus is coded by multiple detectors in a single feature dimension. Multidimensional SDT has been developed to consider these cases in which there are internal representations that correspond with different detectors.[36,37]

Let's consider a simple 2AFC identification experiment in which a single Gabor of one of two orientations is shown to the observer in each trial, and she or he has to identify the orientation of the stimulus. A decision rule may need to consider two Gabor detectors, one for each of the two Gabor orientations. In each trial, a single Gabor stimulus generates one internal response from each of the two detectors. Across trials, the internal representations of the Gabor stimuli can be represented in a 2D space (figure 8.11).

If the two internal response distributions are independent, then the problem is analogous to one-dimensional 2AFC. The two distributions, when projected to an axis

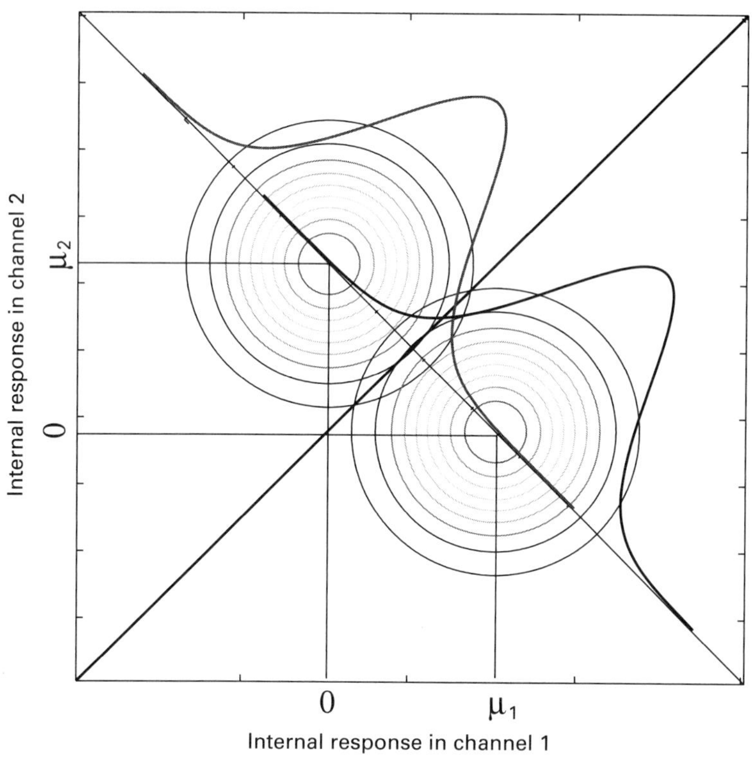

Figure 8.11
Two-alternative identification. Each stimulus generates two internal responses, one from each channel.

orthogonal to the criterion, lead to distributions that correspond to the two shifted distributions of standard SDT. Then, a single linear boundary in the 2D space—the positive diagonal line in figure 8.11—represents the single criterion in standard STD. This projection rule is also equivalent to sampling each channel for the presented stimulus and choosing the response that corresponds to the maximally active channel. In some cases, the responses of the two channels are not orthogonal, in which case SDT application is more complicated.[24,28,35,38]

8.6.2 SDT in 2IFC/2AFC Detection with Uncertainty

Another major development of SDT is the formulation of decision in situations where multiple channels, only one of which is actually relevant, contribute to the response, yet the observers cannot perfectly isolate the relevant channel. This failure to isolate a single channel for decision is called an uncertainty problem. The observer is uncertain which channel to focus on, and so makes a decision based on the response of U task-irrelevant channels and one relevant channel.

Uncertainty arises in several ways. The first is experimenter-controlled (or environmentally controlled) uncertainty. An experiment may include signal stimuli that are drawn from a set of stimuli on each trial. Although one of the detectors or channels will turn out to be relevant for the selected stimulus on a particular trial, the observer does not know in advance which it is and so must monitor the internal responses of a number of detectors to make the decision. For example, the experimenter might present a signal stimulus of one of several spatial frequencies chosen at random over trials. Another situation in which experimenter-controlled uncertainty occurs is when a known signal is displayed in one of a set of randomly selected locations, a situation of signal-known location uncertainty. An example of this is a visual search task in which the observer chooses which interval or which side of the screen contains a tilted Gabor among vertical Gabors. The observer knows that the tilted Gabor is the target. But because the location is not known in advance, the observer must monitor all locations, only one of which will turn out to contain a signal stimulus. This uncertainty has well-known implications for decision accuracy that can be computed in the SDT framework.[39–42]

The second type of uncertainty emerges from the observer. In these cases, the uncertainty is not in the experimental design, but in the neural responses of the observer. Many neural units may potentially be relevant for a perceptual decision, but the observer either does not know which they are or cannot isolate them. The same or related uncertainty computations in the SDT framework may apply in these situations.

We illustrate the computational approach to uncertainty in SDT for a 2AFC/2IFC design. On any given trial, the observer is presented with one known target stimulus chosen from a stimulus set and one distractor stimulus (or interval). Suppose that $(U + 1)$ independent detectors respond to each stimulus. Only one of those detectors is task relevant, but the observer cannot identify in advance which one it is and has to make a decision based on the internal responses of all the detectors that the observer considers broadly task relevant.

Over the two intervals or locations, the observer monitors a total of $2(U+1)$ internal responses, of which only one is associated with the signal stimulus. For this development, we assume that the internal responses of all the $2(U+1)$ detectors are independent and Gaussian distributed. The signal or target stimulus generates an internal response distribution with mean μ_{S+N} and standard deviation σ_{S+N} in the task-relevant detector, while the response distributions of the internal responses of all the other $2U+1$ detectors have mean 0 and standard deviation σ_N.

There are several different decision rules that the observer may use. The optimal rule sums $(U+1)$ internal responses for each of the two stimuli or intervals and chooses the interval or the stimulus with the larger sum as containing the signal. For the sum of the $(U+1)$ random variables for each interval or stimulus, the mean is equal to the sum of the means; the variance is the sum of the variances.

Often, a maximum rule, or *max rule*, is used instead of the summation rule. In many conditions, the max rule is a reasonable approximation[43] to the optimal decision rule. With the maximum rule, the observer compares all the $2(U+1)$ internal responses and chooses the interval or sample with the maximum as the signal location or signal interval. A correct response may arise in two different ways: (1) the internal response of the task-relevant detector to the stimulus containing the signal (with mean μ_{S+N}) is greater than the other $2U+1$ internal responses (each with mean 0), and (2) the internal response of one of the task-irrelevant detectors to the signal is greater than the other $2U+1$ internal responses, including the one from the task-relevant detector for the signal. In the second case, the observer generates the correct response for the wrong reason. The two possibilities are reflected in the two terms of the following equation:

$$P_c = \int_{-\infty}^{+\infty} [g(x-\mu_{S+N},0,\sigma_{S+N})G^{2U+1}(x,0,\sigma_N) + Ug(x,0,\sigma_N)G^{2U}(x,0,\sigma_N)G(x-\mu_{S+N},0,\sigma_{S+N})]dx. \tag{8.16}$$

The maximum rule can also be formulated in a different but equivalent way. The observer could first extract the maximum of the $(U+1)$ internal responses to each interval or stimulus in a trial and decide that the one that contains the greater maximum internal response is generated by the signal. For the $(U+1)$ internal responses generated by non-signals, all with the same probability density function $g(x,0,\sigma_N)$, the probability density function of the maximum is

$$p_1(x \mid \max) = (U+1)g(x,0,\sigma_N)G^U(x,0,\sigma_N). \tag{8.17}$$

For the $(U+1)$ internal responses generated by the interval or the stimulus containing the signal, the probability density function of one of them is $g(x-\mu_{S+N},0,\sigma_{S+N})$; the probability density function of the other U of them is $g(x,0,\sigma_N)$. The probability density function of the maximum internal response is

$$p_2(x \mid \max) = g(x-\mu_{S+N},0,\sigma_{S+N})G^U(x,0,\sigma_N) + Ug(x,0,\sigma_N)G^{U-1}(x,0,\sigma_N)G(x-\mu_{S+N},0,\sigma_{S+N}). \tag{8.18}$$

The probability of making a correct response is equal to the probability that samples from $p_2(x \mid \max)$ are greater than samples from $p_1(x \mid \max)$:

$$P_c = \int p_2(x \mid \max)P_1(x \mid \max)dx = \int_{-\infty}^{+\infty} [g(x-\mu_{S+N},0,\sigma_{S+N})G^{2U+1}(x,0,\sigma_N) + Ug(x,0,\sigma_N)G^{2U}(x,0,\sigma_N)G(x-\mu_{S+N},0,\sigma_{S+N})]dx. \tag{8.19}$$

If $\sigma_S = \sigma_N = \sigma$, we can define $d' = \mu_{S+N} / \sigma$, and simplify equation 8.19 to

$$P_c = \int_{-\infty}^{+\infty} [g(x-d',0,1)G^{2U+1}(x,0,1) + Ug(x,0,1)G^{2U}(x,0,1)G(x-d',0,1)]dx. \tag{8.20}$$

The maximum distributions of the internal responses of the signal-present and signal-absent intervals are shown in figure 8.12. As the number of irrelevant channels increases, the two distributions get closer and closer.

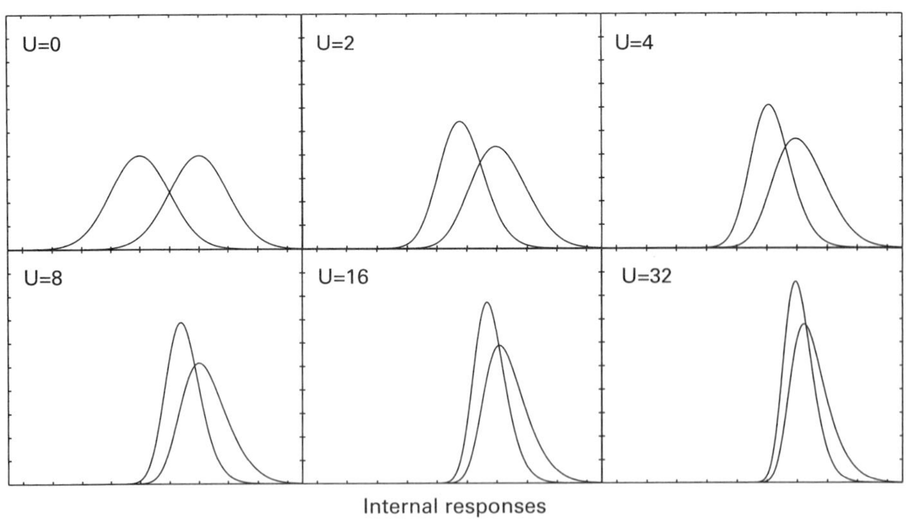

Figure 8.12
Internal response distributions for signal-present (right) and signal-absent intervals, with the number of irrelevant channels (U) indicated at the upper left corner of each panel.

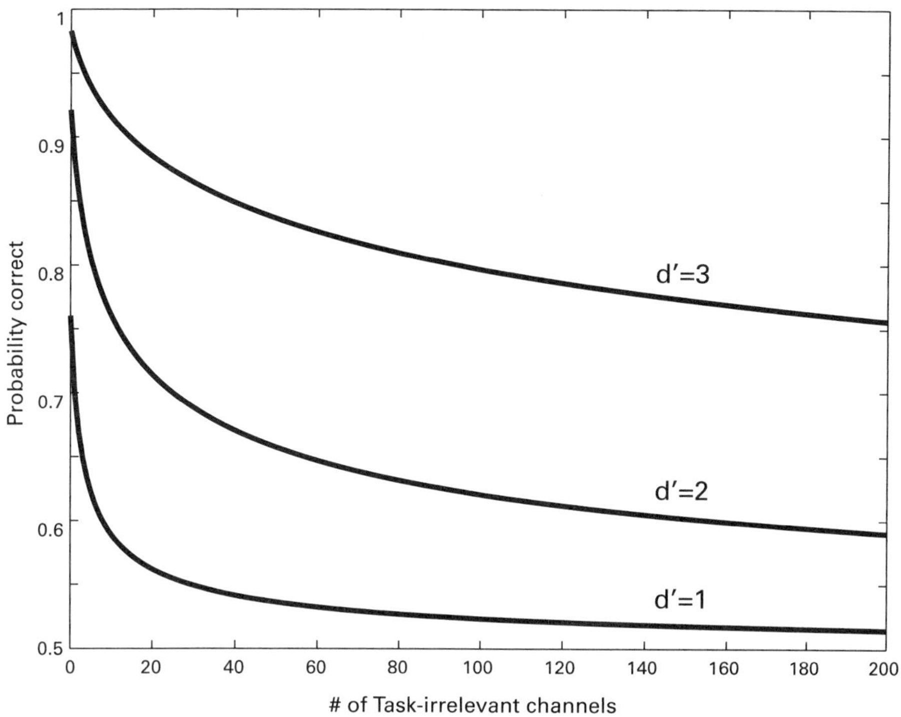

Figure 8.13
Dependence of probability correct on the number of task-irrelevant channels in a 2AFC/2IFC design using the max rule for decision. Three *d′* levels are shown.

Figure 8.13 illustrates how probability correct depends on the number of task-irrelevant channels in a max rule. Uncertainty calculations can be very important in the correct interpretation of the decisions of observers. Adding uncertainty lowers the observed percent correct for the same internal distance between the signal and noise, μ_{S+N}, because all the additional detectors or channels provide more opportunities for a noise sample to exceed the sample from the signal, leading to more opportunities for false alarms.

Uncertainty calculations have been especially important in several areas of research. One such example is the area of attention in which early researchers attributed limited attention processes to reductions in percent correct performance that were instead completely consistent with unlimited and ideal observers given uncertainty considerations of the SDT.[39,41,44,45]

8.6.3 Criterion Noise

SDT has provided investigators with measurements of discrimination sensitivity and bias at the decision level, however SDT assumes fixed decision criteria. If the decision criterion varies from trial to trial, the standard SDT formulation absorbs criterion noise into perceptual noise.[46] However, more than half a century of research has provided a great deal of evidence contradicting the assumption of a static decision criterion, and correct interpretation of sensory processes would ideally segregate criterion noise from perceptual noise. Traditional SDT is quite good at estimating large differences in bias. It provides an excellent approximation as long as decision noise is small relative to the variability in the internal perceptual representations.

Several recent studies[47,48] have suggested a dramatic reassessment of previous research findings from traditional SDT for cases where criterion noise is large. Although extensions of SDT accounting for decision noise have appeared recently, efforts to obtain separate estimates of decision and encoding noise have so far relied on restrictive assumptions about the relationship between various noise components. Recent work has started to make new headway in this classic problem. This new work[49] derives methods capable of independent estimates of the noise components at the decision stage.

8.6.4 Other Issues in SDT

In most applications of SDT, the internal distributions are assumed to be Gaussian. This normality assumption may not be accurate—distributions may depart from normality in certain situations. The predictions of SDT have been computed for a range of plausible distributions[40,44,50] for a number of standard paradigms. For most specific applications, it is possible to derive and carry out SDT-based computations assuming different distributions for the internal representations. Other applications may use bootstrap methods from the data to approximate distributions computationally. The general rationale for the normal or Gaussian distribution assumption in SDT is that the sum of many variables tends to approach a normal distribution—and it seems reasonable that the perceptual variables may result from the integrated activity of many neural processes.

Moreover, SDT by itself does not provide principles for specifying the means and variances or distributions of the internal representations from the variations in the stimulus and the characteristics of the perceptual system. A full understanding of perceptual systems ideally would specify the processes that transform the stimulus to the effective internal representation. Knowing the transformation could specify relationships between many different parametrically varied stimuli. In contrast, SDT must estimate the d' (and slope of the ROC curve) of each condition separately. Modeling the distributions of internal responses to systematically varying stimuli and thereby specifying the transformation from the stimulus to the internal response is the topic of the next chapter on observer models.

8.7 Summary

This chapter provided a basic treatment of the measurement of visual sensitivity or the detection, discrimination, and identification of near-threshold or threshold stimuli. The chapter first reviewed several classical paradigms for measuring perceptual sensitivity. The middle sections of the chapter provided an introduction and explanation of SDT as a framework for understanding detection and discrimination performance and for the separation of measures of sensitivity from measures of criterion or bias. We provided examples and treatments of standard detection paradigms and their extensions to rating experiments that measure ROCs. This section provided an introduction to many of the standard detection, discrimination, and identification paradigms.

We provided a brief overview of the extensions of SDT to multiple-dimensional stimuli, such as stimuli varying in length and orientation for discrimination and for full identification. Situations where stimuli vary in several dimension are considered more generally within the framework of GRT. Finally, we consider several specialized extensions to the SDT such as uncertainty calculations often used in attention research. This chapter should provide the basic theoretical tools for understanding most prominent paradigms for measuring threshold performance.

References

1. Campbell F, Robson J. 1968. Application of Fourier analysis to the visibility of gratings. *J Physiol* 197: 551.
2. Enroth-Cugell C, Robson JG. 1966. The contrast sensitivity of retinal ganglion cells of the cat. *J Physiol* 187: 517.
3. Movshon JA, Thompson ID, Tolhurst DJ. 1978. Spatial and temporal contrast sensitivity of neurones in areas 17 and 18 of the cat's visual cortex. *J Physiol* 283: 101–120.
4. Watson AB. Temporal sensitivity. In: Boff K, Kaufman L, Thomas J, eds. *Handbook of perception and human performance*, Vol. 1. New York: Wiley; 1986, pp. 1–43.
5. Chung STL, Legge GE, Tjan BS. 2002. Spatial-frequency characteristics of letter identification in central and peripheral vision. *Vision Res* 42: 2137–2152.
6. Watson AB, Ahumada AJ. 2005. A standard model for foveal detection of spatial contrast. *J Vis* 5(9): 717–740.
7. Graham NVS. *Visual pattern analyzers*. Oxford: Oxford University Press; 2001.
8. Green DM, Swets JA. *Signal detection theory and psychophysics*, New York: Wiley; 1966.
9. Macmillan NA, Creelman CD. *Detection theory: A user's guide*. Hillsdale, NJ: Lawrence Erlbaum; 2005.

10 Marcum, JA. Statistical theory of target detection by pulsed radar. Project RAND, Douglas Aircraft Company, Inc., RA-15061, Dec. 1947.

11. Peterson W, Birdsall T, Fox W. 1954. The theory of signal detectability. *Transactions of the IRE Professional Group on Information Theory* 4(4): 171–212.
12. Tanner WP, Jr, Swets JA. 1954. A decision-making theory of visual detection. *Psychol Rev* 61: 401–409.
13. Gescheider GA. *Psychophysics: Method and theory*. Hillsdale, NJ: Lawrence Erlbaum; 1976.
14. Kingdom FAA, Prins N. *Psychophysics: A practical introduction*. New York: Academic Press; 2009.
15. DeCarlo LT. 2002. Signal detection theory with finite mixture distributions: Theoretical developments with applications to recognition memory. Psychol Rev 109: 710–721.
16. Cohn TE, Lasley DJ. 1974. Detectability of a luminance increment: Effect of spatial uncertainty. *JOSA* 64: 1715–1719.
17. Eskew RT, Jr, Stromeyer CF, III, Picotte CJ, Kronauer RE. 1991. Detection uncertainty and the facilitation of chromatic detection by luminance contours. *JOSA A* 8: 394–403.

18. Lasley DJ, Cohn T. 1981. Detection of a luminance increment: Effect of temporal uncertainty. *JOSA* 71: 845–850.

19. Wenger MJ, Rasche C. 2006. Perceptual learning in contrast detection: Presence and cost of shifts in response criteria. *Psychon Bull Rev* 13: 656–661.

20. Noreen D. 1981. Optimal decision rules for some common psychophysical paradigms. *Math Psychol Psychophysiol* 13: 237–279.

21. Dai H, Versfeld NJ, Green DM. 1996. The optimum decision rules in the same-different paradigm. *Atten Percept Psychophys* 58: 1–9.

22. Versfeld NJ, Dai H, Green DM. 1996. The optimum decision rules for the oddity task. *Atten Percept Psychophys* 58: 10–21.

23. Petrov AA. 2009. Symmetry-based methodology for decision-rule identification in same-different experiments. *Psychon Bull Rev* 16: 1011–1025.

24. Ashby FG, Townsend JT. 1986. Varieties of perceptual independence. *Psychol Rev* 93: 154–179.

25. Ashby FG. *Multidimensional models of perception and cognition.* Hillsdale, NJ: Lawrence Erlbaum; 1992.

26. Maddox, WT. *Perceptual and decisional separability.* Hillsdale, NJ: Lawrence Erlbaum; 1992.

27. Thomas RD. 1995. Gaussian general recognition theory and perceptual independence. *Psychol Rev* 102: 192–200.

28. Kadlec H, Townsend JT. Signal detection analyses of dimensional interactions. In: Ashby FG, ed. *Multidimensional models of perception and cognition.* Hillsdale, NJ: Lawrence Erlbaum; 1992: pp. 181–227.

29. Kadlec H, Townsend JT. 1992. Implications of marginal and conditional detection parameters for the separabilities and independence of perceptual dimensions. *J Math Psychol* 36: 325–374.

30. Garner WR, Michigan Uo. *The processing of information and structure.* Hillsdale, NJ: Lawrence Erlbaum; 1974.

31. Garner WR. 1976. Interaction of stimulus dimensions in concept and choice processes. *Cognit Psychol* 8: 98–123.

32. Posner MI. 1964. Information reduction in the analysis of sequential tasks. *Psychol Rev* 71: 491–504.

33. Wickens TD, Olzak LA. Three views of association in concurrent detection ratings. In: Ashby FG, ed. *Multidimensional models of perception and cognition.* Hillsdale, NJ: Lawrence Erlbaum; 1992, pp. 229–252.

34. Ashby FG. 2000. A stochastic version of general recognition theory. *J Math Psychol* 44: 310–329.

35. Ashby FG, Gott RE. 1988. Decision rules in the perception and categorization of multidimensional stimuli. *J Exp Psychol Learn Mem Cogn* 14: 33–53.

36. Kadlec H, Townsend J. Signal detection analysis of multidimensional interactions. In: Ashby FG, ed. *Probabilistic Multidimensional Models of Perception and Cognition.* Hillsdale, NJ: Lawrence Erlbaum; 1992, pp. 181–231.

37. Wickens TD. *Elementary signal detection theory.* Oxford: Oxford University Press; 2002.

38. Thomas RD. 1996. Separability and independence of dimensions within the same-different judgment task. *J Math Psychol* 40: 318–341.

39. Sperling G, Dosher B. Strategy and optimization in human information processing. In: Boff KR, Kaufman L, Thomas JP, eds. *Handbook of perception and human performance, Vol. I: Sensory processes and perception.* New York: Wiley; 1986, pp. 2-1–2-65.

40. Shaw ML. 1980. Identifying attentional and decision-making components in information processing. *Attention and Performance* 8: 277–295.

41. Palmer J, Verghese P, Pavel M. 2000. The psychophysics of visual search. *Vision Res* 40: 1227–1268.

42. Palmer J, Ames CT, Lindsey DT. 1993. Measuring the effect of attention on simple visual search. *J Exp Psychol Hum Percept Perform* 19: 108–130.

43. Nolte LW, Jaarsma D. 1967. More on the detection of one of M orthogonal signals. *J Acoust Soc Am* 41: 497–505.

44. Shaw ML. 1982. Attending to multiple sources of information: I. The integration of information in decision making. *Cognit Psychol* 14: 353–409.

45. Eckstein MP. 1998. The lower visual search efficiency for conjunctions is due to noise and not serial attentional processing. *Psychol Sci* 9: 111–118.

46. Wickelgren WA, Norman DA. 1966. Strength models and serial position in short-term recognition memory. *J Math Psychol* 3: 316–347.

47. Benjamin AS, Diaz M, Wee S. 2009. Signal detection with criterion noise: Applications to recognition memory. *Psychol Rev* 116: 84–114.

48. Rosner BS, Kochanski G. 2009. The Law of Categorical Judgment (Corrected) and the interpretation of changes in psychophysical performance. *Psychol Rev* 116: 116–128.

49. Cabrera C, Lu ZL, Dosher B. 2011. Separating decision noise and encoding noise in perceptual decision making. *J Vis* 11: 805.

50. Graham N, Kramer P, Yager D. 1987. Signal-detection models for multidimensional stimuli: Probability distributions and combination rules. *J Math Psychol* 31: 366–409.

9 Observer Models

Observer models specify computations that are sufficient to predict the behavior of an observer for many different input stimuli with a small number of parameters. Observer models specify the transformations leading to the relevant internal representations from the stimulus, and define the decision rules for particular tasks. These models have a remarkable ability to summarize compactly the behavioral outcomes in many conditions and provide a conceptual framework within which to understand the responses of the sensory system. In this chapter, we consider both modern single-channel and multichannel observer models. Each observer in a task can be described by a small number of parameters that fully specify how the stimulus is recoded in an internal response and then subjected to a task-relevant decision. Once these parameters have been estimated from specific experimental tests, it is possible to make predictions about an observer's responses to a wide range of stimuli and paradigms.

9.1 Need for Observer Models

The goal of visual psychophysics is to fully specify the relationship between the stimulus, the internal representation, and the observer's response. A successful theory of perception should include models that specify all of these components, including a transformation from the stimulus inputs to the internal representations, and then a function that takes the internal representation into a response. Chapter 7 on scaling focused on the relationship between perceptual intensity or perceptual qualities and properties of the physical stimulus. Chapter 8 showed how to use signal-detection theory to separate perceptual sensitivity from response bias and subjective criterion. A complete psychophysical theory for any experiment requires detailed specification of the transformations from the input to the output. Signal-detection theory by itself is silent about how the stimulus quality or intensity is related to the internal representation. Scaling infers a quantitative relationship between stimuli, but may or may not specify the transformation from the stimulus to the internal representation, and often fails to provide an explicit model of how the observer generated the scaling responses from the internal representation.

The observer models solve the problem of how to predict an observer's behavioral performance over a wide range of possible stimuli. Even a simple example illustrates the value. Suppose that we want to know the observer's sensitivity to any Gabor pattern. There are hundreds of combinations of contrast and external noise that we might want to predict. How can we make predictions about these hundreds of cases by measuring only a few conditions? One brute-force possibility is to sample the contrast and external noise space with some experimental tests (and hope they are well placed) and then attempt to use interpolation to predict the performance in all the other cases. The observer models provide a sophisticated way to make these predictions from several key observations in a way that also improves our conceptual understanding of how the sensory systems function.

An observer model specifies the transformation from the stimulus to the internal representation and then the transformation to the response. The focus is on specifying the transformation from the stimulus to the internal response. Most observer models then incorporate signal detection or other standard decision module for response generation. Modeling the noise or variability in the internal representation is a key component of any observer model.[1] The noises make the system stochastic—meaning that it involves random variables or probabilities that generate trial-to-trial variability. There are many models in vision; however, only a subset incorporates a significant treatment of all the aspects of observer models.

The simplest observer model must have a component or module that is sensitive to the target pattern or patterns. This is usually called a template. It can be thought of as a detector for the target. The output of the template depends both on the match of the test pattern to the template and the contrast of the pattern. The template response is increased when the match to the template is better and the contrast is higher. To mimic the variability of observers, including the neural and sensory variability in encoding, the observer model must incorporate internal noise sources to generate a noisy, stochastic internal response (figure 9.1). Then, it must have a decision module that maps the noisy internal response into an action, or external response of the observer.

In some applications of observer models, a single template may be applied at the location of the signal stimulus, and the output is a single number that is the basis for subsequent decision. More generally, a template may be applied at many locations, and the output is an array of numbers that drive a decision. Figure 9.1 illustrates template matching by the application (convolution) of a Gabor template at many points in the stimulus. The template is looking for a vertical Gabor pattern. It returns essentially no response over the cloudless sky, stronger responses when the orientation and spatial frequency in the image match the template, and responding most strongly to the vertical edges in the stimulus. The middle panel of figure 9.1 represents a matching value at each spatial location in the image. More complicated observer models may incorporate multiple channels or multiple templates, physiological response nonlinearities and interactions, multiple sources of internal noise, and more complex decision rules.[2,3]

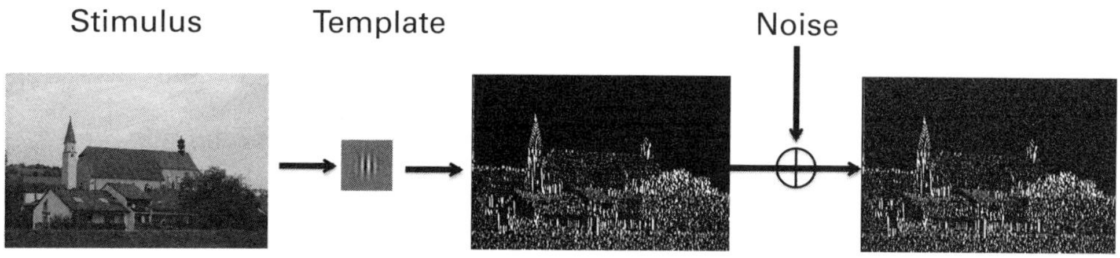

Figure 9.1
An illustration of template matching at many points in the image with additive noise. An input stimulus is convolved with a Gabor template; 30% noise is added to the output.

Once constructed and fully specified, an observer model provides the basis to generate predictions and generalize the results from a small number of experimental tests to a wide range of experimental conditions for a particular observer or class of observers.[4,5] Observer models also provide a theoretical framework to investigate and quantify how the observer changes as a function of the individual's state. An observer state can include such things as adaptation, or training, or attention state, or the observer classification as, for example, dyslexic or amblyopic. The effects of changing the state of the observer are understood by how the model parameters change as a function of state while others remain fixed.[5]

The components of observer models can be specified and tested in a number of ways. Psychophysical paradigms and estimates from neurophysiology or brain imaging may all provide constraints that specify aspects of observer models. In the next section, we describe one observer model and provide illustrations of experiments that are used to specify the model.

9.2 The Perceptual Template Model

Observer models for detection and discrimination have been under development for many years, at least since the 1950s.[6] All of the early models begin with a relevant perceptual template for the signal stimulus, a source of additive internal noise, and a decision module. Other models may include important functions such as nonlinearity, multiplicative noise, and decision uncertainty. A number of observer models have been developed over the past few decades. The most prominent observer models include the linear amplifier model,[7] the induced noise model,[8] the linear amplifier model with decision uncertainty,[9] the induced noise and uncertainty model,[10] and the perceptual template model.[5,11] The perceptual template model (PTM) incorporates and integrates the major components of the previous observer models and has been shown to provide an excellent account of a range of psychophysical data.[5]

The components of the PTM are illustrated in figure 9.2. The PTM includes (1) a perceptual template tuned for the signal stimulus in the target task. For example, the template

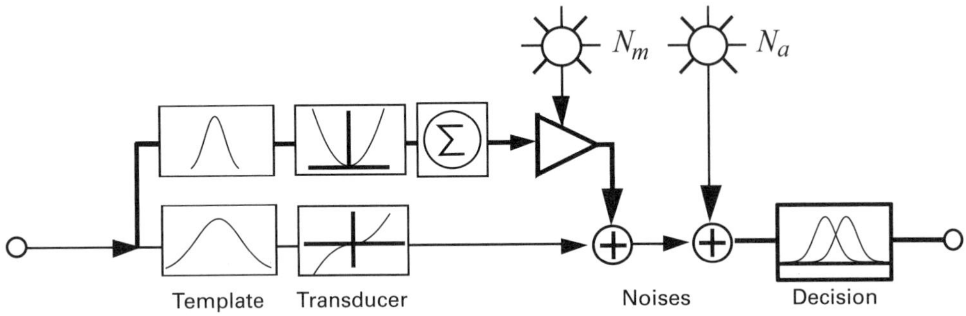

Figure 9.2
The PTM of a human observer (see text for explanation).

may be tuned for an oriented Gabor if that is the signal in a particular experiment. The PTM incorporates (2) a nonlinear transducer function that mimics known nonlinearity in physiological responses and can partially capture the nonlinear relationship between stimulus intensity and the internal response. The PTM also includes two different internal noise sources, (3) multiplicative and (4) additive internal noises, which account for the stochastic or variable nature of the internal representations to the same stimulus. Multiplicative internal noise increases with the contrast energy in the stimulus, and so is related to Weber's law behavior. Additive internal noise limits absolute threshold for very low contrast stimuli. This part of the PTM—the template, nonlinearity, and internal additive and multiplicative noises—specifies the mean internal representation and its variability. The PTM then incorporates a signal-detection module for the specific experimental task that operates on the internal representation(s). The PTM provides a link from the stimulus input to noisy internal representations—a link that is missing in signal-detection theory (SDT).

Here we develop the predictions of the PTM for a two-alternative forced-choice decision task. The PTM, first introduced by Lu and Dosher,[12] processes the input stimuli in two pathways. In the "signal" pathway, input stimuli are processed through a perceptual template that has selectivity for the particular characteristics of the signal stimulus (e.g., spatial frequency, orientation, etc.). This pathway is where the template is used to look for the signal in the stimulus. The response or gain of the template to a signal stimulus is quantified by the model parameter β. The gain of the template to Gaussian white noise is set (normalized) to 1.0. The signal pathway has a nonlinear transducer function [Output = sign(Input)$|$ Input $|^{\gamma_1}$], specified by the model parameter γ_1. This form of nonlinearity is also often used in modeling pattern vision.[13,14]

The output of the second pathway, the "multiplicative noise" pathway, controls the amount of multiplicative internal noise. It also responds to the input stimuli but is normally more broadly tuned than the signal pathway and may integrate (be sensitive to) energy over a broad range of space, time, and features. This is implemented in equations as another perceptual template with gain parameter β_2 to the signal stimulus and 1.0 to white external noise. The output of this "gain-control" template is also submitted to a nonlinear transducer function (Output $= |\text{Input}|^{\gamma_2}$), with parameter γ_2. The variance of multiplicative noise is proportional to the output of the multiplicative noise path. The outputs of the two pathways are combined with an additive internal noise to form the internal representation that is the input to the decision module.

For paradigms where observers discriminate between two stimuli, the final response may be determined by a difference rule; that is, the difference in the response of two different templates to the presented stimulus. If, for example, the observer is discriminating two oriented Gabors of angles +45° or –45° relative to vertical, the PTM compares the response of the two templates to a test stimulus on each trial. One of the templates provides a good match to the signal stimulus, the other mismatches.

In the template detector that matches the signal, the mean of the internal response is $\beta^{\gamma_1} c^{\gamma_1}$, and the total variance of the internal response is

$$\sigma^2_{\text{total 1}} = \sigma^{2\gamma_1}_{\text{ext}} + N^2_{\text{mul}}[\sigma^{2\gamma_2}_{\text{ext}} + (\beta_2 c)^{2\gamma_2}] + \sigma^2_{\text{add}}. \tag{9.1}$$

In the template detector that mismatches the signal (but matches another very different target stimulus), the mean of the internal response is 0, and the total variance of the internal response is

$$\sigma^2_{\text{total 2}} = \sigma^{2\gamma_1}_{\text{ext}} + N^2_{\text{mul}}\sigma^{2\gamma_2}_{\text{ext}} + \sigma^2_{\text{add}}. \tag{9.2}$$

Using the decision module and equations for a two-alternative forced-choice task from chapter 8, the average signal-to-noise ratio (d') in the PTM is

$$d' = \frac{(\beta c)^{\gamma_1}}{\sqrt{\sigma^{2\gamma_1}_{\text{ext}} + N^2_{\text{mul}}\left[\sigma^{2\gamma_2}_{\text{ext}} + \frac{(\beta_2 c)^{2\gamma_2}}{2}\right] + \sigma^2_{\text{add}}}}. \tag{9.3}$$

The numerator or signal part of the equation reflects the difference in response to the test stimulus between the two templates, $(\beta c)^{\gamma_1} - 0$.

Correspondingly, the probability correct is

$$P_c = \int_{-\infty}^{+\infty} g\left\{x - \beta^{\gamma_1} c^{\gamma_1}, 0, \sqrt{\sigma^{2\gamma_1}_{\text{ext}} + N^2_{\text{mul}}[\sigma^{2\gamma_2}_{\text{ext}} + (\beta_2 c)^{2\gamma_2}] + \sigma^2_{\text{add}}}\right\} G\left(x, 0, \sqrt{\sigma^{2\gamma_1}_{\text{ext}} + N^2_{\text{mul}}\sigma^{2\gamma_2}_{\text{ext}} + \sigma^2_{\text{add}}}\right) dx. \tag{9.4}$$

That is, percent correct is determined by the probability that the response of the matching template is greater than the mismatching template, integrated over all possible values.

The equations for the PTM used here are analytic approximations to a fully stochastic model. The PTM formulation uses equations that approximate the full stochastic model by replacing several random variables with their expectations in specifying the noises. The approximation has similar or identical properties to the full stochastic model in most parameter regimes that have been observed so far.[5,15]

The PTM defines the relationship between the physical stimulus, the internal representation, and the response of the observer. It extends SDT by specifying the signal strength and the variability of the internal representations from physical properties of the stimulus. The classical SDT experiments typically correspond to a situation in which external noise is not present ($\sigma^2_{ext} = 0$), while the PTM is designed to account for performance in multiple levels of external noise masking. Once the parameters of the PTM are specified by experimentation, it provides comprehensive predictions of performance under many stimulus conditions, including the condition typically used in SDT without external noise. The properties of the PTM and how to specify them are described in the next section.

9.3 Specifying the PTM Observer Model

9.3.1 Equivalent Input Noise and Multiple Performance Levels

One of the most powerful experimental methods used in the investigation of observer models and human performance is the equivalent input noise method. Originally developed by engineers to study the properties of amplifiers,[16–18] the equivalent input noise method was adapted to measure the perceptual system by Nagaraja in 1964[19] and Pelli and others in the 1980s.[7,20,21] The idea of the equivalent input noise method is that the detection or discrimination of a signal depends upon the signal-to-noise ratio of the internal representation. By adding external noise that the experimenter controls, the other sources of internal noise can be referred to (benchmarked against) an externally controlled or externally measurable quantity—the external noise in the stimulus.

Empirical specification of the PTM often involves testing detection or discrimination of a signal in noise for five to nine different logarithmically spaced external noise levels. One might measure a full parametric set of external noise and signal contrasts. However, this is relatively demanding and also may include many conditions leading to very poor or very good performance. Traditionally, the experiments measure signal contrast threshold as a function of external noise at some performance level, such as 75% correct, on the rising portion of the psychometric function. However, it is essential to characterize the observer at multiple (three or more) performance levels to fully specify an observer model,[11,15] although historically this was rarely done. One good approach is to measure three or more threshold levels for a range of external noises. This subsamples the full stimulus space while providing enough data to constrain the parameter estimates of the model, especially the parameters that estimate the nonlinearity in response.

The contrast threshold for detection or discrimination has a characteristic shape as a function of the contrast of the external noise, as shown in figures 9.3 and 9.4a. At high levels of external noise, the contrast threshold increases directly with the contrast of the

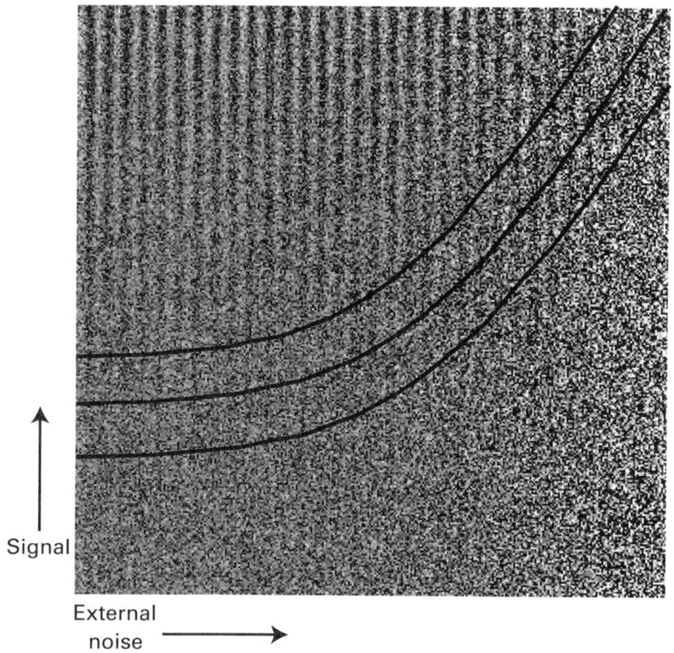

Figure 9.3
Demonstration of the equivalent input noise method. Three images—a vertical sine wave (signal) with increasing contrast from bottom to top, an external noise image with increasing variance from left to right, and an internal noise image with a constant variance—are superimposed. Three contours of equal visibility of the signal grating are traced with the smooth curves. The contour is nearly flat in low external noise conditions and rises with external noise in high external noise conditions. The amplitude of the external noise at the elbow of the contour provides an estimate of the variance of the internal noise.

external noise. In this region of the curve, the external noise is the limiting factor in performance. However, at very low levels of external noise, the contrast threshold for detection or discrimination barely depends upon the external noise. In this region of the curve, the internal noises are the limiting factor in performance. The transition point where external noise starts to control performance is related to the magnitude of the internal noises. By referencing to the external noise, it is possible to peg the scale of the internal noises to a physical quantity.

Frequently in experimental tests, the rising portion of the threshold versus external noise contrast (TvC) functions has a slope of 1.0. In this special case, we can set $\gamma = \gamma_1 = \gamma_2$, and rearrange the fundamental signal-to-noise equation to obtain threshold signal contrast c_τ as a function of external noise contrast σ_{ext} at a given performance criterion (i.e., d'):

$$c_\tau = \left\{ \frac{d'^2[(1+N_{\text{mul}}^2)\sigma_{\text{ext}}^{2\gamma} + \sigma_{\text{add}}^2]}{\beta^{2\gamma} - N_{\text{mul}}^2 \beta_2^{2\gamma} d'^2 / 2} \right\}^{\frac{1}{2\gamma}}. \tag{9.5}$$

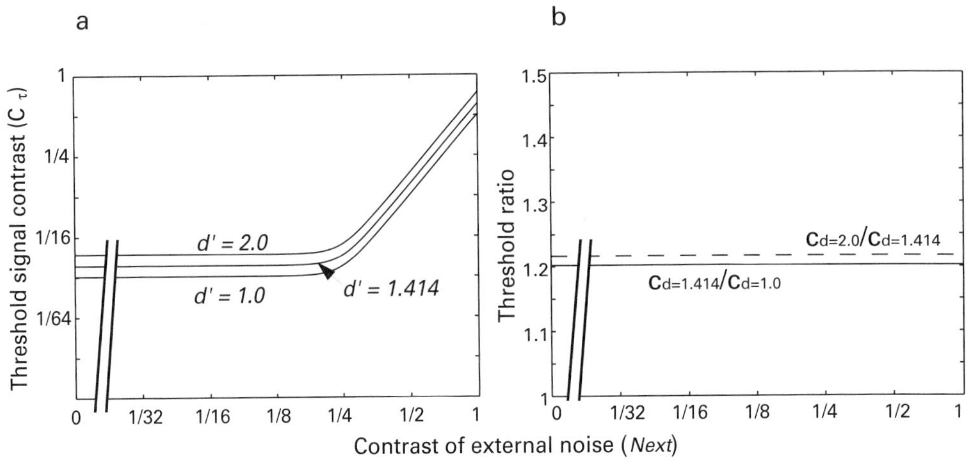

Figure 9.4
Illustration of hypothetical results of a triple-TvC paradigm. (a) TvC functions at three performance levels, corresponding to $d' = 1.0$, 1.414, and 2.0. (b) Threshold ratios between two performance levels in each external noise condition: ratios between thresholds at $d' = 2.0$ and $d' = 1.414$, and ratios between thresholds at $d' = 1.414$ and $d' = 1.0$.

This is the equation for a TvC function at a given threshold level of performance accuracy. It follows directly from this equation that for any given external noise contrast, the ratio between the threshold signal contrasts at two performance levels (corresponding to d_2' and d_1'), is

$$\frac{c_{\tau_2}}{c_{\tau_1}} = \left(\frac{d_2'^2\ \beta^{2\gamma} - N_{\text{mul}}^2 \beta_2^{2\gamma} d_1'^2 / 2}{d_1'^2\ \beta^{2\gamma} - N_{\text{mul}}^2 \beta_2^{2\gamma} d_2'^2 / 2} \right)^{\frac{1}{2\gamma}}. \tag{9.6}$$

The PTM predicts that the ratio of threshold signal contrasts at two criterion performance levels for any given external noise contrast is a nonlinear function of the two corresponding d's and does not depend on the particular external noise level (figure 9.4b).

Measurement of multiple TvC functions at different criterion performance levels provides sets of pairwise tests of these ratio properties. One way is to measure full psychometric functions at each external noise level tested in TvC measurements. In this case, estimating the thresholds for a given accuracy level and graphing these as a function of external noise contrast provides a useful visual display and useful ratio tests as described earlier. Measuring performance at three criterion levels—a three-point proxy for measurement of the full psychometric functions—should be sufficient to estimate the parameters, including the nonlinearity parameters, of a PTM.[11,15] A paradigm that measures the TvC at three threshold levels is sometimes called a triple-TvC paradigm.

Display 9.1 shows a MATLAB program that implements a triple-TvC experiment and a double-pass procedure (see section 9.3.2 for a description of double pass). The experiment measures the contrast thresholds at multiple external noise levels in a two-alternative Gabor orientation identification task. Observers determine the tilt of a Gabor as ±10° from vertical by responding "top tilted right" or "top tilted left." The experiment measures thresholds using the method of constant stimuli and tests seven suitably chosen contrasts for each of eight levels of external noise. Specific random seeds are used in the program to set the random number generators that control all random events in the experiment (see section 4.3.2 of chapter 4), including the randomized trial sequence and the specific samples of external noise. The random seeds of the first four sessions are saved and then used again to reconstruct the exact same sequence of trials and noise displays in sessions five to eight. A double-pass experiment measures the response of the observer to exactly the same stimuli and trial sequence to estimate the variability in the internal response. This experiment corresponds to one described in Lu and Dosher.[5] TvC functions at three performance levels from such an experiment are shown in table 9.1 and figure 9.5.

9.3.2 Estimating Total Internal Noise from Double Pass

In the PTM, several sources contribute to the total noise or variability: external noise, internal multiplicative noise, and internal additive noise. The external noise is specified in physical units. We estimate the internal multiplicative and additive noises in relation to the experimenter-manipulated external noise with the triple-TvC method to fully constrain these parameters of the PTM model.

An alternative and complementary method for estimating internal noise in relation to external noise is the double-pass paradigm. Developed in the auditory domain in the 1950s and 1960s,[22,23] the paradigm capitalizes on the fact that internal noise may lead an observer to make a different response to the same physical signal plus external noise stimulus. To quote Green[23] (p. 397), "On an operational level, internal noise is equivalent to the observation that the same physical stimulus may elicit different responses. In a sense, then, internal noise is the limiting factor in a trial-by-trial prediction of the subject's response." Some accidental features of high external noise may by chance lead to incorrect responses based on random but real stimulus information—but internal noise may cause inconsistencies in response from one copy of an identical trial to another.

The double-pass paradigm requires that exactly the same sequence of stimuli is presented to the observer twice (or sometimes more than twice, for an *n*-pass method).[8,24–26] This means that not just the sequence but exactly the same signal and external noise samples are repeated. The observer's responses can be scored as correct or incorrect. The two copies of the same trials can lead to the same response—response agreement—or the responses can be inconsistent. The two measures, accuracy and agreement, are intrinsically related, because as accuracy goes toward 100%, obviously agreement must increase. Each

Display 9.1

```
%%% Program TvCFunction.m
function TvCFunction (subjID, session)

%% Experimental Module

% Specify the stimulus
p.stimSize = 6;     % image diameter  in visual degrees
p.noises = [0 0.0156 0.0313 0.0625 0.125 0.165 0.250 0.33];
                    % external noise levels
p.contrasts = …
      [0.0237 0.0395 0.0501 0.0639 0.0775 0.0872 0.1027
       0.0237 0.0395 0.0502 0.0640 0.0776 0.0872 0.1027
       0.0238 0.0397 0.0503 0.0642 0.0778 0.0875 0.1031
       0.0249 0.0415 0.0527 0.0672 0.0815 0.0917 0.1079
       0.0343 0.0572 0.0726 0.0925 0.1122 0.1262 0.1486
       0.0433 0.0723 0.0917 0.1169 0.1418 0.1595 0.1878
       0.0644 0.1074 0.1362 0.1737 0.2106 0.2369 0.2790
       0.0847 0.1412 0.1792 0.2284 0.2771 0.3117 0.3670];
p.nFactor = 3;      % noise pixel size
p.sf = 0.9;         % c/d
p.sigma = 1.1;      % degree
p.refAngle = 0;     % gabor reference angle
p.dAngle = 10;      % gabor tilt angle
repeats = 10;       % Number of trials to repeat for
                    % each contrast
p.fixDuration = 0.25;
p.ITI = 1;          % seconds between trials
keys = {'left' 'right' 'esc'};    % allowed response keys

% Compute stimulus parameters
ppd = pi/180 * p.ScreenDistance / p.ScreenHeight * …
      p.ScreenRect(4);              % pixels per degree
m = round(p.stimSize * ppd /p.nFactor) * p.nFactor;
                    % stimulus size in pixels
sf = p.sf / ppd;    % cycles per pixel
sc = round(p.sigma * ppd);         % sigma in pixels
fixXY = [[[-1 1] * 8 0 0] + p.ScreenRect(3) / 2;
        [0 0 [-1 1] * 8] + p.ScreenRect(4) / 2];
nContrast = size(p.contrasts, 2);
                    % number of contrast levels
nNoise = numel(p.noises); % number of external noise levels
nTrials = repeats * nContrast * nNoise;
                    % total number of trials

if nargin < 1, subjID = 's01'; end
if nargin < 2, session = 1; end
```

Display 9.1 (continued)

```
if session > 4
    fileName = sprintf('TvC_rst_%s_%02.0f.mat', subjID, session - 4);
    S = load(fileName, 'p');
    p.randSeed = ClockRandSeed(S.p.randSeed);
        % use seed from session-4 session
else
    p.randSeed = ClockRandSeed;
        % use clock to set seed for the random
        % number generator
end

% procedural gabor allows us to change its parameters
% very quickly
tex = CreateProceduralGabor(windowPtr, m, m, 0, ...
      [1 1 1 0] * 0.5, 1, 0.5);
noiseTex = zeros(1, 4);      % preallocate noise texture

% Initialize a table to set up experimental conditions
p.recLabel = {'trialIndex' 'contrastIndex' 'noiseIndex'...
      'tileRight' 'respCorrect' 'respTime'};
rec = nan(nTrials, length(p.recLabel));
      % matrix rec is made of nTrials x 6 of NaN
rec(:, 1) = 1 : nTrials; % trial number from 1 to nTrials
contrastIndex = repmat(1 : nContrast, [nNoise 1 repeats]);
noiseIndex = repmat((1 : nNoise)', [1 nContrast repeats]);
tileRight = zeros(nNoise, nContrast, repeats) * 2;
      % first set all to 0
tileRight(:, :, 1 : repeats / 2) = 1;
      % change first half to 1
[rec(:, 2) ind] = Shuffle(contrastIndex(:));
      % shuffle contrast index
rec(:, 3) = noiseIndex(ind);
      % shuffle noise index in the same order
rec(:, 4) = tileRight(ind);
      % shuffle interval in the same order

% Prioritize display to optimize display timing
Priority(MaxPriority(windowPtr));

% Start experiment with instructions
str = ['Press left and right arrow keys for top left and '...
      'right tilt responses.\n\n' 'Press SPACE to '...
      'proceed.'];
DrawFormattedText(windowPtr, str, 'center', 'center', 1);
      % Draw Instruction text string centered in window
Screen('Flip', windowPtr);
```

Display 9.1 (continued)

```
        % flip the text image into active buffer
Beeper;

WaitTill('space');        % wait till space bar is pressed
p.start = datestr(now);  % record start time

% Run nTrials trials
for i = 1 : nTrials
    % parameters for this trial
    con = p.contrasts(rec(i, 3), rec(i, 2));
    noise = p.noises(rec(i, 3));
    ang = p.refAngle - sign(rec(i, 4) - 0.5) * p.dAngle;

    % Make 4 independent noise textures
    for j = 1 : 4
        gn = randn(m / p.nFactor);  % Gaussian noise
        while 1
            out = abs(gn) > 3;
            nout = sum(out(:));
            if nout == 0, break; end
            gn(out) = randn(nout, 1);
        end
        gn = gn / 2 * noise + 0.5;  % [0 1]
        gn = Expand(gn, p.nFactor);
        noiseTex(j) = Screen('MakeTexture', windowPtr, ...
            gn, 0, 0, 2);
    end

    if i > 1
       if WaitTill(Secs + p.ITI, 'esc'), break; end
           % wait for ITI
       if strcmp(key, 'esc'), break; end % to stop
    end

    Screen('DrawLines', windowPtr, fixXY, 3, 0); % fixation
    t0 = Screen('Flip', windowPtr);
    Screen('Flip', windowPtr, t0 + p.fixDuration);

     for j = 1 : 2  % 2 noise frames
         Screen('DrawTexture', windowPtr, noiseTex(j));
         Screen('Flip', windowPtr);
    end
     % 1 signal frame
    Screen('DrawTexture', windowPtr, tex, [], [], ang,...
            [], [], [], [], 2, [180 sf sc con 1 0 0 0]);
    Screen('Flip', windowPtr);
```

Display 9.1 (continued)

```
        for j = 3 : 4   % 2 noise frames
            Screen('DrawTexture', windowPtr, noiseTex(j));
            Screen('Flip', windowPtr);
        end
        Screen('Flip', windowPtr);       % turn off last frame

        [key Secs] = WaitTill(keys);     % wait till response
        if iscellstr(key), key = key{1}; end
            % take the first in case of multiple key presses
        if strcmp(key, 'esc'), break; end     % to stop
        rec(i, 5) = strcmp(key, 'right') == rec(i, 4);
           % record correctness
        rec(i, 6) = Secs - t0;                % record respTime
        if rec(i, 5), Beeper; end             % beep if correct
end
p.finish = datestr(now);        % record finish time
fileName = sprintf('TvC_rst_%s_%02.0f.mat', subjID, ...
           session);
save(fileName, 'rec', 'p');     % save results
```

Table 9.1
Threshold versus external noise contrast (TvC) functions at three performance levels (65%, 75% and 85% correct)

N_{ext}(%)	0	3.0	4.5	6.7	10.0	14.9	22.3	33.0
$\tau_{65\%}$(%)	5.1 ± 0.4	5.0 ± 0.7	5.2 ± 0.4	5.2 ± 0.4	5.8 ± 0.5	8.5 ± 0.7	12.5 ± 1.1	16.8 ± 1.3
$\tau_{75\%}$(%)	6.7 ± 0.4	6.9 ± 0.5	6.6 ± 0.3	6.8 ± 0.4	7.8 ± 0.4	11.1 ± 0.6	16.3 ± 0.9	23.9 ± 1.3
$\tau_{85\%}$(%)	8.4 ± 0.4	9.0 ± 0.5	8.2 ± 0.4	8.5 ± 0.4	10.2 ± 0.5	14.0 ± 0.7	20.6 ± 1.0	32.2 ± 1.9

value of the ratio between total internal noise and external noise results in a specific function relating percent correct to percent agreement (see figure 9.6). The variation in accuracy may reflect a manipulation such as signal contrast, for example. For each function, percent correct goes from chance, usually 50%, to very good, near 100%. As percent correct approaches 100%, percent agreement also approaches 100%. As percent correct becomes poor, the agreement in responses between the two copies of the exact same trial depends on whether the errors reflect independent samples of internal noise, and so are not related, or the same sample of external noise that is controlling the error response in the same way.

Figure 9.6 shows several hypothetical functions relating percent correct (P_C) to percent agreement (P_A) for different ratios of the total amount of internal noise and the amount of external noise. The performance for any given condition specified by signal contrast and external noise level (c, N_{ext}) will generate a single (P_C, P_A) point in this graph. Because the experimenter knows the external noise level as well as the signal contrast, it is possible to estimate the total internal noise from the curve upon

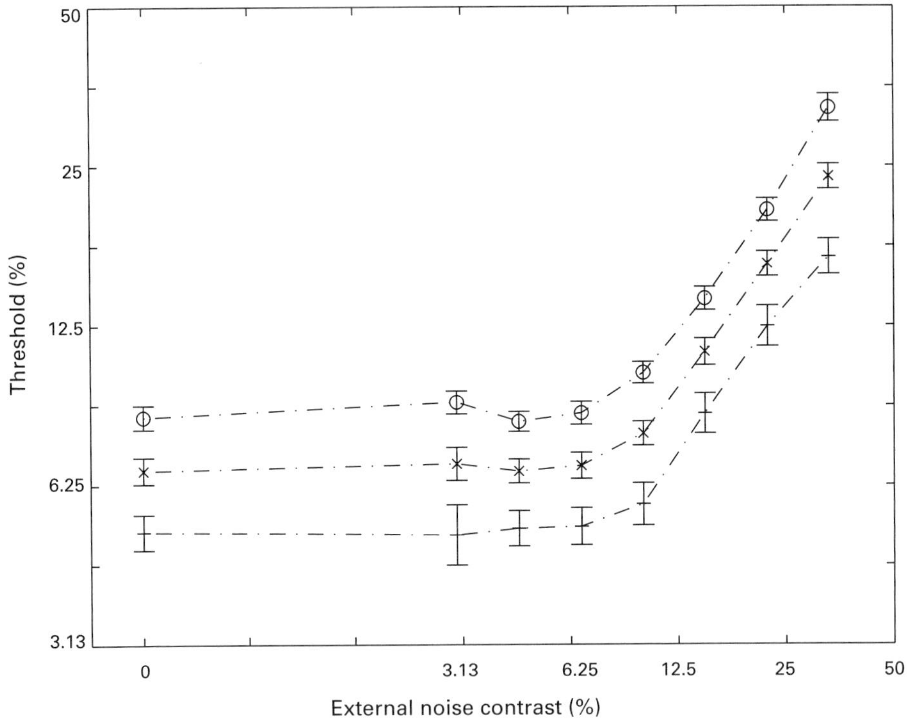

Figure 9.5
TvC functions at three different performance levels (65%, 75%, and 85% correct).

which the data point lies. By varying external noise and signal contrast, one can estimate the total internal noise for many conditions. By plotting estimated total internal noise as a function of the two stimulus variables, contrast and external noise, it is possible to get a sense of the functional relationship between total internal noise and the input stimulus.

The P_C versus P_A double-pass functions for different ratios of total internal noise and external noise are based on SDT, but are otherwise model independent. These generic predictions assume Gaussian signal + noise and noise distributions and are derived for two-alternative forced-choice tasks without bias, where the $\sigma_{\text{total}}^2 = \sigma_{\text{int}}^2 + \sigma_{\text{ext}}^2$. The relevant equations for P_C and P_A are

$$\begin{aligned} P_C &= \{P[(S + N_{\text{ext}1} + N_{\text{int}1a}) > (N_{\text{ext}2} + N_{\text{int}2a})] \\ &\quad + P[(S + N_{\text{ext}1} + N_{\text{int}1b}) > (N_{\text{ext}2} + N_{\text{int}2b})]\}/2.0 \\ &= \int g\left(x - S, 0, \sqrt{\sigma_{\text{ext}1}^2 + \sigma_{\text{int}1}^2}\right) G\left(x, 0, \sqrt{\sigma_{\text{ext}2}^2 + \sigma_{\text{int}2}^2}\right) dx \end{aligned} \tag{9.7}$$

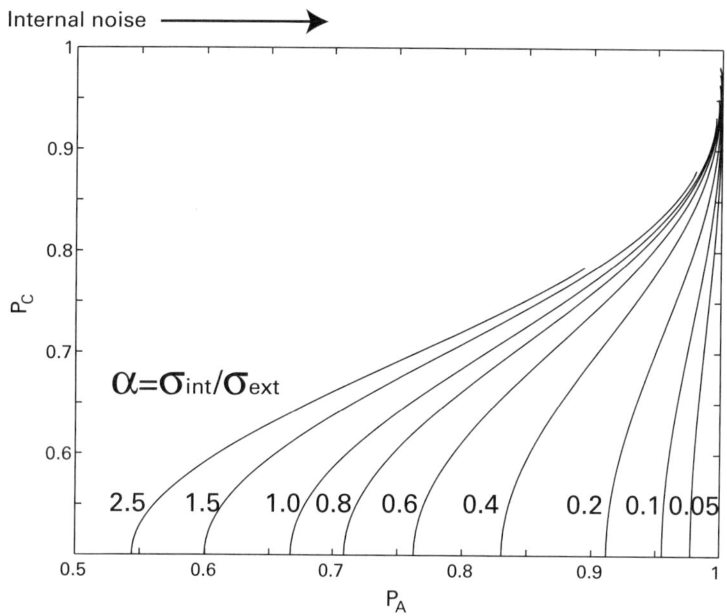

Figure 9.6
Probability correct (P_C) versus probability consistent (P_A) for a range of internal to external noise ratio α's.

$$\begin{aligned} P_A &= P[(S+N_{\text{ext1}}+N_{\text{int1}a})>(N_{\text{ext2}}+N_{\text{int2}a})]P[(S+N_{\text{ext1}}+N_{\text{int1}b})>(N_{\text{ext2}}+N_{\text{int2}b})] \\ &\quad + P[(S+N_{\text{ext1}}+N_{\text{int1}a})<(N_{\text{ext2}}+N_{\text{int2}a})]P[(S+N_{\text{ext1}}+N_{\text{int1}b})<(N_{\text{ext2}}+N_{\text{int2}b})] \\ &= \int_{-\infty}^{+\infty} g(x-S,0,\sqrt{2}\sigma_{\text{ext}})\{G^2(x,0,\sqrt{2}\alpha\sigma_{\text{ext}})+[1-G(x,0,\sqrt{2}\alpha\sigma_{\text{ext}})]^2\}dx. \end{aligned} \tag{9.8}$$

The subscripts 1 and 2 index two intervals/alternatives or two detectors; subscripts a and b index different instances of internal noise.

Unlike this general SDT treatment, the PTM as an observer model further specifies how the internal noises arise and provides an explicit equation for the total internal noise as a function of external noise, and model parameters N_m, N_a, β, and γ. The PTM predicts (P_C, P_A) data points for all of the manipulated conditions in a given experiment through estimating these system parameters. From this, you can estimate the internal noises directly by fitting PTM equations (see sections 10.4.2 and 10.4.4 of chapter 10).

The PTM parameters can be estimated from a triple-TvC experiment. It is also possible to implement both a triple-TvC and double-pass procedures jointly in the same experiment. The data on the relationship between percent correct and percent agreement provide additional complementary constraints for estimating the PTM parameters. Additional empirical constraints can improve the quality of estimation and also provide additional theoretical challenges to test the model.[5]

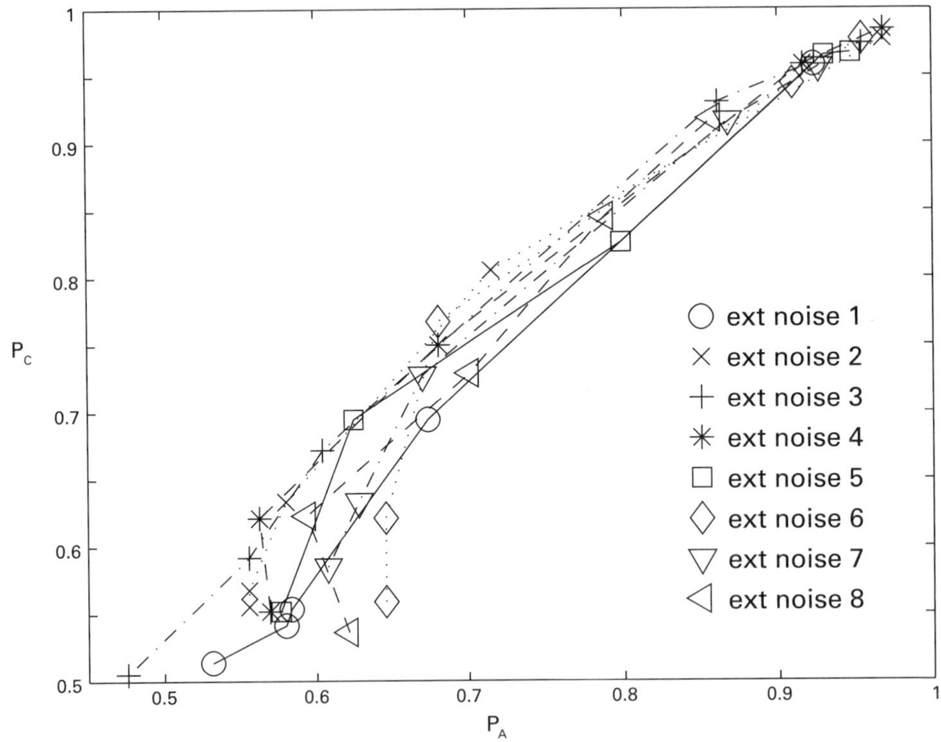

Figure 9.7
Results of a double-pass experiment. Probability correct (P_C) versus probability consistent (P_A) for an average of three observers (based on data reported in Lu and Dosher[5]).

In the example experiment in display 9.1, double pass can be introduced by exactly repeating sessions. In this program, sessions 5–8 repeat the exact sequence of trials, signals, and external noises as sessions 1–4, and sessions 13–16 repeat sessions 9–12. Figure 9.7 shows a resulting graph of percent correct as a function of percent agreement for an average of three observers.

Empirically, most or all of the observed functions relating P_C (percent correct) to P_A (percent agreement) in the literature are quite similar across external noise conditions. This suggests that the ratio of internal to external noise approaches a constant,[5,8] which in turn implies the dominance of multiplicative noise over additive noise, consistent with Weber's law.[27]

9.3.3 Discussion

Observer models provide an important approach to specifying the relationship between the stimulus and the internal representation and the response in a psychophysical task. The

PTM observer model has been applied to data from TvC functions at multiple criteria from many different tasks and manipulations of attention, learning, and other observer states. It also has been applied to several double-pass experiments.[5] The model framework has provided an excellent account of a wide range of data and has done well in comparison with other models of its general class. As a comprehensive model that incorporates the ideas of many earlier models, the success validates this approach to understanding the relationship between the stimulus and the internal representation and decision of the psychophysical observer.

Most external noise experiments in the traditional literature only measured the TvC function at a single performance level. In these cases, a simple linear amplifier model[7] consisting of a single template and additive internal noise, followed by decision, provides an adequate account of performance. However, many other experiments suggest that the simple linear amplifier model cannot account for nonlinearities in the responses of the visual system. These nonlinearities were revealed by the triple-TvC or full psychometric function methods in the previous section. Indeed, in the visual domain, d' is thought to increase as a power function of signal contrast.[13,28–34] The linear amplifier model essentially always fails as soon as even two performance levels are considered jointly—it requires different internal system parameters to account for the data at the different performance levels.

Decision uncertainty (see section 8.6.2 of chapter 8) has been proposed as one explanation for the observed nonlinearities.[9] The idea is that many task-irrelevant or "hidden" channels may contribute to a detection or discrimination decision because the observer is in some way uncertain about the nature of the signal or where the signal is encoded in the visual system. For example, perhaps the observer samples not just the spatial frequency or orientation of the stimulus, but some other orientation and spatial frequency "channels" as well. The uncertainty account presumes that the observer does not know which evidence to consider, and so includes information from irrelevant sources.

Results from double-pass experiments clearly show strong evidence for multiplicative internal noise, which was not incorporated into the linear amplifier model. Models have been proposed to include multiplicative noise[8] and to include multiplicative noise and uncertainty,[10] as well as additive internal noise.

The PTM observer model uses nonlinear transducers and multiplicative noise (that reflects the total energy in the stimulus and not just external noise in the stimulus) instead of incorporating uncertainty and an unknown number of hidden channels. The PTM has outperformed all of the other observer models in accounting for a wide range of experiments.[5]

The formulation of the PTM described in this section is mathematically equivalent to a development in which system nonlinearities are recast as contrast gain control.[35] In contrast gain control, the magnitude of the internal representation is scaled or normalized by a measure of the relevant total contrast energy in the input stimulus.

This contrast gain control form is consistent with a rearrangement of the standard PTM equations.

The PTM as an observer model was developed in the context of external noise masking—the external noises are random. A significant parallel development of observer models occurred in the context of pattern-masking experiments.[13,14,32,36–41] In pattern masking, a pattern stimulus rather than a noise masking stimulus is combined with the signal stimulus that the observer is judging. Although pattern masking and external noise are somewhat different experimental techniques, they both test the same visual system. Models of performance in the two domains should share core properties. The PTM model is functionally very similar to the models developed to account for performance in pattern masking.[14] Many of the other observer models are not. The parallels between the two developments are especially obvious for the gain-control formulation of the PTM.

All of the existing observer models, including the PTM, have been developed to account for the detection or discrimination of orthogonal or nearly orthogonal stimuli—stimuli that differ substantially from one another. To handle the discrimination of very similar stimuli, the PTM must consider close and overlapping templates both of which may respond to the same stimulus, although the response of one template will be stronger than the other. Very closely similar templates will be correlated in their responses to external noise, and so extensions to discrimination of close stimuli may also consider the correlations in response. One such development has been proposed in an elaborated PTM.[42]

9.4 Characterizing the Template

The methods we have discussed so far estimate the response gain of the perceptual template to the stimulus as well as the different sources of perceptual noise, including multiplicative and additive noise, and system nonlinearity. Critical band masking and reverse correlation, or classification images, can also be used to further estimate and specify the properties of the perceptual template. Critical band masking uses masks with specific patterns to discover which features drive the response of the perceptual template. Reverse correlation, or classification images, estimates template sensitivity by categorizing external noise samples by the response they produce to infer which external noise features drive the response. These other masking methods estimate the sensitivity of the perceptual template model to stimulus properties including spatial frequency, orientation, spatial location, and temporal location.

9.4.1 Critical Band Masking

The principle behind all of the methods is rather simple. Only external noise energy at frequencies or orientations to which the template is sensitive, or spatial positions or temporal periods to which the template is sensitive, will impact performance. External noise energy outside the sensitivity of the perceptual template will not influence behavior. The overall response of the perceptual template to a signal stimulus in the PTM is captured by the single parameter β. However, the response of the template can also be specified as a function of any manipulated variable,v, where v could stand for the spatial frequency, orientation, time, or spatial location (or some combination).

The template response to the stimulus along the manipulated variable v is $T_S(v)$, the amplitude of the signal stimulus is $S_S(v)$, and the amplitude of the external noise is $F(v)$. Then the output of the template matching stage through the signal path for the stimulus (signal and the noise) can be expressed as[43]

$$S_1 = \alpha c \int Ts(v) Ss(v) dv \tag{9.9}$$

$$\sigma_{N_1}^2 = \sigma_{\text{ext}}^2 \int T_s^2(v) F^2(v) dv . \tag{9.10}$$

The parameter α is the gain of the template to a signal stimulus relative to external noise.

Similarly, the output of the template in the multiplicative noise pathway to the stimulus is $T_N(v)$. Then, the output of the template for signal and external noise through the multiplicative noise path can be expressed as

$$S_2 = \alpha c \int T_N(v) S_s(v) dv , \tag{9.11}$$

$$\sigma_{N_2}^2 = \sigma_{\text{ext}}^2 \int T_N^2(v) F^2(v) dv . \tag{9.12}$$

Analogous to the simple PTM, the discriminability d' depends upon the output of the signal path compared with the total noise:

$$d' = \frac{S_1^{\gamma_1}}{\sqrt{\sigma_{N_1}^{2\gamma_1} + N_{\text{mul}}^2 \left(\sigma_{N_2}^{2\gamma_2} + \frac{S_2^{2\gamma_2}}{2} \right) + \sigma_{\text{add}}^2}} . \tag{9.13}$$

One example of measuring the sensitivity of the template to a stimulus variable is developed in detail next, followed by a sketch of several other examples. The first example concerns the estimation of the spatial frequency sensitivity of the perceptual template. To measure this, the spatial frequency content of the external noise samples is manipulated. If external noise has spatial frequencies that are outside the template sensitivity, then these will not impact or alter detection or discrimination. Typically, a series of low-pass and high-pass external noise images are created for testing, and the threshold elevation is measured for each kind of noise. Experiments that measure the threshold elevation for different kinds of noise provide different estimates of the template $T_S(f)$. By measuring the threshold as a function of the cutoff spatial frequency (Tvf) functions of the low-pass or high-pass filtered noise, it is possible to estimate the internal noises as well as the profile of the template sensitivity to spatial frequency.

In figure 9.8, we show an example of a series of high-pass and low-pass filters in the spatial frequency domain and examples of external noise through the low-pass and high-pass filters. Figure 9.8a and b depict different spatial frequency cutoffs in a Fourier space representation. (See section 3.3.9 of chapter 3 for a discussion of Fourier representations and the fast Fourier transform and how to program band-pass filters.) Figure 9.8c shows corresponding example noises for the low-pass series, and figure 9.8d shows example noises for the high-pass series. Figure 9.8e shows the estimated thresholds as a function of cutoff for three observers. Figure 9.8f shows the template sensitivities for different spatial frequencies estimated from these threshold profiles. This example is taken from ref. 43.

In figure 9.9, we show another example in which the orientation content of the external noise is filtered around 45° with increasing orientation bandwidths to include energy from more and more orientations, until all orientations are represented. Examples of external noise samples are shown from left to right in figure 9.9a for increasing width of the pass band. Filtered external noise in the orientation domain has been used to estimate the orientation bandwidth of perceptual templates.[35]

In our next example, the spatial footprint of the perceptual template is estimated by adding external noise in different-sized spatial rings that cover different parts of a signal stimulus, which in this example is an oriented Gabor. Testing different combinations of external noise rings estimates the spatial footprint or profile of the template (and tests for interactions between regions of space).[44] If external noise appears in regions that do not drive the template, the noise will not elevate threshold, whereas if the external noise appears in regions that are heavily weighted in the template, there will be significant threshold elevation. Figure 9.10 shows examples of different spatial patterns of external noise rings and the weights on different rings in the spatial template—or the template footprint. The threshold data from which the weights are estimated (16 multipoint psychometric functions) are omitted.

Finally, an analogous manipulation can be carried out in the time domain; external noise is added at various points in time before and after the signal test stimulus. Figure 9.11 shows examples of external noise frames at various points surrounding the signal image frame, from which we can estimate the temporal window of the perceptual template.[45]

In these examples, we have provided sample estimates of the spatial frequency, orientation tuning, spatial footprint, and temporal window of the perceptual template. Each result is associated with a relatively large psychophysical experiment. These estimates of the perceptual template may be related to neurophysiological measures of the tuning properties of visual neurons or visual areas. These same methods have also been used to study how the template may change under manipulations of the state of the observer—for example due to adaptation, or attention, or perceptual learning.[35,44–46]

In an ideal case, the perceptual template could be tested in all these dimensions simultaneously. However, joint measurements of different properties, such as spatial frequency

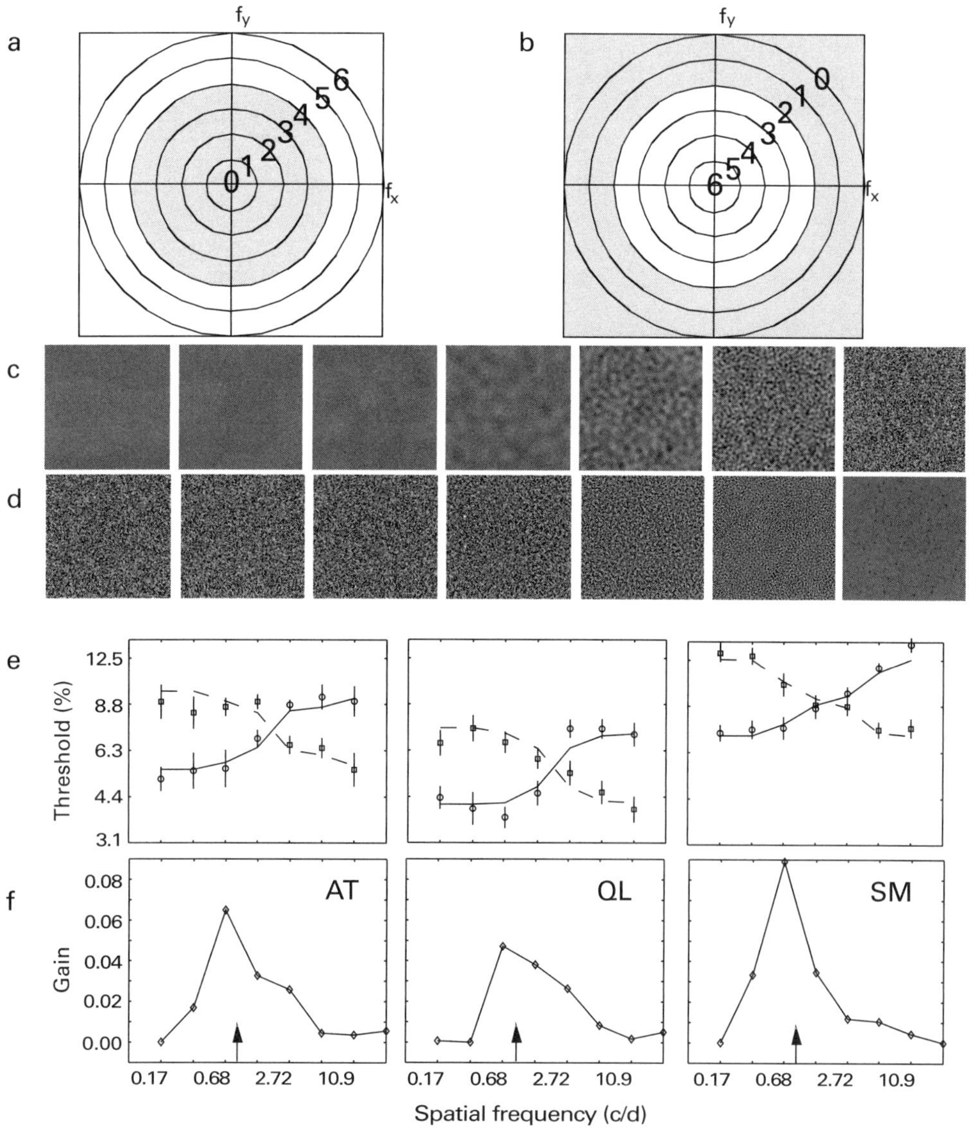

Figure 9.8
Estimating the spatial frequency sensitivity of the template. (a, b) Two-dimensional low-pass and high-pass spatial-frequency filters with seven different passbands. (c, d) From left to right, examples produced by filtering a Gaussian white noise image through the seven filters in (a) and (b). (e) Contrast threshold as a function of pass band of the low-pass and high-pass filters for three observers at 70% correct performance level. The curves were from fits to the PTM. (f) The best-fitting spatial-frequency sensitivity of the perceptual templates. The arrows on the x-axis indicate the center frequency of the signal stimulus (from Lu and Dosher[43]).

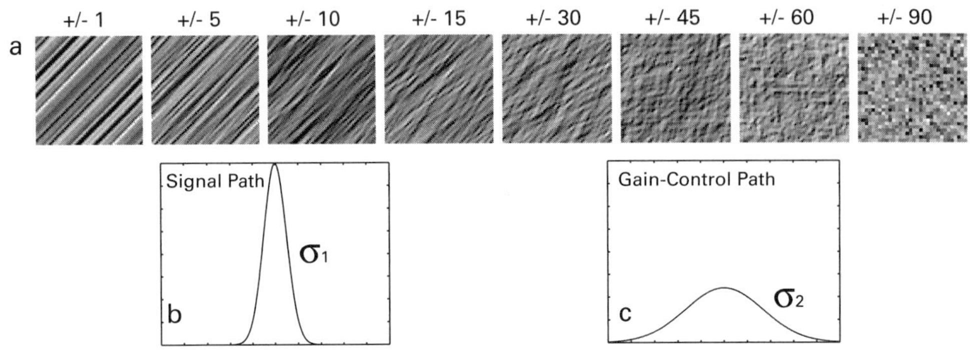

Figure 9.9
Estimating the orientation sensitivity of the template. (a) Orientation-filtered external noise around 45° by filters of different orientation bandwidths. The number above each panel indicates the filter bandwidth in degrees of orientation. (b, c) Estimated perceptual templates in the signal and gain-control paths (after Dao, Lu, and Dosher[35]).

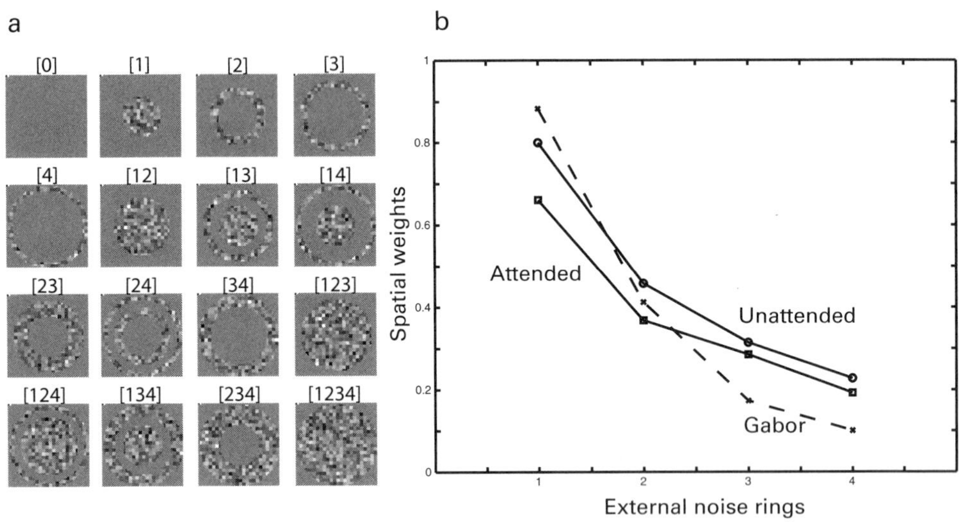

Figure 9.10
Estimating the spatial footprint of the template. (a) Sixteen external noise conditions, consisting of all possible combinations of spatial rings 1, 2, 3, and 4, indicated by the labels above the panels. For example, [124] indicates the combination of these three rings. (b) Estimated weights from the PTM for each of the four rings of external noise in attended and unattended conditions. Also graphed is the proportion root mean square (RMS) contrast for each ring in the signal Gabor stimulus. The spatial profile of both the attended and the unattended condition match the spatial profile of the stimulus (after Dosher, Liu, Blair, and Lu[44]).

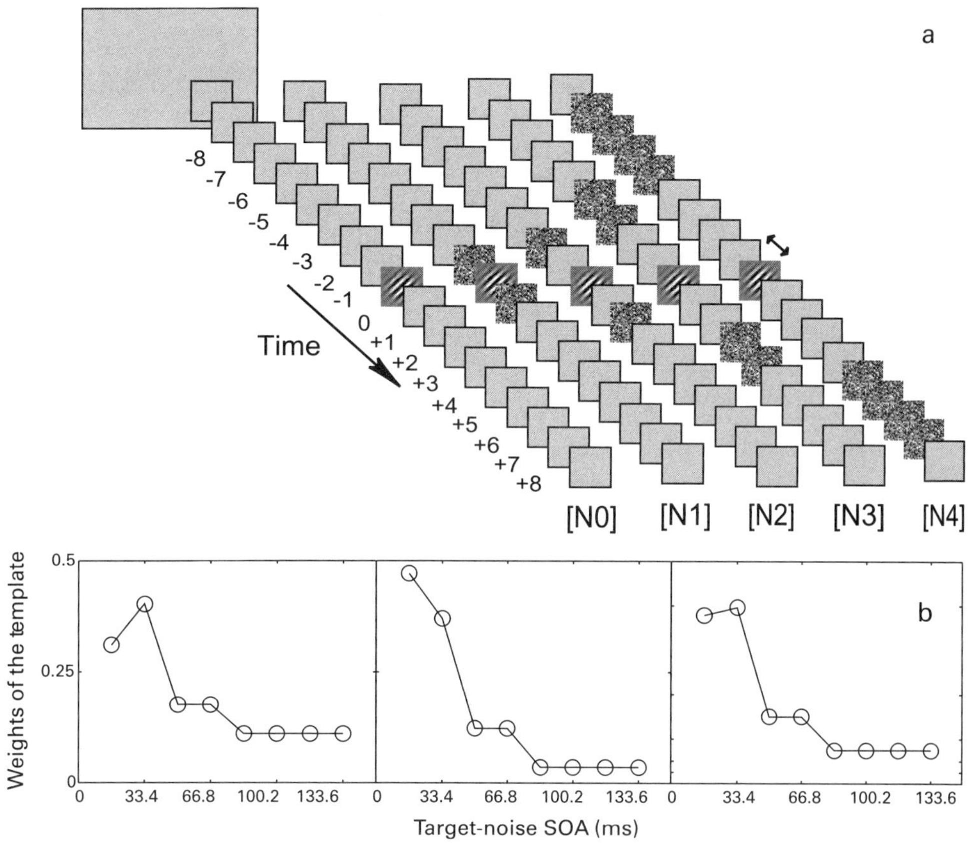

Figure 9.11
Estimating the temporal window of the template. (a) Signal and external noise temporal configurations. The zero noise [N0] and the four non-overlapping basic configurations, [N1], [N2], [N3], and [N4], are shown. Additional mixtures of the basic temporal configurations can also be constructed. (b) Derived temporal characteristics of the perceptual templates from the best fitting PTM (after Lu, Jeon, & Dosher[45]).

and orientation and space and time, are relatively impractical due to the large data collection demands. Investigating one dimension at a time could provide information about optimal positioning for tests in an experiment that simultaneously manipulates multiple dimensions.

Characterizing the detection or discrimination template within the context of the PTM observer framework is essential to correctly understand and estimate all observer properties, including the contributions of internal noises and nonlinearities in performance. The PTM framework allows us to separate perceptual factors from decision factors in the data. This allows the estimation of the underlying perceptual template without contamination by nonlinearity or decision factors.

9.4.2 Classification Images

Another method to visualize the most relevant visual features for perceptual judgments is the classification image method, which was first presented in vision by Ahumada in 1996.[47] Developed from earlier work in audition,[48] the idea is similar to a large multiple regression analysis. It relates the observer's response to stimulus inputs—in this case, the value of the external noise at different locations in the stimulus image. It can also be used to measure sensitivity over time or over space and time.[49] The method uses the correlation of the *decision* of the observer with many noisy stimulus features to infer the weights of those features leading to the decision.[50–52] For example, if an observer is more likely to say "signal" when the image luminance is high (white) rather than low (black) in a particular location, then the template may be "looking for" a white stimulus at that location or at that pixel location. This method is similar to *reverse correlation* methods used for estimating the receptive fields of neurons in visual cortex.[53–55]

One nice example of the method derived raw classification images for a vernier acuity task.[49] Two horizontal bars of 1×4 pixels each, separated by 1 pixel, were presented. In one condition, there was no offset. In another, the left bar is lowered by one pixel below the right bar. Gaussian random external noise was added to each display, and the observer decides whether or not there is an offset. Contrast of the vernier stimuli was controlled to lead to 75% correct decisions.

Trials are classified into four categories, determined by the combination of presence and absence of the offset in the stimulus and the observer's response as "present" or "absent." The classification image is derived by first removing the signal from each image and then subtracting the remaining external noise value for each pixel for trials in which the observer says "offset absent" from trials in which they respond "offset present." The classification image for this task is shown in figure 9.12.

The original applications of the classification image method depend on the assumptions of the linear amplifier model.[56] Later research has investigated different deviations from these assumptions so as to include multiplicative noise[56] and decision uncertainty.[57]

9.5 Multichannel Observer Models

The observer models in the previous sections account for visual detection and discrimination behavior by a single system describing the input–output functions exhibited by observers. A few equations summarize the overall behavior of the observer. The advantage of this approach is that estimating a small number of parameters allows us to predict performance in a wide range of conditions. In addition, good estimation protocols have been developed for this purpose. The framework has been extremely successful in characterizing and understanding changes in observer state because usually only one parameter or a few parameters change from one state to another.[12,35,58]

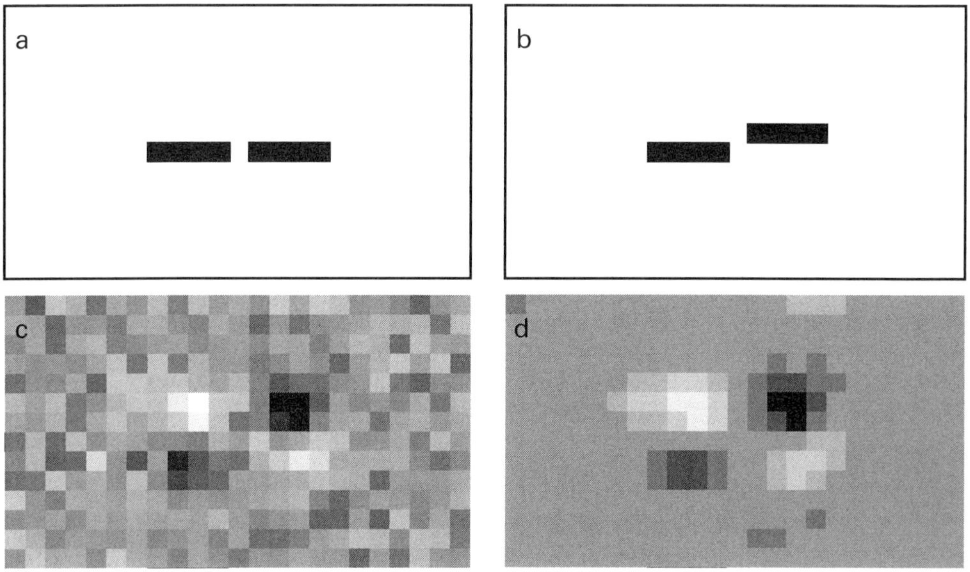

Figure 9.12
Classification images for simple vernier stimuli. (a, b) The vernier stimuli (1 pixel = 1.26 min arc). (c, d) A raw classification image (c) and the same image smoothed and quantized (d), so only weights significantly different from zero are colored differently from the gray background (after Ahumada[47]).

The template in the PTM represents the sensitivity of the whole observer system to the set of stimuli in a particular task. In the visual system, the overall template must be created from neural receptors or visual detectors. The human visual system that embodies the behavior involves the complex interaction of many neural subsystems and many kinds of detectors. In an effort to understand more fully how neural systems produces the observer's behavior, researchers have developed models that include many detectors or channels that correspond more closely with known physiologic properties of the visual system. Any model of this nature is still an approximation of the real neural system. It will be closer to neural reality, but also more complex to specify.[59–65]

Receptors in the visual system are selectively responsive to visual inputs. They are tuned to respond to stimuli with certain properties, such as a particular orientation or spatial frequency. One theoretical approach tries to predict the response to the stimulus from the responses of groups or cells, or *channels*, tuned to somewhat different properties. These are called *multichannel observer models*.

A multichannel observer model combines responses from many channels. Each channel may be thought of as a unit with PTM-like properties—each has tuning, noisiness, gain control, or other nonlinearity. The tuning could be in terms of features such as spatial frequency or orientation but also include tuning in space. Most models assume channels

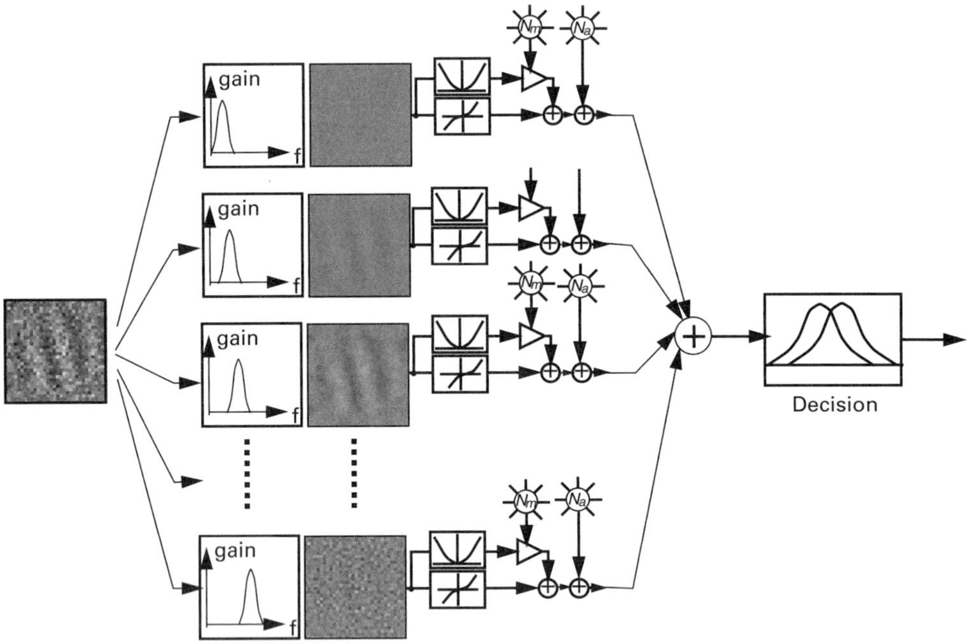

Figure 9.13
A multichannel observer model. The signal plus noise image is shown on the left and is subsequently processed through standard spatial frequency tuned visual channels (shown schematically) to yield the filtered images shown. The image power is passed through the nonlinear transducer functions. Each visual channel is illustrated with its own internal multiplicative and additive noise sources representing processing inefficiencies. Finally, the integrated output of these visual channels is input to a decision process, illustrated at right (after Dosher and Lu[58]).

are duplicated in detectors at many spatial locations. More complex models include systematic variations in channel properties, or their distribution in space, as a function of visual eccentricity.[66]

Figure 9.13 shows a schematic of a hypothetical multichannel observer model.[15,58] A bank of channels or detectors processes a single image. This schematic shows a stimulus image processed through spatial frequency channels, those tuned to very slowly varying low-frequency patterns up to those tuned to very-high-frequency patterns. The images attached to each channel show the stimulus image as filtered by that channel. The most useful information will appear in channels whose spatial frequency tuning matches the spatial frequency content of the signal stimuli. The output of each channel (possibly at many spatial positions) is subject to gain control nonlinearities and internal noises.

The information available in principle to the observer is the output of each of these channels. The observer's job is to integrate all of these pieces of information into a detection or discrimination decision. In this schematic, the decision is made by adding together information from various channels with different weights.

A multichannel observer model illustrated in figure 9.13 has been implemented by Petrov, Dosher, and Lu[67,68] in the context of perceptual learning. The model has successfully accounted for performance for stimuli of different contrasts at several performance levels and is able to reproduce the classical patterns in threshold versus contrast functions observed in external noise experiments.[69]

Different multichannel observer models have different implementations of the channels and their properties and make different assumptions about how evidence is combined, or *pooled* over both space and over channels depending on the task.[36,60,62,70–75] At the theoretical extreme, the channels might be individual neurons or groups of neurons, the properties could be based on reported physiological properties of those neurons, and the model would be a population model consisting of many neurons.[76,77]

In a major effort to specify and test multichannel observer models in visual detection, a large group of visual psychophysicists produced a set of baseline data for foveal detection of representative spatial patterns and used these data as a test bed to evaluate models. This enterprise was called *Modelfest*.[63,64,65,78–80] The Modelfest data set includes contrast threshold measurements from several laboratories for 43 foveal achromatic contrast stimuli (figure 9.14). Included are a series of stimuli with different spatial frequencies to specify the contrast sensitivity function, Gabors of several sizes, Gabors with fixed numbers of cycles and of fixed extent and elongated Gabors of different spatial frequency, and oriented stimuli with different spatial arrangements, edge and line stimuli, several compound pattern stimuli such as a checkerboard, and a few more naturalistic stimuli.

Watson and Ahumada[63] built a framework consisting of a number of stages as a structure for examining multichannel observer models. In the most complete version of the framework, a luminance pattern is converted to contrast, passed through a contrast sensitivity function (CSF), subjected to an oblique effect, cropped by a spatial aperture, and followed by channel analysis. Then the outputs of the channels are pooled together with a so-called Minkowski metric[62,73–75]:

$$c_{\mathrm{T},q} = \left[\sum_{y=1}^{N_y} \sum_{x=1}^{N_x} p_x p_y \left| r_{x,y} \right|^{\beta} \right]^{-\frac{1}{\beta}}. \tag{9.14}$$

This metric specifies the pooling over space for a given channel q. The value c_T is the contrast threshold, p_x and p_y are the width and height of each pixel in degrees, and the $r_{x,y}$ are the pixel values of the new image after preprocessing. For multiple channels, the pooled value is

$$c_\mathrm{T} = \left[\sum_{q=1}^{Q} c_{\mathrm{T},q}^{-\beta} \right]^{-\frac{1}{\beta}}. \tag{9.15}$$

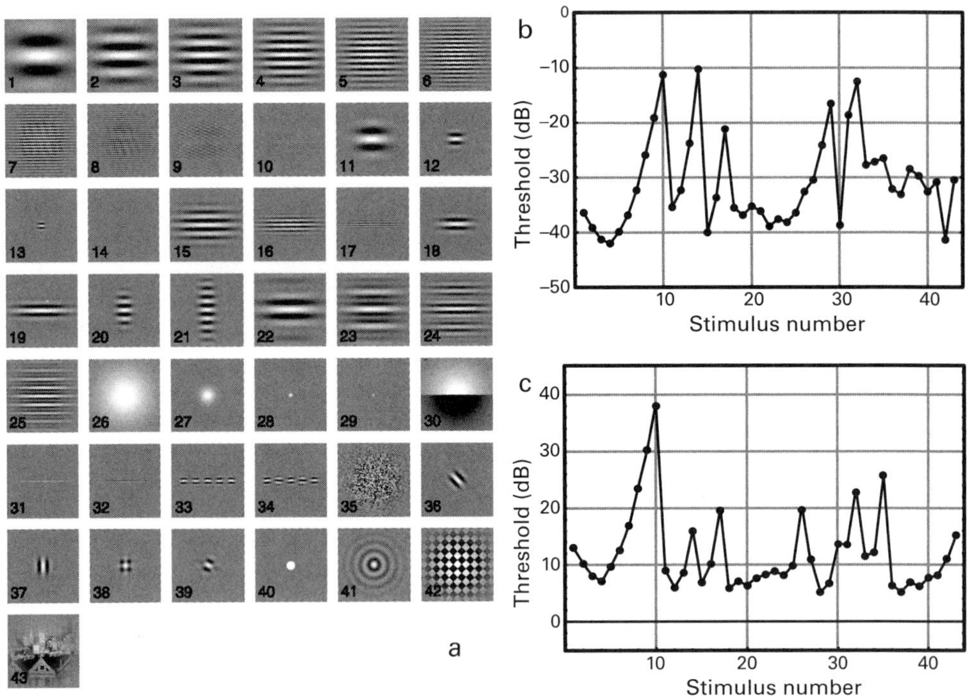

Figure 9.14
(a) Modelfest stimuli. Each is a monochrome image subtending 2.133° × 2.133°. The index numbers have been added for identification and were not present in the stimuli. (b, c) Average Modelfest thresholds in dB (b) and in dBB (c). Each point is the mean of 16 observers, and the error bars indicate ± 2 SE. The dBB is a measure of the contrast energy of a stimulus, normalized by a nominal minimum threshold of 10^{-6} deg^{-2} s^{-1}. Zero dBB is defined so as to approximate the minimum visible contrast energy for a sensitive human observer (after Watson and Ahumada[63]).

A number of variants of this full model did a reasonable job of accounting for the pattern of thresholds over the 43 stimuli.[63] Although the results are excellent, they also illustrate the limited power of even this large data set in its ability to distinguish between different models. This partly reflects the fact that the Modelfest data set does not include psychometric functions, or multiple performance levels, or external noise. Ideal extended tests might involve examining all of these patterns using both external noise and contrast psychometric functions. Another important conclusion from this exercise[63] is that the template models, such as the PTM, could also account for these results without multiple channels, but assuming known templates for a few basic patterns. For example, some researchers have concluded that three basic templates, a spot, a bar, and a grating template, account for the Modelfest[78] data.

9.6 Summary

Observer models aim to provide a strong functional understanding of the internal processes in vision or perception. They provide a quantitative framework that allows us to specify and understand how stimuli are transformed into internal representations and then choice behavior in the mind of the observer. The best models can provide a very compact formulation that predicts behavioral outcomes across many stimulus variations and conditions through the estimation of a few parameters that specify the sensitivity, nonlinearity, and internal noises of the observer. Often, the observer model framework provides key insights into the nature of the processes that determine the internal representation.

This chapter considered a range of modern single-channel observer models, with focus on the PTM. The PTM incorporates many features of earlier models and has proved quite successful. We also discussed several approaches that involve multichannel observer models that to varying degrees approximate the complexity of the human brain processes.

The observer approach and associated paradigms such as the external noise method provide one complementary approach to psychophysical scaling. By measurement of sensitivity or thresholds across a wide range of stimulus contrasts, external noises, and performance levels, observer models derive the internal representation of stimuli. If the goal of scaling is to measure the internal representation, then the observer approach succeeds in specifying that internal representation through the transducer functions and contrast gain control systems.

The observer models, whether single-channel or multichannel, also provide a framework that can be further exploited in order to understand how the system changes when the state of the observer changes due to factors such as attention, learning, or adaptation. The behavioral consequences of these manipulations can often be summarized by a change in one or two parameters in the observer model, and these in turn provide insights and set the stage for many other investigations of the human perceptual systems.

References

1. Sperling G. 1989. Three stages and two systems of visual processing. *Spat Vis* 4: 183–207.

2. Petrov AA, Dosher BA, Lu ZL. 2005. The dynamics of perceptual learning: An incremental reweighting model. *Psychol Rev* 112: 715–743.

3. Bejjanki VR, Beck JM, Lu ZL, Pouget A. 2011. Perceptual learning as improved probabilistic inference in early sensory areas. *Nat Neurosci* 14: 642–648.

4. Pelli DG, Farell B. 1999. Why use noise? *J Opt Soc Am A Opt Image Sci Vis* 16: 647–653.

5. Lu ZL, Dosher BA. 2008. Characterizing observers using external noise and observer models: Assessing internal representations with external noise. *Psychol Rev* 115: 44–82.

6. Barlow HB. 1956. Retinal noise and absolute threshold. *J Opt Soc Am* 1056(46): 634–639.

7. Pelli, DG. 1981. Effects of visual noise. PhD dissertation, Physiology Department, Cambridge University.

8. Burgess AE, Colborne B. 1988. Visual signal detection: IV. Observer inconsistency. *J Opt Soc Am A* 2: 617–627.

9. Pelli DG. 1985. Uncertainty explains many aspects of visual contrast detection and discrimination. *J Opt Soc Am A* 2: 1508–1532.

10. Eckstein MP, Ahumada AJ, Jr, Watson AB. 1997. Visual signal detection in structured backgrounds: II. Effects of contrast gain control, background variations, and white noise. *J Opt Soc Am* 14: 2406–2419.

11. Lu Z-L, Dosher BA. 1999. Characterizing human perceptual inefficiencies with equivalent internal noise. *J Opt Soc Am A Opt Image Sci Vis* 16: 764–778.

12. Lu Z-L, Dosher BA. 1998. External noise distinguishes attention mechanisms. *Vision Res* 38: 1183–1198.

13. Foley JM, Legge GE. 1981. Contrast detection and near-threshold discrimination in human vision. *Vision Res* 21: 1041–1053.

14. Foley JM. 1994. Human luminance pattern-vision mechanisms: Masking experiments require a new model. *J Opt Soc Am A Opt Image Sci Vis* 11: 1710–1719.

15. Dosher BA, Lu Z-L. 1999. Mechanisms of perceptual learning. *Vision Res* 39: 3197–3221.

16. North DO. 1942. The absolute sensitivity of radio receivers. *RCA Review.* 6: 332–344.

17. Friis HT. 1944. Noise figures of radio receivers. *Proceedings of the IRE.* 32: 419–422.

18. Mumford WW, Schelbe EH. *Noise performance factors in communication systems*. Dedham, MA: Horizon House-Microwave Inc.; 1968.

19. Nagaraja NS. 1964. Effect of luminance noise on contrast thresholds. *J Opt Soc Am* 54: 950–955.

20. Burgess AE, Wagner RF, Jennings RJ, Barlow HB. 1981. Efficiency of human visual signal discrimination. *Science* 214: 93–94.

21. Legge GE, Kersten D, Burgess AE. 1987. Contrast discrimination in noise. *J Opt Soc Am A* 4: 391–404.

22. Swets JA, Shipley EF, McKey MJ, Green DM. 1959. Multiple observations of signals in noise. *J Acoust Soc Am* 31: 514–521.

23. Green DM. 1964. Consistency of auditory detection judgments. *Psychol Rev* 71: 392–407.

24. Levi D, Klein S. 2003. Noise provides some new signals about the spatial vision of amblyopes. *J Neurosci* 7: 2522–2526.

25. Gold J, Bennett PJ, Sekuler AB. 1999. Signal but not noise changes with perceptual learning. *Nature* 402: 176–178.

26. Chung STL, Levi DM, Tjan B. 2005. Learning letter identification in peripheral vision. *Vision Res* 45: 1399–1412.

27. Weber EH. *De Pulsu, resorptione, auditu et tactu: Annotationes anatomicae et physiologicae*. Leipzig: CF Koehler; 1834.

28. Cohn TEA. 1974. New hypothesis to explain why the increment threshold exceeds the decrement threshold. *Vision Res* 14: 1277–1279.

29. Leshowitz B, Taub HB, Raab DH. 1968. Visual detection of signals in the presence of continuous and pulsed backgrounds. *Percept Psychophys* 4: 207–213.

30. Nachmias J. 1981. On the psychometric function for contrast detection. *Vision Res* 21: 215–223.

31. Nachmias J, Kocher EC. 1970. Visual detection and discrimination of luminance increments. *J Opt Soc Am* 60: 382–389.

32. Nachmias J, Sansbury RV. 1974. Grating contrast: Discrimination may be better than detection. *Vision Res* 14: 1039–1042.

33. Stromeyer CF, Klein S. 1974. Spatial frequency channels in human vision as asymmetric (edge) mechanisms. *Vision Res* 14: 1409–1420.

34. Tanner WP, Jr. 1961. Physiological implications of psychophysical data. *Ann N Y Acad Sci* 89: 752–765.

35. Dao DY, Lu Z-L, Dosher BA. 2006. Adaptation to sine-wave gratings selectively reduces the contrast gain of the adapted stimuli. *J Vis* 6: 739–759.

36. Fredericksen RE, Hess RF. 1997. Temporal detection in human vision: Dependence on stimulus energy. *J Opt Soc Am* 14: 2557–2569.

37. Gorea A, Sagi D. 2001. Disentangling signal from noise in visual contrast discrimination. *Nat Neurosci* 4: 1146–1150.

38. Klein SA, Levi DM. 1985. Hyperacuity thresholds of 1 sec: Theoretical predictions and empirical validation. *J Opt Soc Am A* 2: 1170–1190.

39. Kontsevich LL, Chen CC, Tyler CW. 2002. Separating the effects of response nonlinearity and internal noise psychophysically. *Vision Res* 42: 1771–1784.

40. Legge GE, Foley JM. 1980. Contrast masking in human vision. *J Opt Soc Am* 70: 1458–1471.

41. Watson AB, Solomon JA. 1997. Model of visual contrast gain control and pattern masking. *J Opt Soc Am* 14: 2379–2391.

42. Jeon ST, Lu Z-L, Dosher BA. 2009. Characterizing perceptual performance at multiple discrimination precisions in external noise. *JOSA A* 26: 43–58.

43. Lu Z-L, Dosher BA. 2001. Characterizing the spatial-frequency sensitivity of perceptual templates. *J Opt Soc Am A Opt Image Sci Vis* 18: 2041–2053.

44. Dosher BA, Liu SH, Blair N, Lu Z-L. 2004. The spatial window of the perceptual template and endogenous attention. *Vision Res* 44: 1257–1271.

45. Lu Z-L, Jeon ST, Dosher BA. 2004. Temporal tuning characteristics of the perceptual template and endogenous cuing of spatial attention. *Vision Res* 44: 1333–1350.

46. Lu Z-L, Dosher BA. 2004. Spatial attention excludes external noise without changing the spatial frequency tuning of the perceptual template. *J Vis* 4(10): 955–966.

47. Ahumada A, Jr. 1996. Perceptual classification images from Vernier acuity masked by noise. [abstract] *Perception* 26: 18.

48. Ahumada AJ, Lovell J. 1971. Stimulus features in signal detection. *J Acoust Soc Am* 49: 1751–1756.

49. Ahumada AJ, Jr. 2002. Classification image weights and internal noise level estimation. *J Vis* 2(1): 121–131.

50. Eckstein MP, Ahumada AJ, Jr. 2002. Classification images: A tool to analyze visual strategies. *J Vis* 2(1): 1x.

51 Abbey, CK, Eckstein, MP. 2002. Classification image analysis: Estimation and statistical inference for two-alternative forced-choice experiments. *J Vis* 2(1): 66–78.

52. Gold JM, Murray RF, Bennett PJ, Sekuler AB. 2000. Deriving behavioural receptive fields for visually completed contours. *Curr Biol* 10: 663–666.

53. Jones JP, Palmer LA. 1987. The two-dimensional spatial structure of simple receptive fields in cat striate cortex. *J Neurophysiol* 58: 1187–1211.

54. Ohzawa I, DeAngelis GC, Freeman RD. 1996. Encoding of binocular disparity by simple cells in the cat's visual cortex. *J Neurophysiol* 75: 1779–1805.

55. Ringach DL, Hawken MJ, Shapley R. 1997. Dynamics of orientation tuning in macaque primary visual cortex. *Nature* 387: 281–284.

56. Murray RF, Bennett PJ, Sekuler AB. 2002. Optimal methods for calculating classification images: Weighted sums. *J Vis* 2(1): 79–104.

57. Tjan BS, Nandy AS. 2006. Classification images with uncertainty. *J Vis* 6(4): 387–413.

58. Dosher BA, Lu Z-L. 1998. Perceptual learning reflects external noise filtering and internal noise reduction through channel reweighting. *Proc Natl Acad Sci USA* 95: 13988–13993.

59. Graham NVS. *Visual pattern analyzers*. Oxford: Oxford University Press; 2001.

60. Wilson HR, Bergen JR. 1979. A four mechanism model for threshold spatial vision. *Vision Res* 19: 19–32.

61. Wilson HR, Humanski R. 1993. Spatial frequency adaptation and contrast gain control. *Vision Res* 33: 1133–1149.

62. Graham N. 1977. Visual detection of aperiodic spatial stimuli by probability summation among narrowband channels. *Vision Res* 17: 637–652.

63. Watson AB, Ahumada AJ. 2005. A standard model for foveal detection of spatial contrast. *J Vis* 5(9): 717–740.

64. Carney T, Klein SA, Tyler CW, Silverstein AD, Beutter B, Levi D, et al. 1999. The development of an image/threshold database for designing and testing human vision models. *Proc SPIE* 3644: 542–551.

65 Walker, L, Klein, S, Carney, T. Modeling the Modelfest data: Decoupling probability summation. *Proceedings of the Optical Society of America Annual Meeting*, Santa Clara, CA: OSA; 1999, pp. SuC5.

66. Najemnik J, Geisler WS. 2005. Optimal eye movement strategies in visual search. *Nature* 434: 387–391.

67. Petrov A, Dosher BA, Lu Z-L. 2005. Perceptual learning through incremental channel reweighting. *Psychol Rev* 112: 715–743.

68. Petrov A, Dosher B, Lu Z-L. 2006. Comparable perceptual learning with and without feedback in non-stationary contexts: Data and model. *Vision Res* 46: 3177–3197.

69. Lu Z-L, Liu J, Dosher B. 2010. Modeling mechanisms of perceptual learning with augmented Hebbian reweighting. *Vision Res* 50: 375–390.

70. Graham N, Sutter A, Venkatesan C. 1993. Spatial-frequency-and orientation-selectivity of simple and complex channels in region segregation. *Vision Res* 33: 1893–1911.

71. Graham N, Sutter A. 1998. Spatial summation in simple (Fourier) and complex (non-Fourier) texture channels. *Vision Res* 38: 231–257.

72. Sutter A, Beck J, Graham N. 1989. Contrast and spatial variables in texture segregation: Testing a simple spatial-frequency channels model. *Atten Percept Psychophys* 46: 312–332.

73. Watson AB. 1979. Probability summation over time. *Vision Res* 19: 515–522.

74. Robson J, Graham N. 1981. Probability summation and regional variation in contrast sensitivity across the visual field. *Vision Res* 21: 409–418.

75. Quick R. 1974. A vector-magnitude model of contrast detection. *Biol Cybern* 16: 65–67.

76. Averbeck BB, Latham PE, Pouget A. 2006. Neural correlations, population coding and computation. *Nat Rev Neurosci* 7: 358–366.

77. Pouget A, Dayan P, Zemel R. 2000. Information processing with population codes. *Nat Rev Neurosci* 1: 125-132.

78. Chen CC, Tyler CW, Rogowitz B, Pappas T. ModelFest: Imaging the underlying channel structure. In: *Human vision, visual processing, and digital display*. Vol. 159. Bellingham, WA: SPIE; 2000, pp. 152–159.

79 du Buf, J. 2005. Modelfest and contrast-interrelation-function data predicted by a retinal model. *Perception* 34 ECVP Abstract Supplement.

80. Klein SA, Tyler CW. 2005. Paradoxical, quasi-ideal, spatial summation in the Modelfest data. *J Vis* 5: 478.

IV EXPERIMENTAL DESIGN, DATA ANALYSIS, AND MODELING

10 Data Analysis and Modeling

This chapter provides a guidebook to the basic issues in quantitative data analysis and modeling. It shows how we test the quality of a model and fit the model to observed data. The quality of a model or theory includes qualitative assessments related to internal consistency, breadth of application, and the ability to make new and useful predictions. Another assessment of the quality of a model is its ability to predict or fit the observed behavioral data in relevant domains quantitatively.

Two criteria for fitting a model to data are considered, a least-squared error criterion and a maximum likelihood criterion, along with methods of estimating the best-fitting parameters of the models. Bootstrap methods are used to estimate the variability of derived data and model parameters. Several methods of comparing and selecting between models are considered. The chapter uses typical psychophysical testing situations to illustrate several standard applications of these methods of data analysis and modeling.

10.1 What Is a Model?

Informal and formal models serve an important role in the development of scientific theories. An informal model is an idea or set of ideas about the mechanisms that underlie some phenomenon—an ability or performance. A formal model is one that has been developed and quantified to the point where precise predictions about that phenomenon can be generated and tested. Developing a formal model often clarifies whether the ideas in an informal model have any likelihood of explaining the phenomenon. For example, people knew for many years that movement of objects depends on the mass of the object and the force applied on it—Newton's second law quantified this as a testable relationship between mass, force, and acceleration. People had observed planet motion and planetary orbits for many centuries. Newton's law of universal gravitation provided a formalism that allowed scientists to test many ideas about the orbits of the planets.

One common meaning of the word *model* is a surrogate or stand-in—often small—representation of something real. It is meant to allow the researcher to explore the more general theory. Just as with a model train or a model plane, certain details may be glossed over in a model of the sensory system. However, as vision or auditory scientists, we are more concerned with the inner workings than with the superficial aspects of a model. The construction of a model often serves two slightly different purposes. A model is sometimes seen as a "description or analogy used to help visualize something . . . that cannot be directly observed." Or, a model is "a system of postulates, data and inferences presented as a mathematical description."[1] The first statement highlights the heuristic value of a model. Construction of models can help us visualize or think through problems. The second statement emphasizes the formal aspects of a model. Formalizing a model of a phenomenon or process will allow us to test precisely how good a model we have.

Quantitative modeling serves several scientific functions. The development of a quantitative model from a set of qualitative ideas (a qualitative theory) can provide clear tests of those ideas (see chapter 12). This is especially important when the qualitative theory is complex. The fit of a quantitative model can be evaluated to see how well it captures all the details of the pattern of performance. This establishes to what degree the model provides an accurate or a complete account of the target phenomenon.

A quantitative model also allows the estimation of specific parameters that summarize performance. This is often critically important in external validation of the model. The primary technique for external validation of a model is to manipulate a relevant aspect of the stimuli or the procedures to see if this results in a sensible change in the values estimated for parameters. Attempts at validation often form the most powerful tests of a model. For example, the introduction of high spatial frequency external noise in an image should disproportionately limit the contrast sensitivity function in parameters that specify high-frequency performance. If the model is substantially correct, then validation tests will support both the model and our interpretation of the model's parameters.

Finally, a successful quantitative model allows the researcher to summarize existing observations efficiently and to make predictions about likely performance under new conditions. Quantitative testing can motivate modifications of the existing model and lead to the creation of a better theory.

10.2 Least Squares in Modeling

One of the often-used methods of assessing a quantitative model uses the least-squares methods of fitting and estimation. It seeks to find a model and a set of estimated parameters for the model that provide a good fit to the data. The fidelity criterion, or function defining a good fit, is that the model with optimized parameters minimizes the sum of squared errors. This is equivalent to minimizing the root mean squared error, which is analogous to minimizing the variance about the predictions.

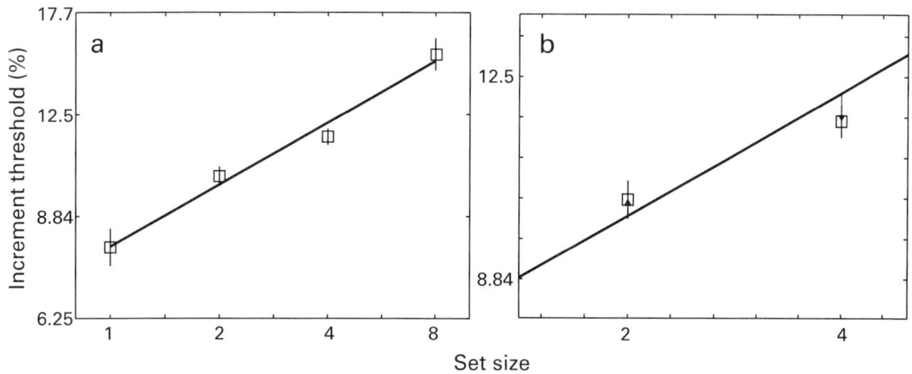

Figure 10.1
A linear model of log contrast increment threshold as a function of log set size in a hypothetical visual search experiment. (a) Contrast increment threshold is plotted as a function of set size. The squares indicate the mean thresholds of four observers; and the solid line is the best-fitting linear regression line on this log–log plot. Error bars are standard errors of the mean. (b) A close-up of a subset of the data (set sizes 2 and 4) that shows the prediction error (indicated by the arrows) for each point.

10.2.1 An Example: A Linear Model

One of the simplest models assumes a linear relationship between behavior and one or more predictors. Here, we consider fitting a linear model with least-squares criteria. One common example from visual search paradigms measures the contrast increment threshold for detecting a target over a range of set sizes.[2] The data from a hypothetical experiment are shown in figure 10.1. The relationship between the log of the increment threshold and the log of the search set size is often modeled as a linear function:

$$\log \Delta\hat{c}(m) = a + b \log m, \tag{10.1}$$

where m is the set size in the visual search display, and $\Delta\hat{c}(m)$ is the predicted contrast increment threshold at set size m. The y-intercept in this case is defined at $\log m = 0$, which corresponds to a set size of 1, and the intercept a is $\log \Delta\hat{c}(1)$, or the contrast increment threshold for set size 1. Then, b is the slope of the contrast increment threshold in relation to $\log m$. The value a is the intercept and b is the slope on a log–log plot. The notation $\log \Delta\hat{c}(m)$ ("log delta-c hat") stands for the predicted log contrast increment threshold. The number of items in a display, m, is manipulated by the experimenter.

Linear models are far simpler in form than most quantitative models but nonetheless illustrate the general principles of fitting models to data.

10.2.2 Model Parameters and Prediction Error

When judging the quantitative adequacy of a model, we examine how well the model fits actual data and whether there are systematic deviations of the model predictions from the data. The model should be concise, parsimonious, and testable and should explain the behavior over a representative range of situations.

The model is fit to the data by choosing values for the slope b and the intercept a. The model specifies that $\log \Delta c$ increases linearly with $\log m$, or the log of the set size. The model does not specify the values of a and b. The values of a and b are chosen to provide a good fit to the data and are called *free parameters*. A good fit between a model and data is intuitively related to how visually close the model predictions are to the data. However, formal model evaluation requires a precise definition of the quality of fit—a fidelity criterion. One common fidelity criterion is the sum of squared errors.

Figure 10.1a shows a linear model and data for set sizes between 1 and 8. Figure 10.1b shows an enlarged graph of the data, which makes clearer the fact that experimental data almost always differ to some degree from the values that the model predicts. Even if the model is correct, deviations between the actual data and the model will arise for reasons such as sampling variability in the data. An "error" refers to the deviation between an observed data value and the model prediction for that data point. Errors are shown as arrows in figure 10.1b. The sum of squared errors, or SSE, is the sum of the squared deviations for all data points:

$$\mathrm{SSE} - \sum_{i=1}^{n} (y_i^{\text{predicted}} - y_i^{\text{observed}})^2 . \tag{10.2}$$

In this equation, an observed data value is y_i^{observed} and a predicted value for that data point is $y_i^{\text{predicted}}$, and n is the number of observed data values. Model parameters are selected to minimize the measure of error, here SSE.

To find a close correspondence between model predictions and observed data points and a low SSE, we need to find a good selection of values for the free parameters, the slope and intercept. A poor selection of parameter values will lead to a poor fit of the model to the data and a high SSE. The variation in error, or SSE, as a function of different parameter values is the *error surface* or *error function* for the model given the data.

Figure 10.2 shows a three-dimensional (3D) projection of the variation in SSE as a function of the slope and intercept parameters. The parameter values that best fit the data are those that minimize the SSE function. Finding the best parameter values corresponds to finding the minimum of the error surface. The best estimates for this data set are $a = -2.53$, and $b = 0.303$. In this case, the error surface is fairly flat near the optimal parameter values, so similar parameter values might be nearly as good in fitting the data. Another fact that is visible in these graphs is the interrelationship between two estimated parameters: A lower intercept can compensate for a too-high slope as seen in the 3D graph.

The best-fitting model and parameters correspond to the minimum of an error surface. From this two-parameter surface it might seem that finding the minimum would be obvious. However, for more complicated models, the full calculation of a multidimensional error surface is prohibitive and difficult to visualize. For the simplest case of a linear model,

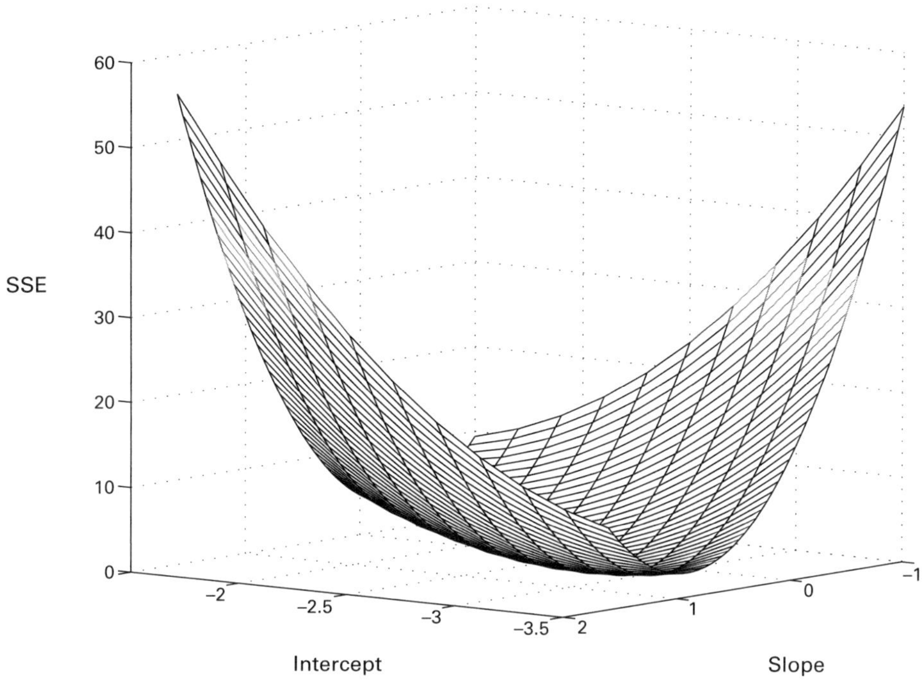

Figure 10.2
The error surface of the linear model of visual search contrast thresholds as a function of the intercept and slope parameters.

there is a direct formula for finding the best parameter values in a linear regression.[3] However, in general the solution for any arbitrary model is not so straightforward, and the minimum must be found by searching the parameter space at many points. There are now many different computational approaches to minimize error functions, including so-called gradient descent methods, grid-search methods, genetic algorithms, and many more. Regardless of the precise search algorithm that is used, the goal is the same—to find the values of parameters that optimize the fit of a particular model to the observed data.

Visual impressions concerning the quality of the fit of the model to the data can be quantified by measures of goodness of fit. For the fidelity criterion of sum of squared errors, there are three forms that provide a quantitative measure of the goodness of fit of the model: the SSE and the related measures root mean squared error (RMSE), and r^2.

The SSE defines an error surface from which we estimate the best fit of a model to a set of data with the best values of the model parameters. The SSE provides one measure of the goodness of fit of a model. Because the best-fitting model has minimized the SSE, other less well fitting models will have higher SSE. Still, it is difficult to have an intuitive interpretation of the SSE as a measure of quality of fit. This is because SSE, which is expressed in units of squared deviations, depends not just on the size of the prediction errors but also on n, the number of data points.

For this reason, it is sometimes useful to convert SSE into the more interpretable measure of RMSE, which is the square root of the average squared error:

$$\mathrm{RMSE} = \sqrt{\mathrm{SSE}/n} = \sqrt{\sum (y_i^{\mathrm{predicted}} - y_i^{\mathrm{observed}})^2 / n}. \tag{10.3}$$

The RMSE is in the same units as the measured performance. For example, it could be in response times or in log percent contrast. You can see from the formula that the RMSE is similar to a standard deviation formula, except that it measures the difference from the predicted value rather than the difference from the mean value. It is quite easy to understand and visualize.

Another very useful measure of quality of fit is r^2, or the proportion of variance accounted for by the model:

$$r^2 = 1.0 - \frac{\sum (y_i^{\mathrm{predicted}} - y_i^{\mathrm{observed}})^2}{\sum (y_i^{\mathrm{observed}} - \overline{y}^{\mathrm{observed}})^2} = \frac{\sum (y_i^{\mathrm{observed}} - \overline{y}^{\mathrm{observed}})^2 - \sum (y_i^{\mathrm{predicted}} - y_i^{\mathrm{observed}})^2}{\sum (y_i^{\mathrm{observed}} - \overline{y}^{\mathrm{observed}})^2}, \tag{10.4}$$

where $\overline{y}^{\mathrm{observed}}$ is the mean of the observed data values. The denominator of this equation is the sum of squared deviations between the observations and the mean value, or $\sum (y_i^{\mathrm{observed}} - \overline{y}^{\mathrm{observed}})^2$. This quantity is the total variability in the observed data. The numerator is the variability that can be accounted for by the model—the prediction errors, or $\sum (y_i^{\mathrm{predicted}} - y_i^{\mathrm{observed}})^2$ subtracted from the total sum of squared errors, or $\sum (y_i^{\mathrm{observed}} - \overline{y}^{\mathrm{observed}})^2 - \sum (y_i^{\mathrm{predicted}} - y_i^{\mathrm{observed}})^2$.

Because r^2 is the proportion of variance in the observed data accounted for by a given model, larger values of r^2 represent better fits of the model, and the r^2 is 1 if the model fits the data perfectly. The r^2 formula compares the variance accounted for by the model to the variance accounted for using the mean as the best prediction. If the mean is a good predictor because there is little variation in the data itself, then even a model with small differences between predicted and observed values may correspond with lower absolute r^2 values, and conversely higher r^2 values are possible if the data have major variations to predict.

It is also good practice to examine the fit of the model to the data visually by graphing or listing the observed data values and the predicted data values. Sometimes the r^2 goodness of fit can seem relatively high even when the model has significant deviations or seems to have a mediocre fit to the data. This is because models that capture even gross trends in the data can capture quite a bit of the variance when compared with using the mean value as the best predictor. And sometimes there may be systematic misfits of the model to the data even when the model fit is truly excellent. Sometimes small but systematic deviations can be useful in developing new variants of the same model.

In the example that we provided of the linear model for the relationship between log threshold increment and log set size in visual search, the minimum SSE is 0.0075, the RMSE is 0.043, and the r^2 is 0.9840. Overall, this is a quite good fit of the linear model to this relationship.

10.2.3 Estimating Variability of Model Parameters

Fitting the best model to the data provides the best estimate of the parameters for a particular data set. However, it is often useful to get a sense of the variability in the estimates of the parameter values themselves. The observed data are only one sample (of a given size) from the "true" or underlying distribution of values that might have been observed for a given experiment. We often use computational resampling to estimate the variability in estimated parameters using some approximation to the underlying distribution from which the samples are drawn.

One way to estimate the variability of model parameters is a bootstrapping procedure.[4,5] Bootstrapping estimates some property of a parameter or other estimator (i.e., the variance) by measuring that parameter or estimator from an approximate distribution. There are several ways in which the approximate distribution may be defined. Usually, bootstrapping constructs a large number of *resampled* data sets of equal size to the original observed data set through sampling *with replacement* from the original data. The idea is that the set of all individual trials of data provides the best estimate of the true distribution of measured values for the experiment. Then, the parameter is estimated from the new resampled data set. This is carried out many times, often 1000 or 10,000 times, depending on the application. From the fits to the set of all resampled data sets, one can estimate the variability of the model parameters, and with a large number of resamples one can estimate the confidence intervals or even the distribution properties of the parameters.

One important assumption of the method is that each observed data value is independent and identically distributed. Sampling with replacement provides new generated data sets in which some original observations may be sampled more than once and others not at all. If the original sample is small, it may actually be preferable to make an approximating assumption, such as assuming that the distribution is Gaussian with the same mean and variance as the measured distribution in that condition, and perform a theoretical resampling from the hypothetical distribution. A number of other corrections or approximations have been proposed in special cases.[4,5]

Consider the linear model example with least-squared error fitting introduced earlier in the chapter. Assume that you measured the aggregate linear relationship between log contrast increment and log set size in 50 observers. Then, you might estimate the variability in the slope and intercept of this function by taking, say, 1000 resamples of size 50 with replacement from the original set of data for your 50 observers, and then compute the linear relationship for each resample (figure 10.3). From these data, you could compute the mean and the standard deviation of the estimated slope and intercept parameters. You

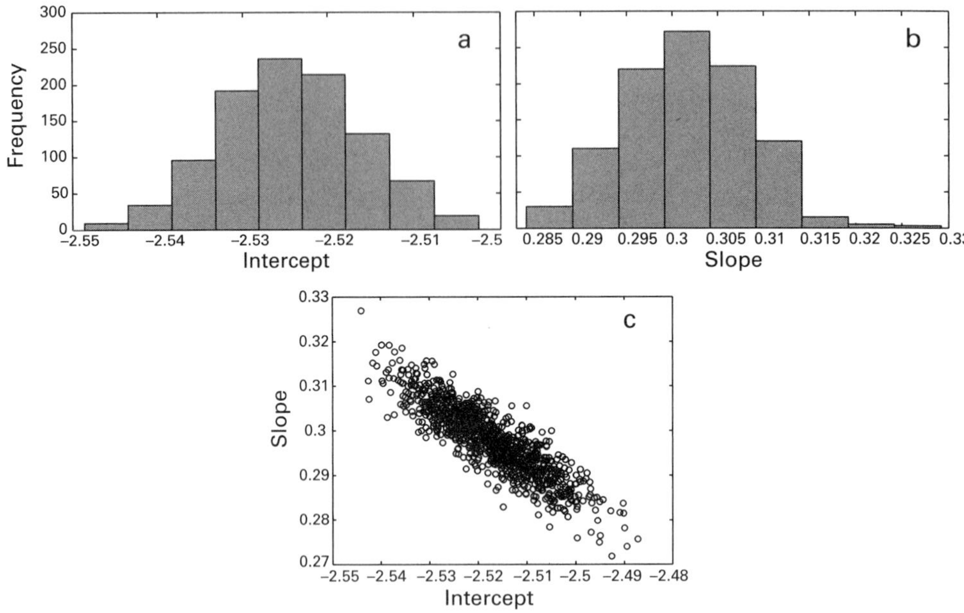

Figure 10.3
Bootstrap evaluation of the parameter variability in the linear model for visual search. (a, b) Distribution of the intercept and slope. (c) Scatterplot of slope versus intercept.

could also look at the distribution of the intercept and of the slope to estimate the confidence intervals that contain the middle 90% of the values for each (figure 10.3a and b). Finally, you could graph the slope and intercept for each resample as a point in two-dimensional (2D) space; over resampled data, this would reveal the correlation between the two estimated parameter values (figure 10.3c).

10.2.4 Model Selection

In science, sometimes we are lucky to find even one reasonably successful model of a complex phenomenon. Even if a model appears to be doing a good job of summarizing data over a range of observations, it is often important to consider substantive variations of that model in order to fine-tune the details. And occasionally, a model that is quite different may be compatible with that same data, and new and different experiments must help us choose between alternative models.

When alternative models are compared, one strategy is to find situations for which the models make very different predictions, which allows a qualitative comparison of the models. For example, the serial search model predicts that reaction time (RT) should increase linearly as set size in the display increases, whereas a parallel feature search model predicts that increasing set size should have (almost) no effect. This difference in the predictions is so straightforward that simply by graphing the results of an experiment, it may be possible to tell which model provides a better account. But another strategy is to evaluate the success of the two models when they are fit quantitatively.[6] Quantitative comparisons are required when two models make fairly similar but slightly different quantitative predictions.

When quantitatively comparing two models, the model with a higher r^2 (or lower SSE or RMSE) is in some loose sense preferable—however the parsimony or simplicity of the model and the statistical reliability of any difference must be considered as well. Usually, we would expect a model with more free parameters to provide a better fit to the data. If the two models are very different in form, however, direct comparisons of r^2 can be misleading.

A simple comparison of two models is most straightforward when one of the models is a nested or special submodel of another. This occurs when the two models are identical except that certain parameters are dropped or held constant in the nested submodel. We refer to the submodel as the reduced model and the original model as the fuller model. Here, the r^2 for the reduced model must by definition be less than or equal to the r^2 for the fuller model. The fuller model can still estimate parameter values that comply with the reduced model, or improve the fit with the additional parameters. In this special case of nested models, it is possible to test whether the improvement in fit provided by the additional free parameters represents a statistically reliable improvement and is unlikely to have occurred by chance. One way to ask this question is to compare the goodness of fit of the fuller model and the reduced model using a nested-F test:

$$F(\mathrm{df}_1, \mathrm{df}_2) = \frac{(r_{\mathrm{full}}^2 - r_{\mathrm{reduced}}^2) / \mathrm{df}_1}{(1 - r_{\mathrm{full}}^2) / \mathrm{df}_2}. \tag{10.5}$$

Here, $\mathrm{df}_1 = k_{\mathrm{full}} - k_{\mathrm{reduced}}$, $\mathrm{df}_2 = N - k_{\mathrm{full}}$, k_{full} is the number of parameters of the full model, k_{reduced} is the number of parameters of the reduced model, and N is the number of predicted data points. An F test compares variances to see if the variance in the numerator statistically exceeds that in the denominator. An examination of the formula shows this is a test of whether the difference in the r^2s (essentially the difference between the SSE) between the fuller and the reduced model *per added free parameter* is greater than would be expected from the error per degree of freedom in the data given the fuller model.

In our visual search example, consider the relationship of contrast increment thresholds as a function of set size. One model of visual search predicts that all items are processed in parallel and any reductions in performance—increases in threshold—are the result of having multiple sources of false alarms. Some authors[7] have argued that in this case, the slope of the log threshold function should be 0.33. We can test this hypothesis for the sample data by using the nested-F test to compare a model in which the slope is free to vary with a submodel in which the slope is set to 0.33. The r^2 of the full model is 0.9840 and of the reduced model is 0.9763. This leads to a value of F of 0.9625, with degrees of freedom 1 and 2. By comparing these to an F distribution, we see that the reduced and full models provide statistically equivalent fits to the data ($p > 0.40$). We conclude that the slope is consistent with the predicted slope of 0.33 of the parallel search model.

10.3 Maximum Likelihood in Modeling

One alternative to the least-squares model estimation and testing uses a maximum likelihood framework. The maximum likelihood approach to fitting and testing models seeks to find a model and a set of estimated parameters to provide a good fit to the data. The fidelity criterion for a good fit is to maximize the likelihood that the data were sampled from the model.

10.3.1 A Simple Example

An experiment that provides a simple example of maximum likelihood methods is the estimation of a single probability—in this case the estimated probability of detection for a simple stimulus. The task is a two-interval forced-choice detection in which the experimenter has selected a Gabor of a particular contrast. The Gabor occurs in only one of the intervals, and the observer indicates which interval. There are 100 trials, and the observer has produced m correct responses, and $n-m$ or $100-m$ incorrect responses. The process by which the observer makes a correct response can be modeled as a binomial process, with a probability p of making a correct response in each trial. Each process, like a coin toss, is assumed to be independent and drawn from the same distribution.

For a given value of p, the likelihood of observing m correct responses out of n trials is[8]

$$\text{Likelihood} = \frac{n!}{m!(n-m)!} p^m (1-p)^{n-m}. \tag{10.6}$$

10.3.2 Maximum Likelihood Estimation of Models

In this example, the goal is to estimate the underlying probability p in the binomial process that best describes the data of m correct responses in n trials. The maximum likelihood procedure searches for the parameter value that maximizes the likelihood of the observed data. Because many search functions, such as `fminsearch` in MATLAB, are written to find the *minimum* of a cost function, we usually minimize –1 times the log of the likelihood, or $L = -\log(\text{likelihood})$. Minimizing L maximizes the likelihood.

The maximum likelihood calculation uses the underlying binomial process to compute the likelihood. For the given example, $m = 75$ and $n = 100$, we can show that the intuitive estimate of $p = 0.75$ minimizes L and so maximizes the likelihood. This is illustrated in figure 10.4, which graphs the likelihood as a function of the value of parameter p on the left and the $L = -\log(\text{likelihood})$ on the right. This simple example is a model

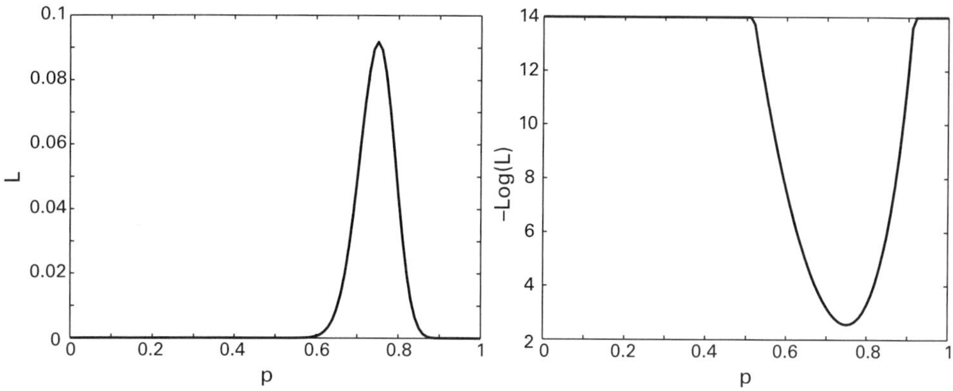

Figure 10.4
Likelihood and –log(likelihood) as functions of parameter p.

with only a single parameter p, corresponding with a one-dimensional minimization surface.

Bootstrap resampling of the trial data can be used to estimate the variability of p in the maximum likelihood procedure. By generating 1000 resampled estimates of p, we can see its distribution and estimate the variability in the estimated parameter. In this example, the standard deviation of the estimated probability of detection is about 0.044, which is very close to the formula value for the binomial at 0.043.

10.3.3 Model selection: G^2, Akaike Information Criterion, and Bayesian Information Criterion

Now suppose that the Gabor detection experiment is performed over a period of several days, for thousands of trials. To evaluate the perceptual learning over this practice period, we compare the performance in the first 100 trials and the last 100 trials. The question is whether the performance at the end of practice differs from that at the beginning of practice—or whether the two observed proportions differ significantly from one another. A more general, fuller model allows for a change in the underlying probability p_i with practice. A reduced model assumes that the two values of p are the same, $p_{\text{early}} = p_{\text{late}} = p$, consistent with no effect of learning on performance.

The likelihood formula from the binomial that combines evidence from both conditions is

$$\text{Likelihood} = \prod \frac{n_i!}{m_i!(n_i - m_i)!} p_i^{m_i} (1 - p_i)^{n_i - m_i} . \tag{10.7}$$

The subscript i refers to the conditions, here early or late in practice, and the product Π refers to the product over the two experimental conditions. This product becomes a sum in the $L = -\log(\text{likelihood})$ formulation that is minimized by a search function in optimizing the selection of parameters.

To answer whether the training has an effect on performance, we can compare nested models. That is, one model is a submodel of the other in which the parameters p_i have been equated. In the case of the maximum likelihood calculation, the nested test is a likelihood ratio test G^2:

$$G^2(\text{df}) = 2.0 \times \log\left(\frac{\text{max likelihood}_{\text{full}}}{\text{max likelihood}_{\text{reduced}}}\right), \tag{10.8}$$

with $\text{df} = k_{\text{full}} - k_{\text{reduced}}$. This G^2 distribution of the nested likelihood ratio test is well approximated by a χ^2 ("chi-squared") distribution with the same degrees of freedom. In the programs and the subsequent examples, we simply refer to this as a χ^2 test.

In our example, the observed values are $m_{\text{early}} = 75$ and $m_{\text{late}} = 90$. The nested χ^2 test has a value of 8.007 and a probability of 0.0017, so we conclude that the fuller model provides a statistically better fit to the joint data and that perceptual learning has occurred in the experiment.

So far, we have discussed model selection only between models in a nested structure. In general, it may be useful to compare models that are not related to one another in this way.

One approach for the comparison of non-nested models penalizes models with more free parameters. Models with more free parameters should, all things being equal, be better able to account for the data. For this reason, a direct comparison of the likelihood of the data given the respective models should discount or adjust for the number of free parameters in assessing the quality of fit. Several different measures have been developed to assist in the selection of models in non-nested model comparisons.

The Akaike Information Criterion (AIC) is a measure of relative goodness of fit that penalizes the model for the number of free parameters.[9] The AIC is defined as

$$\text{AIC}_{\text{model}} = 2k_{\text{model}} - 2\log(\text{max likelihood}_{\text{model}}), \tag{10.9}$$

where k_{model} is the number of parameters in that model. The AIC does not provide an intuitively interpretable measure of the goodness of fit of the model. It produces a number that characterizes the model for a given data set. It is designed to be useful specifically in comparing several models for the same data. We compute the AIC value for the different models that are being compared and choose the model with the minimum AIC value. In the current example, the AIC for the model with two independent probabilities, one for early and one for late in training, is 12.83, and the AIC for the single-probability model is 18.84. This model choice is consistent with the results of the nested model test in which we concluded that the two-parameter model was needed to fit the data.

Another commonly used metric for comparing the fits of different models is the Bayesian Information Criterion, or BIC.[10] The BIC is related to the AIC. It is also based on the likelihood function, and like the AIC, the purpose is to compare models to one another. The BIC is defined as

$$BIC_{\text{model}} = k_{\text{model}} \log(n) - 2\log(\text{max likelihood}_{\text{model}}). \tag{10.10}$$

The parameter k_{model} is the number of parameters of that model, and n is the number of data points. In this simple example, the number of data points, n, is 2, for the two observed proportions. (It is not the sample size of either proportion. These sample sizes are implicitly incorporated in the likelihood values.)

The BIC imposes a heavier penalty on the number of model parameters than the AIC. Here, too, the point is to choose the best model as that with the lower BIC. In our example, the BIC of the full model is 10.21 and of the reduced model is 17.53.

10.3.4 Model Fitting and Selection with Bayesian Methods

New advances in methods for comparing and selecting models have been developed using Bayesian logic that incorporates prior probabilities as well as the likelihood of observing the data given different models. New Bayesian technologies are being developed as alternative methods for model selection as well. The χ^2 test for nested model selection and the more general AIC and BIC indexes that use penalties for free parameters to allow comparison between quite different models are replaced by a comparison of (the ratio between) the (marginal) likelihoods of the data given one or another model—the so-called Bayes factor. The basic logic of Bayesian inference and incorporation of Bayesian logic into adaptive methods is treated in chapter 11. These newer methods are of increasing interest in the field today.[11–16]

10.4 Examples of Fitting Models to Data

In this section, we provide a series of examples of model estimation, model selection, and estimation of parameter variability. The examples are designed to cover the methods described more abstractly in the previous section. Sample code for the model estimation and fitting is provided for several common problems and experiments in psychophysics. At the same time, we also show some examples of model comparison and selection.

10.4.1 Modeling Receiver Operating Characteristic Functions

The receiver operating characteristic (ROC) function measures the relationship between hits and false alarms in a condition. The purpose of the ROC analysis in signal-detection theory is to estimate the variances of signal-present and signal-absent internal response distributions, or signal + noise and noise distributions, to provide an accurate estimate of the mean difference between the two distributions, and to estimate criteria corresponding with different levels of confidence.[17–19] The theory and purpose underlying the ROC curve and analysis was considered in section 8.4.3 of chapter 8, where we described the theory but not the details of how data are analyzed.

The data for an ROC analysis often come from a rating experiment in which, for example, an observer indicates both a Yes/No (Present/Absent) response but also the confidence associated with that response. In the example of section 8.3.4, this led to six ratings, Absent-3, Absent-2, Absent-1, Present-1, Present-2, and Present-3. If we run an experiment with 150 signal-present trials (S + N) and 150 signal-absent trials (N), the observations are the frequencies of each of the six response categories in the two test conditions. It is these frequencies that are the input to an ROC analysis. These frequencies are first converted into cumulative frequencies of hits and false alarms, and then to the corresponding proportions of hits and false alarms. In this case, the cumulative frequencies of hits and false alarms start with Absent-3, then consider Absent-3 or Absent-2, and then Absent-3 or Absent-2 or Absent-1 (all the "no" responses), and so forth.

Figure 8.3 of chapter 8 shows examples of hit and false alarm probabilities graphed as an ROC curve on probability–probability axes and on z–z axes. The next sections consider in more detail how to go about analyzing such data.

In a rating procedure, the $z_i(\text{Hit})$ and $z_i(\text{FA})$ in category i are functions of the distance between the signal and noise distributions d, criterion c_i, the standard deviation of the noise distribution σ_{N}, and the standard deviation of the signal plus noise distribution $\sigma_{\text{S+N}}$:

$$\begin{aligned} z_i(\text{Hit}) &= G^{-1}[1 - G(c_i, d, \sigma_{\text{S+N}})] \\ z_i(\text{FA}) &= G^{-1}[1 - G(c_i, 0, \sigma_{\text{N}})]. \end{aligned} \tag{10.11}$$

As usual in signal-detection theory analyses, G is the cumulative Gaussian—the proportion of the distribution up to criterion c_i, and G^{-1} is the inverse cumulative Gaussian that returns a z-score from a proportion for the standard normal distribution. Because $\sigma_{\text{S+N}}$ and σ_{N} can only be known up to their ratio, we typically set σ_{N} to 1.0 and set the mean of the noise distribution to 0. These parameters set the scale. For a rating experiment with n categories, the model has $n + 1$ parameters: d, $\sigma_{\text{S+N}}$, and criteria c_i ($i = 1$, to $n - 1$). The next task in the analysis is to fit this functional form to observed data.

In this example, we choose to fit the observed z-scores for hits and false alarms, z_i (Hits) and z_i (FA), for the n different rating categories. We use a least squares procedure to fit the model in equation 10.11 to the empirical ROC function. The cost function is defined as

$$L = \sum_{i=1}^{n-1} [z_i^{\text{observed}}(\text{Hit}) - z_i^{\text{predicted}}(\text{Hit})]^2 + \sum_{i=1}^{n-1} [z_i^{\text{observed}}(\text{FA}) - z_i^{\text{predicted}}(\text{FA})]^2 . \tag{10.12}$$

Display 10.1a–c shows a program `fittingROC1.m` that performs this ROC analysis. This program takes as input the frequencies of each confidence rating for the target-

Display 10.1a

```
%%% Program fittingROC1.m
function ROC = fittingROC1(Srating, Nrating)

freqHits(1)  = Srating(6);
freqFA(1)  = Nrating(6);
for i = 2:6
    freqHits(i)  = freqHits(i-1)  + Srating(7-i);
    freqFA(i)  = freqFA(i-1)  + Nrating(7-i);
end

pHits = freqHits(1:5)/freqHits(6);
pFA = freqFA(1:5)/freqFA(6);
zHits = norminv(pHits);
zFA = norminv(pFA);
data = [zHits; zFA];

guess = [1 1.5 -2 -1 0 1 2];
options = optimset('fminsearch');
ROC = fminsearch('ROC1costfunc', guess, options, data);

% compute r2
zHits_observed =data(1,:);
zFA_observed = data(2, :);
d = ROC(1);
sigma = ROC(2);
criteria = ROC(3:7);
zHits_predicted = norminv(1-normcdf(criteria, d, sigma));
zFA_predicted  = norminv(1- normcdf(criteria, 0, 1));
mean_zHits = mean(zHits_observed);
mean_zPA = mean(zFA_observed);
r2 = 1- 0.5*sum((zHits_observed - zHits_predicted).^2)...
      /sum((zHits_observed-mean_zHits).^2) - ...
      0.5*sum((zFA_observed -zFA_predicted).^2) ...
      /sum((zFA_observed-mean_zPA).^2)
```

present and the target-absent trials, `Srating` and `Nrating`. The program computes cumulative frequencies and cumulative proportions and the corresponding z-scores for hits and false alarms associated with each rating criterion. A MATLAB search function `fminsearch` is called to find parameter values that minimize a cost function, `ROC-1costfunc`. The cost function computes the predicted values corresponding to each data point for given parameter values and returns the fidelity score L. The `fminsearch` algorithm uses the simplex search (modified gradient descent) method[20] to optimize the selection of parameters to find the best-fitting model. The program returns the best estimates of d', σ_{S+N}, and the $n-1$ criteria $c_{i=1:n-1}$. It also returns r^2 as a summary of the goodness of fit (see equation 10.13).

Display 10.1b

```
%%% Program ROC1costfunc.m
function L = ROC1costfunc(guess, data)

zHits_observed =data(1,:);
zFA_observed = data(2, :);
d = guess(1);
sigma = guess(2);
criteria = guess(3:7);
zHits_predicted =norminv(1-normcdf(criteria, d, sigma));
zFA_predicted   = norminv(1- normcdf(criteria, 0, 1));
L = sum((zHits_observed - zHits_predicted).^2 + ...
      (zFA_observed - zFA_predicted).^2);
```

Display 10.1c

```
>> Srating = [ 7     15     27     35     20     46 ];
>> Nrating = [ 5     36     68     38      4      1 ];
>> ROC = fittingROC1(Srating, Nrating)
r2 = 0.9999
ROC -   1.494   1.994   2.484   1.831   0.5791 -0.6112 -1.842
```

In ROC curves, both the x- and y-values (false alarms and hits) are estimated from data. For this reason, the fidelity criterion in the cost function that is minimized sums the deviations from predicted values in both x and y. This is different from the principles of fitting a simple regression model, where it is assumed that one value is given or manipulated and the second, observed value is being predicted.

The degrees of freedom for the fit of the model to the data is df $= 2 \times (n - 1) - (n + 1) = n - 3$. In MATLAB, `fminsearch.m` is typically used to minimize the cost function.

The goodness of fit is described with r^2:

$$r^2 = 1 - \frac{0.5\sum_{i=1}^{n-1}[z_i^{\text{observed}}(\text{Hit}) - z_i^{\text{predicted}}(\text{Hit})]^2}{\sum_{i=1}^{n-1}\{z_i^{\text{observed}}(\text{Hit}) - \text{mean}[z^{\text{observed}}(\text{Hit})]\}^2} - \frac{0.5\sum_{i=1}^{n-1}[z_i^{\text{observed}}(\text{FA}) - z_i^{\text{predicted}}(\text{FA})]^2}{\sum_{i=1}^{n-1}\{z_i^{\text{observed}}(\text{FA}) - \text{mean}[z^{\text{observed}}(\text{FA})]\}^2}. \tag{10.13}$$

One important question is whether the simplifying assumption of equal variance fits the ROC curve; that is, whether $\sigma_{S+N} = \sigma_N = 1$. If the variances are equal, the signal-detection model is a reduced model of the fuller model where σ_{S+N} is a free parameter used to fit the data. Display 10.2a–c shows the function program `fittingROC2.m` that

Display 10.2a

```
%%% Program fittingROC2.m
function ROC = fittingROC2(Srating, Nrating)

freqHits(1)  = Srating(6);
freqFA(1)  = Nrating(6);
for i = 2:6
    freqHits(i)  = freqHits(i-1)  + Srating(7-i);
    freqFA(i)  = freqFA(i-1)  + Nrating(7-i);
end

pHits  = freqHits(1:5)/freqHits(6);
pFA  = freqFA(1:5)/freqFA(6);
zHits  = norminv(pHits);
zFA  = norminv(pFA);
data  = [zHits; zFA];

guess  = [1 1.5 -2 -1 0 1];
options  = optimset('fminsearch');
ROC  = fminsearch('ROC2costfunc', guess, options, data);

% compute r2
zHits_observed =data(1,:);
zFA_observed = data(2, :);
d = ROC(1);
criteria  = ROC(2:6);
zHits_predicted =norminv(1-normcdf(criteria, d, 1));
zFA_predicted   = norminv(1- normcdf(criteria, 0, 1));
mean_zHits  = mean(zHits_observed);
mean_zPA  = mean(zFA_observed);
r2  = 1- 0.5*sum((zHits_observed -zHits_predicted).^2)...
     /sum((zHits_observed-mean_zHits).^2) - ...
     0.5*sum((zFA_observed - zFA_predicted).^2) ...
     /sum((zFA_observed-mean_zPA).^2)
```

Display 10.2b

```
%%% Program ROC2costfunc.m
function L = ROC2costfunc(guess, data)

zHits_observed =data(1,:);
zFA_observed = data(2, :);
d = guess(1);
criteria  = guess(2:6);
zHits_predicted =norminv(1-normcdf(criteria, d, 1));
zFA_predicted   = norminv(1- normcdf(criteria, 0, 1));
L = sum((zHits_observed - zHits_predicted).^2 + ...
      (zFA_observed - zFA_predicted).^2);
```

Display 10.2c

```
>> Srating = [ 7     15     27     35     20     46 ];
>> Nrating = [ 5     36     68     38      4      1 ];
>> ROC=fittingROC2(Srating, Nrating)
r2 =  0.8489
ROC =   1.102   2.043   1.546   0.6131   -0.0032   -1.209
```

reduces the model by eliminating σ_{S+N} as a free parameter. As in `fittingROC1.m`, it takes frequency data as input and returns the best estimates of d', the $n-1$ criteria $c_{i=1:n-1}$, and the r^2.

One way to ask the question about equal variances is to compare the goodness of fit of the fuller model and the reduced model using a nested-F test:

$$F(\text{df}_1, \text{df}_2) = \frac{(r_{\text{full}}^2 - r_{\text{reduced}}^2)/\text{df}_1}{(1 - r_{\text{full}}^2)/\text{df}_2}, \tag{10.14}$$

where $\text{df}_1 = k_{\text{full}} - k_{\text{reduced}}$, $\text{df}_2 = N - k_{\text{full}}$, and N is the number of predicted data points. The fuller model has parameters d, σ_{S+N}, and $n-1$ criteria c_i, or $n+1$ parameters, while the reduced model has only d, and $n-1$ criteria c_i, or n parameters. In our example, k_{full} = 7 parameters, k_{reduced}= 6 parameters, df_1 = 1 (the number of parameters difference between the models), and df_2 = 5. In figure 8.3 and table 8.3 (section 8.3.4 of chapter 8), we simulated data from a situation in which $\sigma_{S+N} > \sigma_N = 1$. Fitting the fuller model to the data using `fittingROC1.m` led to an r^2 of 0.9999. Fitting the reduced model to the data using `fittingROC2.m` led to an r^2 of 0.8489. When we apply the F test for nested models to these data, we find $F(1,5)$ = 7550.0, $p < 5\times10^{-9}$. Consistent with the true situation—that is, the simulated situation—we reject the hypothesis that $\sigma_{S+N} = \sigma_N = 1$.

In this example, we chose to use bootstrap methods to generate variance estimates for the parameters. The bootstrap method resamples individual experimental trials and responses in a bootstrap procedure. Display 10.3a and b shows a program `ROCstd.m` that takes the original rating frequency data as input. It creates many new ROC data sets that are then subjected to tabulation and fitting. It computes the standard deviation of each parameter of the best-fitting model from the resampled data sets to provide an estimate of parameter variability. The program outputs the variance estimates for d', σ_{S+N}, and the $n-1$ criteria $c_{i=1:n-1}$ (display 10.3b).

10.4.2 Modeling Psychometric Functions

The psychometric function relates human performance accuracy to some parameter of variation of the stimulus. Typically, psychometric functions relate detection or discrimination accuracy to contrast, stimulus intensity, angular difference, and so forth. In section

Display 10.3a

```
%%% Program ROCstd.m
function ROCstd = ROCstd(Srating0, Nrating0)

% reconstruct single trial data (sorted by rating)
sorted_S_trials = [1*ones(Srating0(1),1)
    2*ones(Srating0(2),1)
    3*ones(Srating0(3),1)
    4*ones(Srating0(4),1)
    5*ones(Srating0(5),1)
    6*ones(Srating0(6),1)];

sorted_N_trials = [1*ones(Nrating0(1),1)
    2*ones(Nrating0(2),1)
    3*ones(Nrating0(3),1)
    4*ones(Nrating0(4),1)
    5*ones(Nrating0(5),1)
    6*ones(Nrating0(6),1)];

ROC0 = [];
for i=1:1000
    resampled_S_trials=[];
    resampled_N_trials=[];
    for j=1:150    % sample 300 trials from the sorted
                   % single trial data with replacement
        index1 = floor(150*rand(1))+1;
        index2 = floor(150*rand(1))+1;
        resampled_S_trials = [resampled_S_trials; …
            sorted_S_trials(index1)];
        resampled_N_trials = [resampled_N_trials; …
            sorted_N_trials(index1)];
    end

    Srating = [];
    Nrating = [];
    for j=1:6      % compute Srating and Nrating from the
                   % resampled data
        Srating = [Srating; …
              length(find(resampled_S_trials ==j))];
        Nrating = [Nrating; …
              length(find(resampled_N_trials ==j))];
    end
    Srating(Srating==0)=1;
    Nrating(Nrating==0)=1;

    ROC = fittingROC1(Srating, Nrating);
    ROC0=[ROC0; ROC];
end

ROCstd=std(ROC0);
```

Display 10.3b

```
>> Srating0 = [ 7      15      27      35      20      46 ];
>> Nrating0 = [ 5      36      68      38       4       1 ];
>> ROCstd=ROCstd(Srating0, Nrating0)
ROCstd = 0.1901  0.1485  0.0426  0.1646  0.2830  0.1155  0.2244
```

2.2 of chapter 2, we showed an example that used the method of constant stimuli and two-interval forced-choice to measure a 7-point contrast psychometric function for detecting a sine-wave grating. The purpose of measuring a psychometric function is to find the best estimate of the threshold and to measure the slope of the psychometric function. Together, the threshold, the slope, and therefore the shape provide important information about the visual system.

The data for psychometric functions often come from the method of constant stimuli that tests a predefined set of six to nine stimulus values spanning from near chance to maximal performance. Section 4.3.2 of chapter 4 provided an example of measuring a contrast psychometric function with 7 contrasts and 100 trials per contrast in a two-interval forced-choice (2IFC) task, leading to a tabulation of the number of correct and incorrect trials for each contrast (see table 4.1 of chapter 4).

Psychometric functions are often modeled by the Weibull function[21]:

$$P(c) = \xi + (1 - \xi - \lambda)(1 - e^{-(c/\tau)^{\eta}}). \tag{10.15}$$

In this formula, c is the signal contrast, τ is the threshold, η is the slope of the psychometric function, ξ represents the chance performance level, and λ represents the observer's "lapse rate," which sets a maximum accuracy below 100%. In a 2IFC task, ξ is set to 0.5; λ is often set to a value less than 0.05.[21]

The theoretical basis and implications of fitting Weibull functions to psychometric data have been debated.[22,23] A number of other similarly shaped functions have been used as well.[24] A more theoretically driven approach is to specify the functional form of the psychometric functions based on the internal responses of the observer. For example, we can use the PTM observer model to specify d' as a function of stimulus contrast and external noise and use SDT to relate d' to percent correct, as in section 9.2 in chapter 9.[25,26]

In this section, we consider model estimation using the Weibull, which is one of the most popular descriptive psychometric functions. The principles of fitting any of the other functional forms would be analogous. The maximum likelihood procedure[8] is typically used to fit psychometric functions. As described earlier, this procedure maximizes the likelihood of the observed data given the probabilities predicted by the model or function with a particular set of parameter values. We assume that the observed data in each condition are drawn from a binomial distribution. So, the likelihood is a function of the predicted probability for a given contrast condition, p_i, the total number of trials, n_i, and the number of correct trials, m_i, in each experimental condition i:

$$\text{Likelihood} = \prod \frac{n_i!}{m_i!(n_i - m_i)!} p_i^{m_i} (1 - p_i)^{n_i - m_i} . \tag{10.16}$$

The product Π runs across all the experimental conditions i for an observer.

We maximize the likelihood by minimizing the cost function, $L = -\log(\text{likelihood})$. Minimizing L maximizes the likelihood. Taking the log changes multiplication to summation. To avoid values of infinity, model probabilities 0 and 1 are adjusted up or down slightly by $1/2n_i$.[19]

Display 10.4a–c provides a sample program `Weibull.m`. The program reads in a matrix with a column for each condition, each with three rows with the contrast (or other variable) and the numbers of correct and incorrect trials. It sets the chance performance level to 0.5 and the lapse rate to 0.05 and provides starting values or guesses for the threshold τ (tau) and slope η (eta) parameters of the psychometric function. It returns the best estimates of τ and η, estimates of thresholds at selected levels of performance (see later), and the resulting likelihood fidelity criterion for the best fit, `maxloglikelihood`.

Display 10.4a

```
%%% Program Weibull.m
function [tau, eta, thresholds, maxloglikelihood] = Weibull(PF)

data = PF;          % PF is the measured psychometric function.
                    % First row: contrasts;
                    % Second row: # of correct trials;
                    % Third row: # of incorrect trials.

guess = [0.02 3];   % tau eta
options = optimset('fminsearch');
[psy, minus_maxloglikelihood] = fminsearch('Weibullcostfunc', ...
      guess, options, data);
maxloglikelihood = - minus_maxloglikelihood;

%compute contrast thresholds at three performance levels
p = [0.65 0.75 0.85];
tau = psy(1);
eta = psy(2);
xi = 0.50;
lamda = 0.02;
thresholds   = ((-1)*log((1-lamda-p)./(1-lamda- ...
               xi))).^(1/eta)*tau;
```

Display 10.4b

```
%%% Program Weibullcostfunc.m
function L = Weibullcostfunc(guess, data)

tau = guess(1);
eta = guess(2);
Nconditions = size(data, 2); % # of stimulus conditions
xi = 0.50;
lamda = 0.02;
L=0;

for i=1:Nconditions
    p = xi + (1 - xi - lamda)*(1-exp(-(data(1,i)/tau).^eta));
        % Eq. 10.15
    if (p < 1/2/(data(2, i)+data(3, i)))
        % putting lower and upper boundaries on p
        p = 1/2/(data(2, i)+data(3, i));
    elseif (p> 1-1/2/(data(2, i)+data(3, i)))
        p = 1- 1/2/(data(2, i)+data(3, i));
    end
    L = L - (data(2, i)*log(p) + data(3, i)*log(1-p));
end
```

Display 10.4c

```
>> PF = [.0025   .0040   .0063   .0100   .0159   .0252   .0400
           52      53      59      74      95      97      98
           48      47      41      26       5       3      2];
>> [tau, eta, thresholds, maxLoglikelihood]=Weibull(PF)
tau = 0.0112, eta = 2.84, thresholds = .0079  .0101  .0123
maxLoglikelihood = -306.7750
```

Threshold signal contrast at a given performance level can be calculated from the best-fitting model to the psychometric function (figure 10.5). The model for the psychometric function is used to interpolate a contrast level associated with some probability correct, p. For a Weibull function, the threshold at performance level p can be solved by rearranging terms:

$$c_p = \left(-\log\frac{1-\lambda-p}{1-\lambda-\xi}\right)^{\frac{1}{\eta}}\tau. \tag{10.17}$$

This formula can be used to estimate a 75% threshold, or an 85% threshold, or any other target value. For the data in section 4.3.2 of chapter 4, the best-fitting Weibull has set

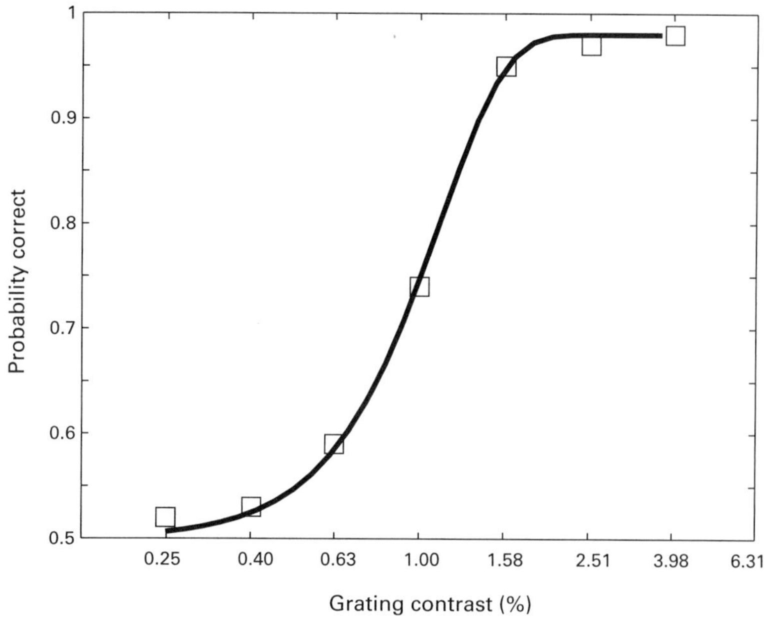

Figure 10.5
A psychometric function with a best-fitting Weibull model. Square symbols are measured data.

parameters of $\lambda = 0.02$, $\xi = 0.50$, and estimated parameters of $\tau = 0.0112$, and $\eta = 2.84$, and the 65%, 75%, and 85% thresholds were estimated as 0.79%, 1.01%, and 1.23% contrasts.

Bootstrap techniques can be used to estimate the variability of estimated parameters of the Weibull function, but also can be used to estimate the variability of other derived values, such as the three threshold estimates.[27,28] Resampling can be performed literally by resampling trials. However, in many cases we simply assume that the number of correct responses has a binomial distribution with a single event probability $P_i = p_i^{\text{observed}}$ for each experimental condition. Display 10.5a and b shows a program `Weibullstd.m` that uses this latter form of bootstrap to estimate the standard deviations of the parameters of the Weibull and of the estimated thresholds. Finally, the Weibull function is fit to each of the resampled psychometric functions, and then the standard deviations of the parameters and the thresholds at multiple performance levels are calculated.

In many experiments, psychometric functions are collected for different conditions and compared. Sometimes we may want to compare psychometric functions from two or more conditions or wish to constrain psychometric functions from several different conditions

Display 10.5a

```
%%% Program Weibullstd.m
function [tau_std, eta_std, eta_iqr, thresholds_std] = …
      Weibullstd(PF0)

contrasts = PF0(1,:);   % PF0 is the measured psychometric
                        % function. First row: contrasts;
                        % Second row: # of correct trials
                        % Third row: # of incorrect trials.
totalTrials = PF0(2, :) + PF0(3, :);
p = PF0(2, :) ./ totalTrials;
Nconditions = size(PF0, 2);

tau0 = []; eta0 = []; thresholds0=[];
for i = 1 : 2000
    resampled_Ncorrect = [ ];
    for j = 1 : Nconditions
        resampled_Ncorrect = [resampled_Ncorrect …
            binornd(totalTrials(j), p(j))];
    end
    resmapled_Nincorrect = totalTrials - …
            resampled_Ncorrect;

    resampled_PF = [contrasts; resampled_Ncorrect;…
            resmapled_Nincorrect];

   [tau, eta, thresholds, maxLoglikelihood] = …
            Weibull(resampled_PF);
   if eta > 5
       resampled_PF
   end

   tau0 = [tau0; tau];
   eta0 = [eta0; eta];
   thresholds0 = [thresholds0; thresholds];
end

tau_std = std(tau0);
eta_std = std(eta0);
eta_iqr = iqr(eta0);
thresholds_std = std(thresholds0);
```

Display 10.5b

```
>> PF0 = [.0025   .0040   .0063   .0100   .0159   .0252   .0400
            52      53      59      74      95      97      98
            48      47      41      26       5       3      2];
>> [tau_std, eta_std, eta_iqr, thresholds_std] =
Weibullstd(PF0)
tau_std = 7.520e-04, eta_std = 4.0349, eta_iqr = 0.8940
thresholds_std = 0.00091  0.00074  0.00089
```

to share one or more parameters (e.g., refs. 29 and 30). We can compare different (nested) fits of two or more psychometric functions using χ^2statistics:

$$\chi^2(\mathrm{df}) = 2.0 \times \log\left(\frac{\text{max likelihood}_{\text{full}}}{\text{max likelihood}_{\text{reduced}}}\right), \tag{10.18}$$

where df $= k_{\text{full}} - k_{\text{reduced}}$.

For example, although psychometric functions at different external noise levels have different thresholds (location parameters), they may have the same slope parameters, and these shared parameter values may be important for testing different observer models.

Display 10.6a–c shows the sample program `Weibullcomp.m` that tests for equivalent slope parameters by nested model comparisons of full and reduced models. The program takes the data from three psychometric functions collected at different levels of external noise. It fits a fuller model with parameters τ and η for each of the three external noise conditions, for a total of six free parameters. It also fits a reduced model that shares the slope parameter η, which has a total of four free parameters. For these sample data, the $\chi^2 = 0.5683$, and $p = 0.2473$, suggesting that the reduced model with equivalent slope gives a good account of the data from the three levels of external noise.

In this example, the likelihood is directly computed from the underlying binomial process generating the data, or at least the binomial provides a good approximation to the actual process. In other cases, where likelihood is not directly available, we often use either least squares or weighted least squares. Likelihood can be computed for continuous measurements if one is able to specify predicted distributions. More often, weighted least squares criteria would be selected as an approximation to a maximum likelihood model solution.[31]

10.4.3 Modeling Contrast Sensitivity Functions

The contrast sensitivity function is a characterization of the sensitivity of the visual system to different spatial frequencies. It is a major building block for models of the visual observer and important in many practical applications.[24,32,33] The contrast sensitivity function, $S(f)$, represents sensitivity (1/threshold) as a function of grating frequency of the

Display 10.6a

```
%%% Program Weibullcomp.m
function [chi2, p] = Weibullcomp(PFS)
        % PFS contains the measured psychometric
        % functions. Every three rows of PFS is a
        % psychometric function

NPF = size(PFS, 1)/3;

% full model fit
fullmodel_maxloglikelihood=0;
for i = 1: NPF
        data =  PFS( (i-1)*3+1 : (i*3), :);
        guess = [0.02 3];    % tau eta
        options = optimset('fminsearch');
        [psy, minus_maxloglikelihood] = …
              fminsearch('Weibullcostfunc', guess, …
              options, data);
        maxloglikelihood = - minus_maxloglikelihood;

        fullmodel maxloglikelihood = …
              fullmodel_maxloglikelihood …
              + maxloglikelihood;
end

% reduced model: same slope across all psychometric
% functions
data = PFS;
guess = [0.02*ones(1, NPF) 3];
options = optimset('fminsearch');
[psy, minus_maxloglikelihood] = fminsearch('Weibullcomp_
costfunc', guess, options, data);
reducedmodel_maxloglikelihood = - minus_maxloglikelihood;

chi2 = 2*(fullmodel_maxloglikelihood - …
       reducedmodel_maxloglikelihood);
p = chi2cdf(chi2, NPF-1);
```

Display 10.6b

```
%%% Program Weibullcomp_costfunc.m
function L = Weibullcomp_costfunc(guess, data)

NPF = size(data, 1)/3;
Nconditions = size(data, 2);    % # of stimulus conditions
xi = 0.50;
lamda = 0.02;
L=0;
for n=1:NPF
    tau = guess(n);
    eta = guess(NPF+1);
    data1 =  data( (n-1)*3+1 : (n*3), :);

    for i=1:Nconditions
        p = xi + (1 - xi - lamda) * (1-exp(- ...
            (data1(1,i)/tau).^eta));  % Eq. 10.15
        if (p < 1/2/(data1(2, i)+data1(3, i)))
            % putting lower and upper boundaries on p
            p = 1/2/(data1(2, i)+data1(3, i));
        elseif (p> 1-1/2/(data1(2, i)+data1(3, i)))
            p = 1- 1/2/(data1(2, i)+data1(3, i));
        end
        L = L - (data1(2, i)*log(p) + ...
            data1(3, i)*log(1-p));
    end
end
```

Display 10.6c

```
>> PFS = [.0025  .0040  .0063  .0100  .0159  .0252  .0400
             58     51     55     59     81     97     97
             42     49     45     41     19      3      3
          .0025  .0040  .0063  .0100  .0159  .0252  .0400
             54     58     56     75     93     99     97
             46     42     44     25      7      1      3
          .0100  .0159  .0252  .0400  .0635  .1008  .1600
             50     51     51     69     90     97     96
             50     49     49     31     10      3      4  ]
>> [chi2, p] = Weibullcomp(PFS)
chi2 = 0.5683, p = 0.2473
```

sine wave. In sections 2.3 of chapter 2 and 4.3.3 of chapter 4, we showed an example of a contrast sensitivity function estimated from an experiment.

We choose to test nine spatial frequencies from coarse to fine. For each spatial frequency, we chose seven contrast levels. On each trial, the computer selects a particular sine pattern and a particular contrast level to show to the observer. For each combination of frequency and contrast, we might measure 100 trials—so this experiment requires 9 times the number of trials as the example experiment in section 2.2 of chapter 2. The contrast sensitivity function was estimated by graphing the inverse of the 75% correct threshold as a function of spatial frequency. The contrast sensitivity function is usually analyzed by estimating a best-fitting summary function to these data.

Based on a comprehensive review of nine functional forms yielding similar shapes that have been used to describe empirical contrast sensitivity functions (CSFs) in normal vision, Watson and Ahumada[24] concluded that all provide a roughly equivalent description of the CSFs in a shared Modelfest data set (see section 9.5 of chapter 9).

Qualitatively, the CSF has a shape similar to a band-pass filter. That is, it has a peak at a mid-level spatial frequency and attenuated sensitivity at both high and low spatial frequencies relative to the peak. Here, we have chosen to describe the CSF as the *truncated log-parabola* (see figure 10.3). This model has four parameters: (1) the peak gain (sensitivity) $\gamma_{\max}$; (2) the peak spatial frequency $f_{\max}$; (3) the bandwidth β, and (4) δ, the truncation level at low spatial frequencies.

The *log-parabola* form of the contrast sensitivity function, $\log_{10}[S'(f)]$ is

$$\log_{10}[S'(f)] = \log_{10}(\gamma_{\max}) - \log_{10}(2)\left(\frac{\log_{10}(f) - \log_{10}(f_{max})}{\log_{10}(2\beta)/2}\right)^2 \tag{10.19}$$

The truncated log-parabola is defined as:

$$\log_{10}[S(f)] = \begin{cases} \log_{10}(\gamma_{\max}) - \log_{10}(2)\left(\dfrac{\log_{10}(f) - \log_{10}(f_{max})}{\log_{10}(2\beta)/2}\right)^2, & \textit{if } f \ge f_{\max} \\ \log_{10}(\gamma_{\max}) - \delta, \quad \textit{if } f < f_{\max} \ \& \ S(f) < \gamma_{\max} - \delta \end{cases} \tag{10.20}$$

The advantage of this four-parameter description, especially in relation to other three-parameter models,[34] is that it accounts for asymmetry between low and high frequencies in the shape of the CSF. It typically provides a good fit to both individual CSFs and aggregate CSFs for groups of observers. Compared to other five-parameter models, the four-parameter truncated log-parabola is more compact and provides essentially equivalent shape predictions.

A least squares method is typically used to fit the functional form to the CSF. The cost function is defined as:

$$L = \sum_f \{\log_{10}[S^{\text{observed}}(f)] - \log_{10}[S^{\text{predicted}}(f)]\}^2 . \tag{10.21}$$

As usual, the quality of the fit is gauged by:

$$r^2 = 1 - \frac{\sum_f \{\log_{10}[S^{\text{observed}}(f)] - \log_{10}[S^{\text{predicted}}(f)]\}^2}{\sum_f \{\log_{10}[S^{\text{observed}}(f)] - \text{mean}(\log_{10}[S^{\text{predicted}}(f)])\}^2}. \quad (10.22)$$

Display 10.7a–c shows the program for analyzing the CSF, `fittingCSF.m`. The data from a measured CSF experiment are entered into the program as a matrix with two rows. The first row contains the spatial frequency of each test condition. The second row is the sensitivity for those conditions. The program has selected initial starting guesses for the parameters [`Gmax Fmax beta delta`], or [$\gamma_{\max}$ $f_{\max}$ β δ]. It returns the best estimates of these parameters and the r^2. With an r^2 of 0.9884, the fit to the data is excellent (figure 10.6).

Here, we illustrate a somewhat different resampling procedure for estimating the standard deviations of the model parameters—a *Monte Carlo resampling* method. In resampling, each sensitivity that is measured, $S(f)^{\text{Observed}}$, is assumed to be drawn from a normal distribution that is estimated from the psychometric function for each spatial frequency by the method in section 10.2.3 of chapter 10. We resample new hypothetical $S(f)^{\text{Observed}}$ for each spatial frequency, and refit the truncated log-parabola model to each set of resampled data. Another possible approach to resampling, not shown here, is to use a bootstrap procedure on the raw data to construct new psychometric functions and new CSFs based on the bootstrapped data sets.

The number of resampled data sets in this example is 1000. Display 10.8a and b shows the sample program `CSFstd.m` using this procedure to estimate the variances of the fitted parameters. This program takes as input the CSF data and the estimated standard deviation

Display 10.7a

```
%%% Program fittingCSF.m
function [Gmax, Fmax, beta, delta, r2] = fittingCSF(CSF)
data = CSF;     % CSF is the measured contrast sensitivity
                % function. First row: spatial frequency;
                % Second row: sensitivity.
guess = [200 1.0 8 0.5];     % Gmax, Fmax, beta, delta
options = optimset('fminsearch');
[guess, L] = fminsearch('CSFcostfunc', guess, options, ...
        data);
meanS = mean(log10(CSF(2, :)));
r2 = 1 - L/sum((log10(data(2, :)) - meanS).^2);

Gmax = guess(1);
Fmax = guess(2);
beta = guess(3);
delta = guess(4);
```

Display 10.7b

```
%%% Program CSFcostfunc.m
function L = CSFcostfunc(guess, data)

Gmax = guess(1);
Fmax = guess(2);
beta = guess(3);
delta = guess(4);
f = data(1, :);
S_observed = log10(data(2, :));
Nfrequencies = size(data, 2);
        % # of spatial frequency conditions

S_predicted = [];
for i=1:Nfrequencies
    S = log10(Gmax) - log10(2) * ((log10(f(i)) - …
        log10(Fmax))/(log10(2*beta)/2))^2;
    if (f(i) >= Fmax)
        S = S;
    elseif (f(i) < Fmax & S < log10(Gmax) - delta)
        S = log10(Gmax) - delta;
    end
    S_predicted = [ S_predicted S ];
end

L = sum( (S_observed - S_predicted).^2 );
```

Display 10.7c

```
>> CSF = …
     [.125   .25   .50    1.00   2.00   4.00   8.00  16.0  32.0
      70.1  144.7  202.2  204.6  189.1  183.5  98.7  46.0  21.1];
>> [Gmax, Fmax, beta, delta, r2] = fittingCSF(CSF)
Gmax = 222.8, Fmax = 1.168, beta = 17.58, delta = 0.4777
r2 = 0.9884
```

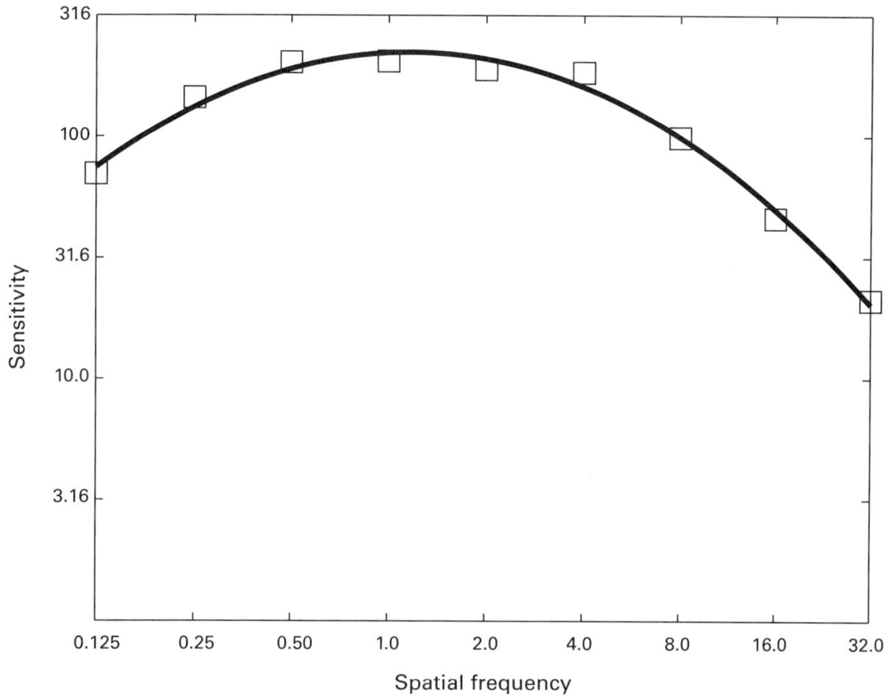

Figure 10.6
A contrast sensitivity function with the truncated log-parabola model fit. The symbols represent measured data, or 1/threshold from psychometric functions for tests of different spatial frequencies.

of sensitivity at each spatial frequency. It outputs the standard deviations of the estimated parameters of the CSF model fit. For the data in section 4.3.3 of chapter 4, the best-fitting CSF model parameters are γ_{max} = 222.8 ± 7.3, f_{max} = 1.17 ± 0.04, β = 17.6 ± 1.2, and δ = 0.48 ± 0.08.

One practical application of the comparison of several CSFs is in the study of amblyopia, where the CSF of the amblyopic eye is compared to the CSF of the "fellow" or non-amblyopic eye.[35,36] This comparison is sometimes carried out using analysis of variance on the two sets of empirical CSFs (figure 10.7). An alternative (and we believe better) approach is to test whether the same values of the four parameters of the truncated log-parabola model can be used jointly to fit the two CSFs. Display 10.9a–c shows a program that compares a model in which each CSF has four parameters, for a total of eight parameters, to a situation where the two CSFs share the same four-parameter description. If the two fits are statistically equivalent, we infer that the two CSFs are equivalent, otherwise we infer that they are different. For this purpose, with least-squares model fitting, the nested-F test (equation 10.14) yields $F(4,10) = 58.62$, $p < 6.63 \times 10^{-7}$, indicating that the two CSFs are statistically different.

Display 10.8a

```
%%% Program CSFstd.m
function [Gmax_std, Fmax_std, beta_std, delta_std] = ...
      CSFstd(CSF0)

Nfrequencies = size(CSF0, 2)
Gmax0 = []; Fmax0 = []; beta0=[]; delta0=[];
for i = 1 : 1000
  CSF=[CSF0(1,:); CSF0(2, :) + CSF0(3, :).*randn(1, ...
       Nfrequencies)];

  [Gmax, Fmax, beta, delta, r2] = fittingCSF(CSF);
  Gmax0 = [Gmax0; Gmax];
  Fmax0 = [Fmax0; Fmax];
  beta0 = [beta0; beta];
  delta0 = [delta0; delta];
end

Gmax_std = std(Gmax0);
Fmax_std = std(Fmax0);
beta_std = std(beta0);
delta_std = std(delta0);
```

Display 10.8b

```
>> CSF0= ...
     [.125  .25  .50   1.00  2.00  4.00  8.00 16.0 32.0
      70.1 144.7 202.2 204.6 189.1 183.5 98.7 46.0 21.1
       4.9  10.1  11.6  12.2  12.9  12.5  7.6  3.2  1.4];
>> [Gmax_std, Fmax_std, beta_std, delta_std] = CSFstd(CSF0)
Gmax_std = 7.3, Fmax_std = 0.043, beta_std = 1.2, delta_std =
0.085
```

Using the same logic, we can evaluate other intermediate models in which some subset of the four parameters is shared between the two empirical CSF measurements. This statistical comparison is valid as long as the *reduced* model is nested within the *fuller* model—in which the reduced model equates certain parameters. For example, an eight-parameter model could be compared to a seven-parameter model in which the center frequency parameter f_{max} is constrained to be identical for the two empirical CSFs.

10.4.4 Modeling Threshold versus External Noise Contrast Functions

Threshold versus external noise contrast (TvC) functions play a central role in defining observer models in psychophysics.[25] A TvC function relates the contrast threshold in a

Display 10.9a

```
%%% Program CSFcomp.m
function [F, df1, df2,  p] = CSFcomp(CSFs)
                          % CSFs contains the measured CSFs

NCSF = size(CSFs, 1)/2; % compute # of CSFs for comparison

% full model fit
fullmodel_L=0;
for i = 1: NCSF
        data =  CSFs( (i-1)*2+1 : (i*2), :);
        guess = [200 1.0 8 0.5]; % Gmax, Fmax, beta, delta
        options = optimset('fminsearch');
        [guess, L] = fminsearch('CSFcostfunc', guess, ...
                     options, data);
        fullmodel_L = fullmodel_L + L;
end

%reduced model: same parameters for all CSFs
data = CSFs;
guess = [200 1.0 8 0.5];    % Gmax, Fmax, beta, delta
options = optimset('fminsearch');
[guess, L] = fminsearch('CSFcomp_costfunc', guess, ...
        options, data);
reducedmodel_L = L;

meanLogS = mean(mean(log10(CSFs(2:2:(2*NCSF), :))));
SS = sum(sum((log10(CSFs(2:2:(2*NCSF), :)) - ...
        meanLogS).^2));
r2_full = 1 - fullmodel_L/SS;
r2_reduced = 1 - reducedmodel_L/SS;
df1 = 4*(NCSF-1);
df2 = size(CSFs, 2)*NCSF - 4*NCSF;
F = (r2_full - r2_reduced)/df1 / ((1 - r2_full)/df2);
p = 1 - fcdf(F, df1, df2);
```

Display 10.9b

```
%%% Program CSFcomp_costfunc.m
function L = CSFcomp_costfunc(guess, data)

NCSF = size(data, 1)/2; % compute # of CSFs for comparison
Nfrequencies = size(data, 2); % # of stimulus conditions
Gmax = guess(1);
Fmax = guess(2);
beta = guess(3);
delta = guess(4);

L=0;
for n = 1 : NCSF
    f = data((n-1)*2+1, :);
    S_observed = log10(data(2*n, :));
    S_predicted = [];
    for i=1:Nfrequencies
        S = log10(Gmax) - log10(2) * ((log10(f(i)) - …
            log10(Fmax))/(log10(2*beta)/2))^2;
        if (f(i) >= Fmax)
            S = S;
        elseif (f(i) < Fmax & S < log10(Gmax) - delta)
            S = log10(Gmax) - delta;
        end
        S_predicted = [ S_predicted S ];
    end
    L = L+sum( (S_observed - S_predicted).^2 );
end
```

Display 10.9c

```
>> CSFs = …
     [.125   .25   .50    1.00   2.00   4.00   8.00 16.0 32.0
      70.1 144.7 202.2 204.6 189.1  183.5 98.7 46.0 21.1
      .125   .25   .50    1.00   2.00   4.00   8.00 16.0 32.0
      60.1 110.7 160.2 150.6 110.1   45.5 19.0 10.0  2.1];
>>  [F, df1, df2, p] = CSFcomp(CSFs)
F = 58.62, df1 = 4,  df2 = 10, p = 6.6347e-07
```

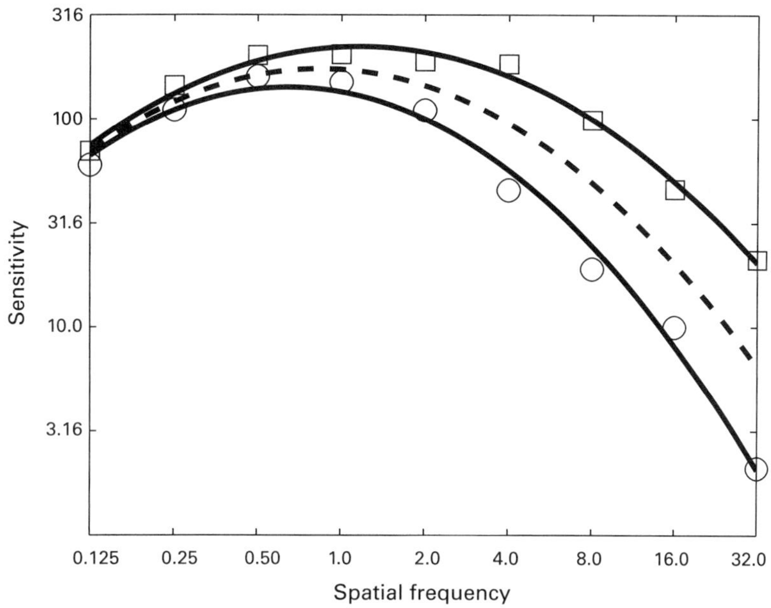

Figure 10.7
Comparing two contrast sensitivity functions. The solid curves represent predictions of the full model; the dotted curve represents the predictions of the reduced model assuming no difference between the two functions. The two data sets are shown as square and circle symbols, respectively.

detection or discrimination task to the amount of external noise added to the signal stimulus. In section 9.3.1 of chapter 9, we showed empirical TvC functions estimated from 8-point contrast psychometric functions measured at each of eight external noise levels. The TvC is traditionally graphed on log–log axes. Taking the log contrast threshold helps to equalize the variability of the thresholds across a large range of measured values.

Here, we use a function from the perceptual template model (PTM) to describe the TvC data. Fitting a set of TvCs at different criterion threshold levels for a single condition requires four parameters: the internal multiplicative and additive equivalent noises, N_{m} and N_{a}, the gain response of the template to a signal stimulus, β, and the nonlinearity parameter, γ. The PTM model and its parameters were extensively discussed in section 9.3.1 of chapter 9. The PTM model predictions are fit to empirical TvC functions through least squares methods, using the predicted log contrast as:

$$\log(c_{\tau}) = \frac{1}{2\gamma}\left\{2\log(d') + \log[(1 + N_{\text{mul}}^2)\sigma_{\text{ext}}^{2\gamma} + \sigma_{\text{add}}^2] - \log\left(1 - \frac{N_{\text{mul}}^2 d'^2}{2}\right)\right\} - \log(\beta). \tag{10.23}$$

See the derivation of this functional form in section 9.2 of chapter 9.

Display 10.10a–c shows the program `fittingTvC.m` that fits this equation to observed TvC data (figure 10.8). A matrix of values containing the levels of external noise and the

Display 10.10a

```
%%% Program fittingTvC.m
function [beta, gamma, Nm, Sa, r2] = fittingTvC(TvC)

data = TvC;
NperformanceLevels = size(TvC, 1) -1;
NnoiseLevels = size(TvC, 2) - 1;
guess = [4 2 0.1 0.01];    % beta, gamma, Nm, Sa
options = optimset('fminsearch');
[guess, L] = fminsearch('TvCcostfunc', guess, options, ...
     data);
beta = guess(1);
gamma = guess(2);
Nm = guess(3);
Sa = guess(4);
Log_observedThresholds = ...
    log(TvC(2:(NperformanceLevels+1), 2:(NnoiseLevels+1)));
meanT = mean(mean(Log_observedThresholds));
r2 = 1 - L/sum(sum((Log_observedThresholds - meanT).^2));
```

Display 10.10b

```
%%% Program TvCcostfunc.m
function L = TvCcostfunc(guess, data)

beta = guess(1);
gamma = guess(2);
Nm = guess(3);
Sa = guess(4);
NperformanceLevels = size(data, 1) -1;
NnoiseLevels = size(data, 2) - 1;
Next = data(1, 2:(NnoiseLevels+1));
dprime = norminv(data(2:(NperformanceLevels+1), 1)) - ...
      norminv( 1- data(2:(NperformanceLevels+1), 1));

L=0;
for i = 1:NperformanceLevels
    Log_observedThresholds = ...
      log(data(i+1, 2:(NnoiseLevels+1)));
    d = dprime(i);
    Log_predictedThresholds = 1 / (2*gamma) * (2*log(d') ...
      + log((1+Nm^2) * Next.^(2*gamma) + Sa^2) - ...
      log(1-Nm.^2*d.^2/2)) - log(beta);
    L = L+sum((Log_observedThresholds -...
      Log_predictedThresholds).^2);
end
```

Display 10.10c

```
>>   TvC=...
   [0    0.000 0.030 0.045 0.067 0.100 0.149 0.223 0.330
    0.65 0.051 0.050 0.052 0.052 0.058 0.085 0.125 0.168
    0.75 0.067 0.069 0.066 0.068 0.078 0.111 0.163 0.239
    0.85 0.084 0.090 0.082 0.085 0.102 0.140 0.206 0.322]
>> [beta, gamma, Nm, Sa, r2] = fittingTvC(TvC)
beta = 1.635, gamma = 1.96, Nm = 0.135,   Sa = 0.0095
r2 =   0.9946
```

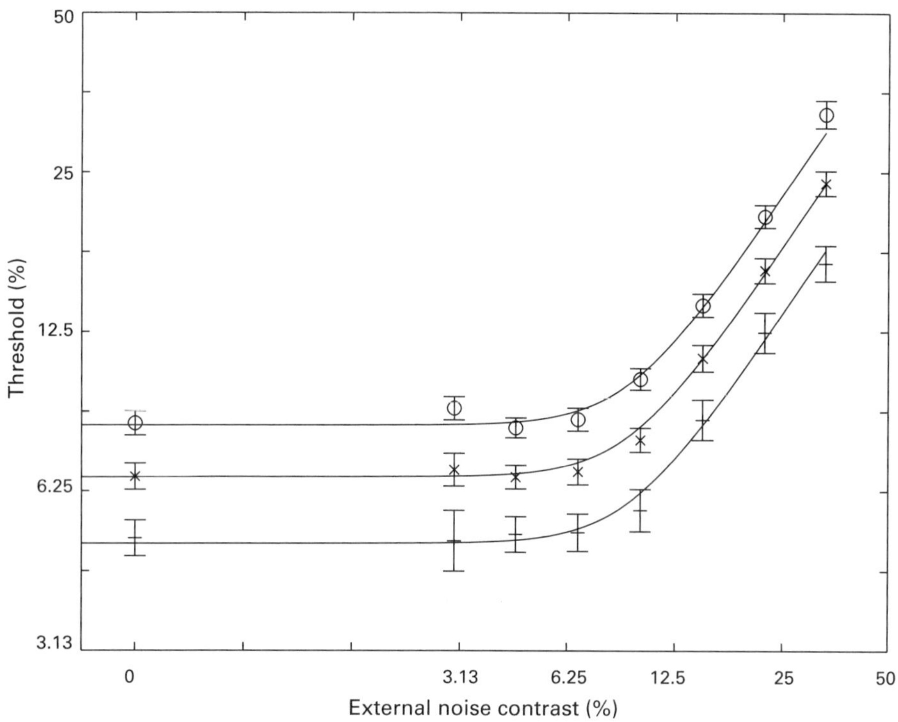

Figure 10.8
TvC functions showing thresholds as a function of external noise contrast. Symbols show the data with error bars estimated by bootstrap methods; smooth curves represent the best-fitting PTM observer model.

thresholds at three performance levels are input to the program. In this sample experiment, there are eight levels of external noise. The input data includes the 8 × 4 data matrix in which the first row is the external noise contrast and the next three rows are the estimated thresholds at 0.65, 0.75, and 0.85 proportion correct accuracies. The input matrix is however 9 × 4, which includes a left-most column that contains the proportion correct targets for the thresholds and a placeholder 0 in the external noise row. The program assumes starting values for and then estimates and returns the best values for the PTM parameters [`beta gamma Nm Sa`] along with the r^2. The fit of this PTM model to the triple-TvC data is excellent, with $r^2 = 0.9946$.

Fitting contrast thresholds in the log form, $\log(c_\tau)$, roughly equates the variability of the different contrasts. This approximates a weighted least squares solution and so approximates maximum likelihood estimation.[37]

As in the previous examples, it is possible to use bootstrapped resampling to estimate the variability of the four parameters of the PTM model. Display 10.11a and b shows the program `TvCstd.m` that carries out this resampling analysis. The input to the program is

Display 10.11a

```
%%% Program TvCstd.m
function [beta_std, gamma_std, Nm_std, Sa_std] = TvCstd(TvC0)

data = TvC0;
NperformanceLevels = (size(TvC0, 1) -1)/2;
NnoiseLevels = size(TvC0, 2) - 1;
beta0 = []; gamma0 = []; Nm0=[]; Sa0=[];

for n = 1 : 1000
    TvC = TvC0(1, :);
   for i = 1 : NperformanceLevels
       TvC = [TvC; TvC0((i+1), 1) ...
         TvC0((i+1), 2:(NnoiseLevels+1)) + ...
         TvC0((NperformanceLevels+i+1), 2:(NnoiseLevels+1)) ...
         .*randn(1, NnoiseLevels)];
   end
   [beta, gamma, Nm, Sa, r2] = fittingTvC(TvC);

   beta0 = [beta0; beta];
   gamma0 = [gamma0; gamma];
   Nm0 = [Nm0; Nm];
   Sa0 = [Sa0; Sa];
end
beta_std = std(beta0);
gamma_std = std(gamma0);
Nm_std = std(Nm0);
Sa_std = std(Sa0);
```

Display 10.11b

```
>>  TvC0 = …
 [ 0    0.000 0.030 0.045 0.067 0.100 0.149 0.223 0.330
  0.65 0.051 0.050 0.052 0.052 0.058 0.085 0.125 0.168
  0.75 0.067 0.069 0.066 0.068 0.078 0.111 0.163 0.239
  0.85 0.084 0.090 0.082 0.085 0.102 0.140 0.206 0.322
  0.65   .0044 .0066 .0040 .0042 .0053 .0074 .0109 .0130
  0.75   .0043 .0049 .0034 .0039 .0041 .0064 .0089 .0129
  0.85   .0044 .0045 .0035 .0043 .0048 .0071 .0102 .0191]
>>  [beta_std, gamma_std, Nm_std, Sa_std] = TvCstd(TvC0)
beta_std = 0.066, gamma_std = 0.42, Nm_std = 0.17,
Sa_std = 0.0044
```

the contrast thresholds in the different external noise conditions, the three levels of contrast thresholds, and the corresponding standard deviations of the threshold estimations, input in a format analogous to fitting the TvC. The output includes the PTM parameter estimates and their standard deviations.

Nested-*F* tests can compare the TvCs from two or more conditions using fuller or more reduced models. The nested model structure can be used to identify which of several parameters change when the state of the observer changes. For example, nested model testing has been used to identify mechanisms of attention and perceptual learning.[29,38–40]

Consider an example from an experiment on perceptual learning[39] in which observers were trained for 10 sessions, nearly 14,400 trials, on an orientation discrimination task in the lower right visual field, while performing a letter/number identification task in a rapid stream of characters in the center of the display. The orientation task was tested in eight different levels of external noise from 0 contrast to 0.33 contrast, all intermixed. Contrast thresholds were estimated using staircase procedures (see chapter 11) at two accuracy levels, 70.7% and 79.4%. In figure 10.9, we show the TvC functions from every two practice sessions to illustrate the effects of perceptual learning. The TvCs at both performance levels were fit simultaneously with PTM equations.

Display 10.12a–c provides the program `TvCcomp.m` to analyze and test the PTM models of TvC data at two accuracy criteria. The input consists of five sets of two TvC functions. The data are input as in the last example as a matrix in which the first column is a placeholder 0 in the first row and indicates the target accuracy level in the other rows, and the first row lists the external noise contrasts, and subsequent rows have the thresholds for the lower and higher accuracy staircases for each of 5 days. The search algorithm `fminsearch` is used to find best estimates of the parameters of several different PTM model variations.

Perceptual learning may improve performance through several mechanisms expressed in the PTM model. Learning may reduce the impact of external noise through improved filtering, it may reduce internal additive noise, or it may reduce internal multiplicative

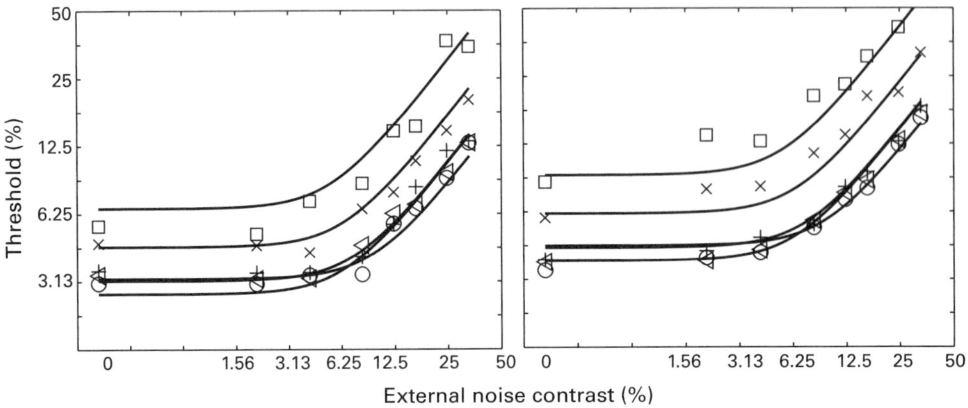

Figure 10.9
TvC functions in a perceptual learning experiment, fit with a PTM model. The left panel and right panels show thresholds at two different accuracy criteria, and each curve represents 2 days of data collection. Practice improved external noise filtering and reduced internal additive noise (data from Dosher and Lu[39]).

noise—each associated with a different kind of change in the TvC functions following practice. Reducing the impact of external noise improves contrast thresholds in high external noise. Reducing additive internal noise improves contrast thresholds in low external noise. Reducing internal multiplicative noise improves contrast thresholds in both low and high noise, but by a different amount (in log contrast threshold) depending upon the performance threshold level. Perceptual learning is expressed in the PTM equations for the TvC, then, by added parameters (<1) that multiplicatively reduce the impact of these different noises after training: $A_f^2 N_{ext}^2$ replaces N_{ext}^2 after training, $A_a^2 N_a^2$ replaces A_a^2 after training, and $A_m^2 N_m^2$ replaces N_m^2 after training—if performance has improved in all these ways.

Nested model tests allowed us to infer that perceptual learning reduced the impact of external noise ($A_f^2 N_{ext}^2$) and reduced additive internal noise ($A_a^2 N_a^2$), but did not alter internal multiplicative noise ($A_m^2 N_m^2 = N_m^2$). The two factors of improvement A_f^2 and A_a^2 were both needed to account for the data, but they were not necessarily the same. This is typical of observed changes due to perceptual learning, where one or the other, but usually both factors are needed to explain improved TvC performance—but the factors may be affected partially independently.[41]

This example is meant to illustrate how the framework of a quantitative model, combined with model comparison, can assist the researcher in summarizing and quantifying differences between conditions.

10.4.5 Modeling Speed–Accuracy Trade-off Functions

Speed–accuracy trade-off (SAT) functions measure the time course of visual discrimination—or some other process—by measuring the relationship of performance accuracy to

Display 10.12a

```
%%% Program TvCcomp.m
function [beta, gamma, Nm, Sa, Aa, Af, r2_full, r2_reduced, …
       F, df1, df2,  p] = TvCcomp(TvCs, Nstates)
         % TvCs contains the measured TvCs

NperformanceLevels = (size(TvCs, 1) - 1)/Nstates;
NnoiseLevels = size(TvCs, 2) - 1;

% full model fit
data = [Nstates zeros(1, NnoiseLevels); TvCs];
        % include info on # of states and # of performance
        % levels
guess = [4 2 0.1 0.01 0.50*ones(1, Nstates - 1) …
    0.50*ones(1, Nstates - 1) ]; % beta gamma Nm Sa Aa's Af's
options = optimset('fminsearch');
[guess, fullmodel_L] = fminsearch('TvCcomp_costfunc', guess,…
    options, data);

beta = guess(1);
gamma = guess(2);
Nm = guess(3);
Sa = guess(4);
Aa = [1 guess(5 : (5 + Nstates - 2))];
Af = [1 guess((5+Nstates-1) : (5 + 2* Nstates -3))];

%reduced model: same parameters for all TvCs
data=TvCs;
guess = [4 2 0.1 0.01];    % Gmax, Fmax, beta, delta
options = optimset('fminsearch');
[guess, reducedmodel_L] = fminsearch('TvCcostfunc', guess,…
      options, data);
meanLogS = mean(mean(log(TvCs(2:(NperformanceLevels*Nstates),…
      2:NnoiseLevels))));
SS = sum(sum((log(TvCs(2:(NperformanceLevels*Nstates), …
      2:NnoiseLevels)) - meanLogS).^2));
r2_full = 1 - fullmodel_L/SS;
r2_reduced = 1 - reducedmodel_L/SS;
df1 = 2*(Nstates-1);
df2 = Nstates*NperformanceLevels*NnoiseLevels - 4 - …
      2*(Nstates-1);
F = (r2_full - r2_reduced)/df1 / ((1 - r2_full)/df2);
p = 1 - fcdf(F, df1, df2);
```

Display 10.12b

```
%%% Program TvCcomp_costfunc.m
function L = TvCcomp_costfunc(guess, data)

Nstates = data(1, 1);
NperformanceLevels = (size(data, 1) - 2)/Nstates;
beta = guess(1);
gamma = guess(2);
Nm = guess(3);
Sa = guess(4);
Aa = [1 guess(5 : (5 + Nstates - 2))];
Af = [1 guess((5+Nstates-1) : (5 + 2* Nstates -3))];
NperformanceLevels = (size(data, 1) - 2)/Nstates;
NnoiseLevels = size(data, 2) - 1;
Next = data(2, 2:(NnoiseLevels+1));
L=0;

for i = 1 : Nstates
    for j = 1 : NperformanceLevels
        d  = ...
          norminv(data((i-1)*NperformanceLevels+2+j, 1)) - ...
          norminv( 1- data((i-1)*NperformanceLevels+2+j, 1));
        Log_observedThresholds = ...
          log(data((i-1)*NperformanceLevels+2+j, ...
          2:(NnoiseLevels+1)));
        Log_predictedThresholds = 1 / (2*gamma) * (2*log(d')...
          + log((1+Nm^2) * (Af(i)*Next).^(2*gamma) + ...
          (Aa(i)*Sa)^2) - log(1-Nm.^2*d.^2/2)) - log(beta);
        L = L+sum ((Log_observedThresholds - ...
          Log_predictedThresholds).^2);
    end
end
```

processing time (see section 6.3.1 of chapter 6). Usually, visual discrimination improves with added processing time for short times, and then levels off when more processing time ceases to yield new information. Originally used in the measurement of memory retrieval for verbal materials,[42,43] the time course of processing has also been measured for several cases of visual discriminations.[44–48]

As in the CSF example earlier, the rapid early improvements and asymptotic form of the SAT function can be fit by a number of functional forms.[42,49–53] Here we consider the exponential approach to a limit as one of the best descriptive equations for the functional form of the SAT function:

$$d'(t) = \lambda\left(1 - e^{-\beta(t-\delta)}\right), \tag{10.24}$$

Display 10.12c

```
>>   TvCs = ...
[0    0.0000 0.0205 0.0410 0.0820 0.1230 0.1640 0.2460 0.3281
.707 0.0553 0.0510 0.0712 0.0854 0.1458 0.1530 0.3669 0.3455
.794 0.0847 0.1363 0.1281 0.2041 0.2300 0.3059 0.4111 0.5137
.707 0.0462 0.0453 0.0421 0.0658 0.0779 0.1073 0.1461 0.2007
.794 0.0588 0.0786 0.0808 0.1134 0.1369 0.2038 0.2119 0.3176
.707 0.0349 0.0342 0.0343 0.0403 0.0561 0.0823 0.1190 0.1312
.794 0.0382 0.0418 0.0479 0.0548 0.0804 0.0937 0.1279 0.1840
.707 0.0334 0.0321 0.0323 0.0454 0.0628 0.0689 0.0953 0.1292
.794 0.0374 0.0378 0.0423 0.0572 0.0758 0.0862 0.1299 0.1707
.707 0.0307 0.0305 0.0333 0.0336 0.0565 0.0661 0.0899 0.1289
.794 0.0343 0.0389 0.0411 0.0530 0.0706 0.0794 0.1238 0.1631]
>>   [beta, gamma, Nm, Sa, Aa, Af, r2_full, r2_reduced, F, ...
      df1, df2,  p]   = TvCcomp(TvCs, 5)
beta = 0.9050, gamma = 1.3138, Nm = 0.1759, Sa = 0.0225
Aa = 1.0000     0.5971     0.3774     0.3161     0.3876
Af = 1.0000     0.5662     0.3508     0.3493     0.2793
r2_full = 0.9490, r2_reduced = 0.4896
F = 76.5759, df1 = 8, df2 = 68, p = 0
```

where $d'(t)$ is the d' discriminability at processing time t, λ is the highest or asymptotic d' level achieved at long processing time, β is the exponential rate of accumulation of information over time, and δ is the minimum processing time, or intercept, below which d' is zero. The exponential is a good approximation to the form of many measured SAT functions in different domains.[43–46,50,51,54,55]

In section 6.3.1 of chapter 6, we described and programmed an experiment in which an observer is asked to search for a C among O's and an O among C's, and we tested display sizes of 4 and 12 items appearing around an annulus at the 4.12° of visual angle away from fixation (after ref. 45). The observer decides whether the display contains a C in one case and an O in the other case.

The search display is briefly presented (100 ms), but processing the visual information and making a decision requires further processing time. The time course of visual search is measured in a speed–accuracy paradigm by interrupting the decision at different times and having the observer respond as quickly as possible with whatever information is available. Here we used seven interruption points ranging from 0.05 s to 2 s following onset of the search display. For the earliest point of interruption, performance accuracy is at chance, while asymptotic performance at very long processing time is about 2.10 and 1.10 d' for display set size 4 and 12, respectively. Performance is strongly affected by the display size, with better performance for only four elements, and reduced performance for 12 elements (figure 10.10). We use a least squares method to fit the SAT function and

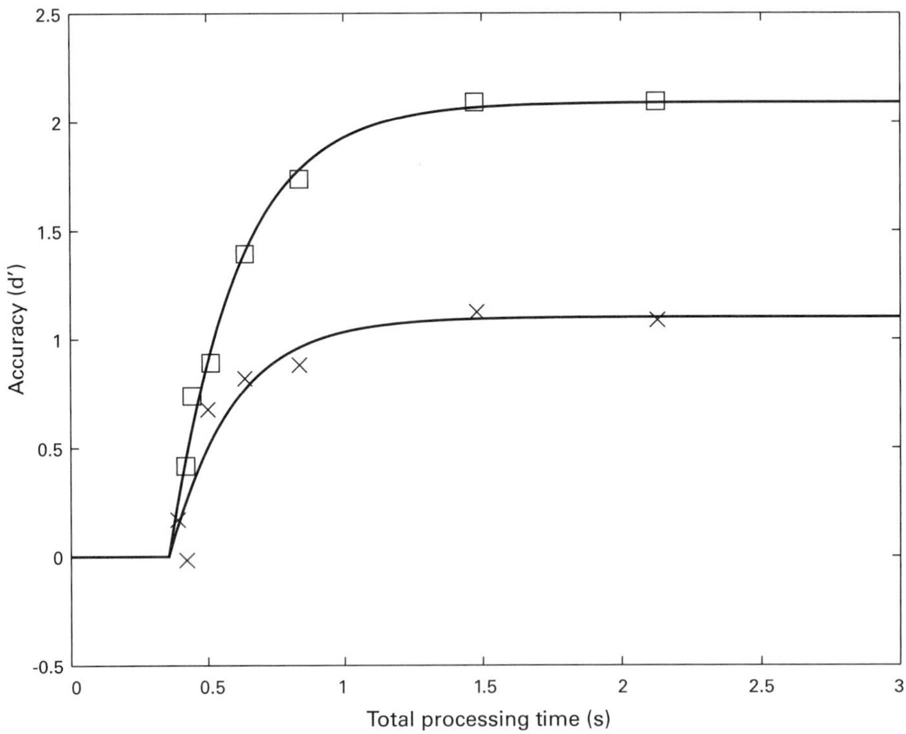

Figure 10.10
SAT functions with the best fitting exponential model. Symbols are measured data at seven different interruption times for display sizes of 4 (top curve, square) and 12 (bottom curve, x). The best-fitting exponential functions differ in asymptotic accuracy, but neither rate nor intercept (data after Dosher, Han, and Lu[45]).

summarize the quality of fit with r^2. Nested model tests indicate that the two display sizes differ in asymptotic levels, and only modestly in rate, but not in intercept.

Display 10.13a–d shows a program used for model comparison and fitting of SAT data, `fittingSATs.m`. The input to the program is a matrix of the SAT data in two conditions, for set size 4 and 12, for 7 points of interruption. The first two rows have the total processing time (lag plus latency) and d' of the set size 4 at the seven interruption times, while the third and fourth rows have total processing time and d' for set size 12. The output is the best-fitting parameters of the reduced model, the r^2 of the fuller model, the r^2 of a reduced model, and the value of the F test for the nested models with degrees of freedom, and the resulting p-value for the test.

As in other examples, bootstrap resampling could be used to estimate the variability in the parameters of the fits of the descriptive model to the data. In this example, we have shown how to fit the descriptive exponential approach to a limit as the functional form of

Display 10.13a

```
%%% Program fittingSATs.m
function [lamda, beta, delta, r2_full, r2_reduced, F, …
      df1, df2,  p] = fittingSATs(SATs)
        % SATs contains the measured SATs

NSATs = size(SATs, 1)/2;
Ndelays = size(SATs, 2);
data = SATs;

% full model fit
guess = [2*ones(1, NSATs) 4*ones(1, NSATs), …
      0.30*ones(1, NSATs)]; % lamda, beta, delta
options = optimset('fminsearch');
[guess, fullmodel_L] = fminsearch('SATfull_costfunc',…
      guess, options, data);

%reduced model: same delta for all SATs
guess = [3*ones(1, NSATs) 0.25*ones(1, NSATs), 0.20];
        % lamda, beta, delta
options = optimset('fminsearch');
[guess, reducedmodel_L] = fminsearch('SATreduced_costfunc',…
      guess, options, data);
lamda= guess(1:NSATs);
beta = guess((NSATs+1):(2*NSATs));
delta = guess((2*NSATs+1));
meanLogS = mean(mean(SATs(2:2:NSATs, :)));
SS = sum(sum((SATs(2:2:NSATs,:) - meanLogS).^2));
r2_full = 1 - fullmodel_L/SS;
r2_reduced = 1 - reducedmodel_L/SS;
df1 = NSATs-1;
df2 = NSATs*Ndelays - 3*NSATs;
F = (r2_full - r2_reduced)/df1 / ((1 - r2_full)/df2);
p = 1 - fcdf(F, df1, df2);
```

Display 10.13b

```
%%% Program SATfull_costfunc.m
function L = SATfull_costfunc(guess, data)

NSATs = size(data, 1)/2;
Ndelays = size(data, 2);
lamda= guess(1:NSATs);
beta = guess((NSATs+1):(2*NSATs));
delta = guess((2*NSATs+1):(3*NSATs));

L=0;
for i = 1:NSATs
    t = data((i-1)*2+1, :);
    Observed_dp = data((i-1)*2+2, :);
    Predicted_dp = lamda(i) .* ( 1 - exp(-beta(i) …
        * ( t - delta(i) ) ) );
    L = L+sum ((Observed_dp - Predicted_dp).^2);
end
```

Display 10.13c

```
%%% SATreduced_costfunc.m
function L = SATreduced_costfunc(guess, data)

NSATs = size(data, 1)/2;
Ndelays = size(data, 2);
lamda= guess(1:NSATs);
beta = guess((NSATs+1):(2*NSATs));
delta = guess((2*NSATs+1));

L=0;
for i = 1:NSATs
    t = data((i-1)*2+1, :);
    Observed_dp = data((i-1)*2+2, :);
    Predicted_dp = lamda(i) .* ( 1 - exp(-beta(i) …
       * ( t - delta ) ) );
    L = L+sum ((Observed_dp - Predicted_dp).^2);
end
```

Display 10.13d

```
>>   SATs = …
   [0.420   0.443   0.511   0.638   0.837   1.474   2.124
    0.392   0.424   0.500   0.636   0.836   1.480   2.128
    0.418   0.740   0.892   1.393   1.738   2.091   2.094
    0.171  -0.016   0.677   0.820   0.882   1.124   1.087]
>> [lamda, beta, delta, r2_full, r2_reduced, F, df1, df2,  p]
= fittingSATs(SATs)
lamda = 2.093 1.101, beta =  4.007  4.365, delta = 0.3604
r2_full = 0.9506, r2_reduced = 0.9443
F = 1.0279,  df1 = 1, df2 = 8, p =  0.3403
```

the SAT, or time-course functions. Other functional forms of very similar shape have been derived for process models of visual search.[45,46] Yet other functional forms have been derived from general models of decision, such as the diffusion model.[56] All of these functional forms are quite similar to the exponential and can be used to test the process models that they are designed to capture.

10.5 Summary

In this chapter, we have focused on classical methods of model fitting, parameter estimation, estimation of parameter variability, and model comparison based on least-squares and maximum likelihood methods. The initial part of the chapter provided a general introduction and intuitive analysis of these methods, and the second part of the chapter provided example applications and programs that are central to typical experiments in psychophysics. The methods and the examples treated in this chapter should provide the reader with the tools to develop their own programs to carry out model testing and estimation for their applications.

References

1. Grolier. *New Webster's dictionary and thesaurus of the English language*. New York: Grolier; 1992.

2. Palmer J, Verghese P, Pavel M. 2000. The psychophysics of visual search. *Vision Res* 40(10): 1227–1268.

3. Graybill FA, Iyer HK. *Regression analysis*. Belmont, CA: Duxbury Press; 1994.

4. Efron B, Tibshirani RJ. *An introduction to the bootstrap*. Monographs on statistics & applied probability. Boca Raton: Chapman & Hall/CRC Press; 1994.

5. Davison AC, Hinkley DV. *Bootstrap methods and their application*. Cambridge, UK: Cambridge University Press; 1997.

6. Myung JI, Pitt MA. 2004. Model comparison methods. *Methods Enzymol* 383: 351–366.

7. Palmer J, Ames CT, Lindsey DT. 1993. Measuring the effect of attention on simple visual search. *J Exp Psychol Hum Percept Perform* 19(1): 108–130.

8. Hays WL. *Statistics for the social sciences*, Vol. 410. New York: Holt, Rinehart and Winston; 1973.

9. Akaike H. 1974. A new look at the statistical model identification. *IEEE Transactions on Automatic Control* 19(6): 716–723.

10. Schwarz G. 1978. Estimating the dimension of a model. *Ann Stat* 6(2): 461–464.

11. Kruschke JK. *Doing Bayesian data analysis: A Tutorial with R and BUGS*. Burlington, MA: Academic Press; 2010.

12. Lee MD. 2008. Three case studies in the Bayesian analysis of cognitive models. *Psychon Bull Rev* 15(1): 1–15.

13. Raftery AE. 1995. Bayesian model selection in social research. *Sociol Methodol* 25: 111–164.

14. Pitt MA, Myung IJ, Zhang S. 2002. Toward a method of selecting among computational models of cognition. *Psychol Rev* 109(3): 472–491.

15. Gelman A, Carlin JB, Stern HS, Rubin DB. *Bayesian data analysis*. Boca Raton, FL: CRC Press; 2004.

16. Shiffrin RM, Lee MD, Kim W, Wagenmakers EJ. 2008. A survey of model evaluation approaches with a tutorial on hierarchical Bayesian methods. *Cogn Sci* 32(8): 1248–1284.

17. Green DM, Swets JA. *Signal detection theory and psychophysics*. New York: Wiley; 1966.

18. Wickens TD. *Elementary signal detection theory*. Oxford: Oxford University Press; 2002.

19. Macmillan NA, Creelman CD. *Detection theory: A user's guide*. Hillsdale, NJ: Lawrence Erlbaum; 2005.

20. Lagarias JC, Reeds JA, Wright MH, Wright PE. 1998. Convergence properties of the Nelder-Mead simplex method in low dimensions. *SIAM J Optim* 9: 112–147.

21. Wichmann FA, Hill NJ. 2001. The psychometric function: I. Fitting, sampling, and goodness of fit. *Atten Percept Psychophys* 63(8): 1293–1313.

22. Mortensen U. 2002. Additive noise, Weibull functions and the approximation of psychometric functions. *Vision Res* 42(20): 2371–2393.

23. Tyler CW, Chen CC. 2000. Signal detection theory in the 2AFC paradigm: Attention, channel uncertainty and probability summation. *Vision Res* 40(22): 3121–3144.

24. Watson AB, Ahumada AJ. 2005. A standard model for foveal detection of spatial contrast. *J Vis* 5(9): 717–740.

25. Lu ZL, Dosher BA. 2008. Characterizing observers using external noise and observer models: Assessing internal representations with external noise. *Psychol Rev* 115(1): 44–82.

26. Lesmes LA, Lu ZL, Tran NT, Dosher BA, Albright TD. 2006. An adaptive method for estimating criterion sensitivity (d') levels in yes/no tasks. *J Vis* 6(6): 1097.

27. Maloney LT. 1990. Confidence intervals for the parameters of psychometric functions. *Atten Percept Psychophys* 47(2): 127–134.

28. Lu Z-L, Dosher BA. 1999. Characterizing human perceptual inefficiencies with equivalent internal noise. *J Opt Soc Am A Opt Image Sci Vis* 16(3): 764–778.

29. Dosher BA, Lu ZL. 2000. Noise exclusion in spatial attention. *Psychol Sci* 11(2): 139–146.

30. Lu ZL, Lesmes LA, Dosher BA. 2002. Spatial attention excludes external noise at the target location. *J Vis* 2(4): 312–323.

31. Wonnacott TH, Wonnacott RJ. *Regression: A second course in statistics*. New York: Wiley; 1981.

32. Campbell FW, Robson JG. 1968. Application of Fourier analysis to the visibility of gratings. *J Physiol* 197(3): 551–566.

33. Enroth-Cugell C, Robson JG. 1966. The contrast sensitivity of retinal ganglion cells of the cat. *J Physiol* 187(3): 517–552.

34. Rohaly AM, Owsley C. 1993. Modeling the contrast-sensitivity functions of older adults. *JOSA A* 10(7): 1591–1599.

35. Zhou Y, Huang C, Xu P, Tao L, Qiu Z, Li X, Lu Z-L. 2006. Perceptual learning improves contrast sensitivity and visual acuity in adults with anisometropic amblyopia. *Vision Res* 46(5): 739–750.

36. Huang CB, Zhou Y, Lu ZL. 2008. Broad bandwidth of perceptual learning in the visual system of adults with anisometropic amblyopia. *Proc Natl Acad Sci USA* 105(10): 4068–4073.

37. Busemeyer JR, Diederich A. *Cognitive modeling*. Thousand Oaks, CA: Sage Publications; 2009.

38. Lu Z-L, Dosher BA. 1998. External noise distinguishes attention mechanisms. *Vision Res* 38(9): 1183–1198.

39. Dosher BA, Lu Z-L. 1998. Perceptual learning reflects external noise filtering and internal noise reduction through channel reweighting. *Proc Natl Acad Sci USA* 95: 13988–13993.

40. Dosher BA, Lu Z-L. 1999. Mechanisms of perceptual learning. *Vision Res* 39(19): 3197–3221.

41. Lu ZL, Dosher BA. 2009. Mechanisms of perceptual learning. *Learning & Perception*. 1(1): 19–36.

42. Reed AV. 1973. Speed-accuracy trade-off in recognition memory. *Science* 181(4099): 574–576.

43. Dosher BA. 1976. The retrieval of sentences from memory: A speed-accuracy study. *Cognit Psychol* 8(3): 291–310.

44. McElree B, Carrasco M. 1999. The temporal dynamics of visual search: Evidence for parallel processing in feature and conjunction searches. *J Exp Psychol Hum Percept Perform* 25(6): 1517–1539.

45. Dosher BA, Han S, Lu ZL. 2004. Parallel processing in visual search asymmetry. *J Exp Psychol Hum Percept Perform* 30(1): 3–27.

46. Dosher BA, Han S, Lu ZL. 2010. Information-limited parallel processing in difficult heterogeneous covert visual search. *J Exp Psychol Hum Percept Perform* 36(5): 1128–1144.

47. Carrasco M, McElree B. 2001. Covert attention accelerates the rate of visual information processing. *Proc Natl Acad Sci USA* 98(9): 5363–5367.

48. Carrasco M, McElree B, Denisova K, Giordano AM. 2003. Speed of visual processing increases with eccentricity. *Nat Neurosci* 6(7): 699–700.

49. Reed AV. 1976. List length and the time course of recognition in immediate memory. *Mem Cognit* 4(1): 16–30.

50. Dosher BA. 1979. Empirical approaches to information processing: Speed-accuracy tradeoff functions or reaction time—A reply. *Acta Psychol (Amst)* 43(5): 347–359.

51. Dosher BA. 1981. The effects of delay and interference: A speed-accuracy study. *Cognit Psychol* 13(4): 551–582.

52. Ratcliff R. 1978. A theory of memory retrieval. *Psychol Rev* 85(2): 59–108.

53. Ratcliff R. 1980. A note on modeling accumulation of information when the rate of accumulation changes over time. *J Math Psychol* 21(2): 178–184.

54. McElree B, Dosher BA. 1989. Serial position and set size in short-term memory: The time course of recognition. *J Exp Psychol Gen* 118(4): 346–373.

55. McElree B, Dosher BA. 1993. Serial retrieval processes in the recovery of order information. *J Exp Psychol Gen* 122(3): 291–315.

56. Ratcliff R. 1985. Theoretical interpretations of the speed and accuracy of positive and negative responses. *Psychol Rev* 92(2): 212–225.

11 Adaptive Psychophysical Procedures

Adaptive procedures are developed to reduce the burden of data collection in psychophysics by creating more efficient experimental test designs and methods of estimating either statistics or parameters. In some cases, these adaptive procedures may reduce the amount of testing by as much as 80% to 90%. This chapter begins with a description of classical staircase procedures for estimating the threshold and/or slope of the psychometric function, followed by a description of modern Bayesian adaptive methods for optimizing psychophysical tests. We introduce applications of Bayesian adaptive procedures for the estimation of psychophysically measured functions and surfaces. Each method is accompanied by an illustrative example and sample results and a discussion of the practical requirements of the procedure.

11.1 Need for Adaptive Experimental Testing

Previous chapters have described classical methods in visual psychophysics for measuring important perceptual properties of detection, such as the threshold or the psychometric function. One such method, the method of constant stimuli, requires the measurement of behavioral performance for a number of stimulus values and requires a large number of test trials. Such procedures are relatively expensive in data collection. They also presume a certain a priori knowledge about where to place stimulus values to cover a range of performance levels from chance to maximum performance. The experimental demand is multiplied in situations where we are measuring either a threshold or a psychometric function for multiple groups or different conditions. This is more typical than not in testing different hypotheses about vision. Furthermore, in some circumstances we may not really know in advance which stimulus values to test for a given observer, a special population, or for novel stimulus manipulations.

Adaptive procedures serve a critical function in the measurement of psychophysical properties. They provide procedures and methods that use each new observation to adjust the stimulus values tested, making the most of each new observation. This leads to methods that are more efficient. Good estimates may be achieved with much less testing.

Furthermore, within limits, adaptive procedures are self-ranging and able to select stimulus values to test that are more responsive to the observer.

The adaptive methods are all the more critical when multiple measurements or conditions are required or when measuring overall perceptual functions, such as the contrast sensitivity function. A method of constant stimulus approach may require the measurement of a full psychometric function at each of a number of stimulus conditions or different groups. Without adaptive methods, certain studies simply require too much testing to be feasible. Increasingly in vision science, it is functions, like the contrast sensitivity function or the threshold versus external noise contrast function, that provide important information in the specification of individuals or tasks. Adaptive testing procedures have the potential for an especially large payoff in efficiency, especially for these new applications.

With the advancement of new computer technology and the availability of new algorithms for search and optimization, the area of adaptive testing is making rapid advancements and is likely to be increasingly important in the field. This chapter begins with some of the simpler, classical approaches to adaptive testing and ends with some examples of adaptive measurements of complex visual functions.

11.2 Classical Adaptive Procedures for a Threshold

A number of non-parametric adaptive procedures have been developed to estimate a threshold with a small number of trials. The methods are adaptive in that the stimuli presented to the observer depend on the observer's prior responses. The goal of the methods is to estimate the threshold in as small a number of trials as possible to achieve the level of desired accuracy in the measurement. Distinct procedures estimate thresholds associated with different proportions of a given response (i.e., 50%, 70.7%, etc.) Often, these correspond with percent correct; sometimes they correspond to percent "yes" for the required judgment. The classical adaptive procedures for measuring threshold are non-parametric—they do not make any strong assumptions about the specific shape of the psychometric function. The psychometric function is assumed to be monotonically increasing with the relevant stimulus variable. It is assumed to be stationary or constant throughout the process of measurement. In addition, the response of the observer to each trial is assumed to be independent of other trials. A number of classical adaptive procedures for the estimation of a threshold, including simple staircase methods and more complex adaptive sampling methods, are considered in the following sections.

11.2.1 Up–Down Staircases

Up–down staircases are designed to estimate thresholds at certain fixed probabilities of a correct response by changing the stimulus variable—such as contrast—up or down after integer numbers of correct or error responses. The *truncated staircase* refers to a 1/1, or

1-up and 1-down, staircase that converges on 50% correct (or "yes") responses.[1] The transformed staircases such as 2/1, 3/1, or 4/1 reduce contrast (or increase difficulty) after 2, 3, or 4 consecutive correct responses in a row and increase contrast after every error. They converge to probabilities of 70.7%, 79.4%, and 84.1%, respectively.

A common use of truncated staircases is for the estimation of a point of subjective equality or the point at which a test stimulus subjectively matches a standard stimulus. For example, a truncated staircase might be used to estimate the perceived length of a line ending in arrows by comparing it to a line alone (the Müller-Lyer illusion[2]). Or, the subjectively equal luminance of a patch in a black surround may be estimated by comparing it to a sample patch in a neutral gray surround. The two percepts are equal when one chooses the test over the standard 50% of the time.

Figure 11.1a shows a possible psychometric function for a contrast matching experiment. The psychometric function graphs the percent "yes" for classifying a test patch as having more contrast than a standard patch. The function goes from near zero at the lowest contrast values where the test clearly has less contrast, through the 50% point, and approaches 100% at the highest contrasts that are clearly more than the standard. The 1-up and 1-down (or 1/1) staircase estimates the 50% point in a relatively small number of trials.

The 1/1 adaptive staircase increases the manipulated variable, here the contrast of the test, after every "no" response (when the test stimulus is estimated as lower) and decreases the contrast of the test after every "yes" response. Over a sequence of responses, the staircase tracks the point of subjective equality. Figure 11.1b shows an example of such a staircase that tracks a threshold stimulus value for a simulated observer with the psychometric function in figure 11.1a. Because there is noise or randomness in the observer's

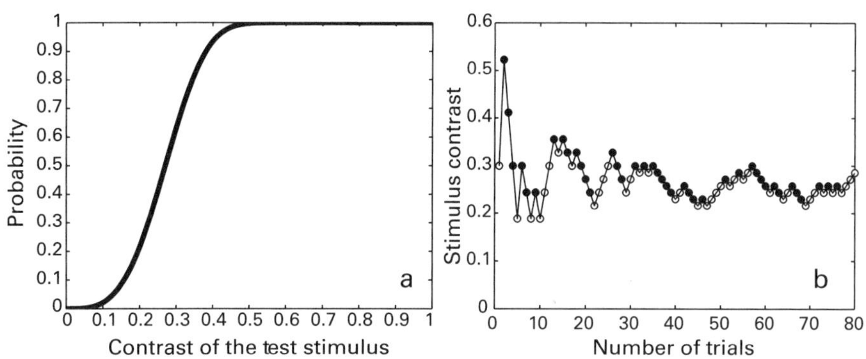

Figure 11.1
A psychometric function and sample staircase for contrast matching. (a) A psychometric function for the matching task. (b) A sample trial sequence of a truncated (1/1) staircase. The filled circles indicate "yes" responses; the unfilled circles indicate "no" responses.

internal response, each run of the staircase leads to a slightly different sequence of stimuli and responses.

In a 2-down, 1-up (or 2/1) staircase, the contrast or signal level is decreased after every new set of two consecutive correct responses and is increased after every incorrect response.[3] When the staircase converges, the performance level p can be computed by equating the probability that the staircase goes up and down:

$$\begin{aligned}&(1-p)+p(1-p)=p^2\\&p=\sqrt{2}/2=0.707,\end{aligned} \tag{11.1}$$

corresponding to 70.7% correct.

In a 3-down, 1-up (or 3/1) staircase, the signal level is decreased after every new set of three consecutive correct responses, and the signal level is increased after each error. When the staircase converges, the performance level p again can be computed by equating the probability of the staircase going up or down:

$$\begin{aligned}&(1-p)+p(1-p)+p^2(1-p)=p^3\\&p=\sqrt[3]{1/2}=0.794,\end{aligned} \tag{11.2}$$

corresponding to 79.4% correct. A similar computation for the 4-down, 1-up (or 4/1) staircase leads to a performance level of 84.1%.

Figure 11.2a shows a psychometric function of percent correct versus stimulus contrast starting at 50% correct for zero or small contrasts and increasing to 100% at high contrasts, a typical psychometric function for two-alternative forced-choice discrimination. The psychometric function in this example is typical for the discrimination of the orientation of a ±15° Gabor in the periphery. Sample trial sequences for 2/1, 3/1, and 4/1 staircases are shown in Figures 11.2b-d.

All up–down staircases begin at a *starting value*—a value specified by the experimenter—and increase or decrease the next presented stimulus value by a *step-size*. Following the suggestions of Levitt, we illustrated trial runs in which the step-size is decreased by half after the first, third, seventh reversals, and so on. A reversal occurs if the stimulus value moves up when it was last moved down, or vice versa. All staircase procedures have a *stop rule*, a rule that determines when to stop the estimation. The stop rule might involve testing a certain number of trials, but often the rule itself is also adaptive, and specifies testing until after a certain number of "reversals." The procedures in figures 11.1 and 11.2 are illustrated for a stop rule of 80 trials, but many investigators use a stop rule of a fixed number of reversals (e.g., 10 reversals). A standard practice in psychophysics is to exclude the first three reversals in computing the threshold if the number of total reversals is odd or to exclude the first four reversals if the total number of reversals is even. The estimate of the threshold in the example in figure 11.2 averages the last six endpoints. This eliminates some of the early range-finding trials from the threshold estima-

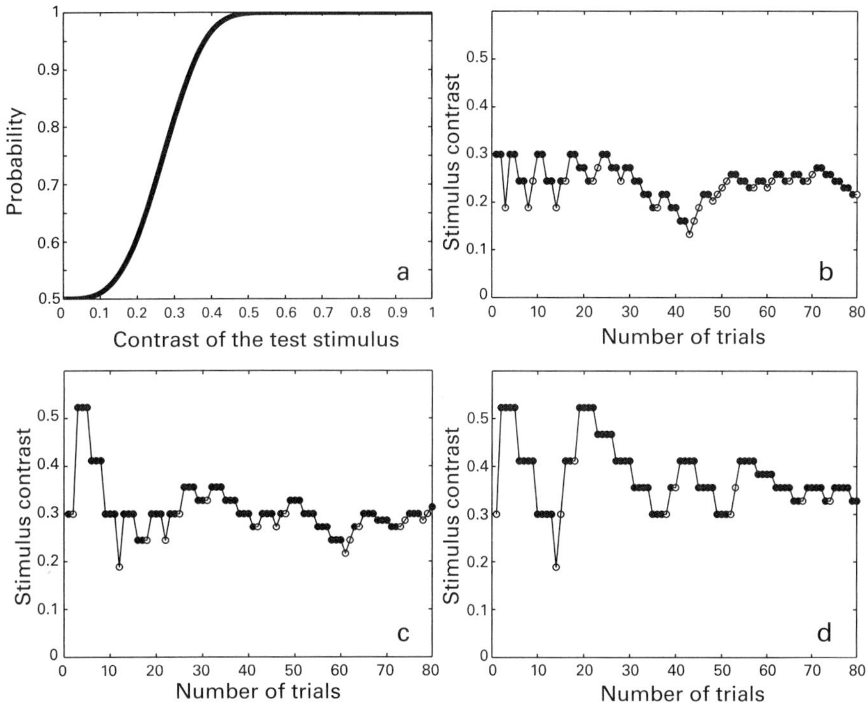

Figure 11.2
Sample transformed staircases for 2/1, 3/1, and 4/1 methods for the same psychometric function tracking 70.7%, 79.4%, and 84.1% correct, respectively. (a) A psychometric function for Gabor orientation discrimination. (b) A sample trial sequence of a 2-down, 1-up staircase. The estimated threshold is 0.263. (c) A sample trial sequence of a 3-down, 1-up staircase. The estimated threshold is 0.309. (d) A sample trial sequence of a 4-down, 1-up staircase. The estimated threshold is 0.356.The filled circles indicate correct responses; the unfilled circles indicate incorrect responses.

tion and guarantees a balanced number of up and down reversal points, which reduces bias in the estimate.

A staircase procedure to estimate a particular threshold is judged in three ways: *bias*, *precision*, and *efficiency*. Bias refers to whether the average threshold estimated by the staircase corresponds to the actual value or is higher or lower than it. Precision refers to the variability of the estimate. The relative efficiency is related to the number of trials required to achieve a given precision. The technical definitions of these criteria are discussed in section 11.2.5. These three aspects of performance depend on how far the starting value is from the actual threshold, the step-size, and the stop rule.

Figure 11.3 (plate 8) illustrates estimates of bias, precision, and typical trial numbers for a 2-up and 1-down staircase for a psychometric function like the one in figure 11.2b.

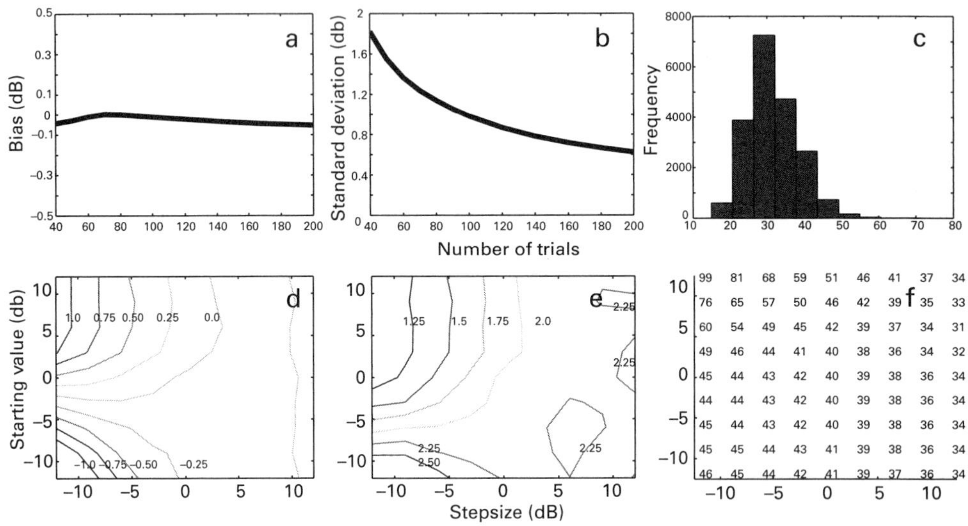

Figure 11.3 (plate 8)
Bias, precision, and sampling demands of a 2/1 staircase. (a) Bias of the estimated threshold as a function of number of trials. (b) Precision of the estimated threshold from these staircases as a function of number of trials. (c) Histogram of the number of trials needed to reach 10 reversals in the staircase. (d) Contour plot of the biases of a 2-down, 1-up staircase as a function of the starting value and step-size. (e) Contour plot of the precision of the 2-down, 1-up staircase as a function of the starting value and step-size. (f) Number of trials needed to obtain 10 reversals in the staircase with 90% probability. Bias dB is referred to 0 at the actual threshold; step-size dB is referred to 0 at the recommended step-size, or 0.5/slope of the psychometric function near threshold.

We used simulation studies to investigate the behavior of these 2/1 staircases with different starting values and step-sizes, ranging from –10 dB below the true threshold to +10 dB above the true threshold. [A decibel is a logarithmic function of the ratio of two amplitude values, $20 \log_{10}(x/y)$; 1 dB has an amplitude ratio of about 1.222.]

Consistent with the early reports of Levitt,[3] the optimal starting value for the staircase is near the actual threshold. Here, we know the "true" threshold because we are simulating the process. In experimental situations, this is an unknown, and experimenters do their best to estimate a reasonable starting value based on available information.

The optimal initial step-size depends on both the threshold and the slope of the psychometric function. It should be large enough to traverse the rising portion of the psychometric function in one or two steps. For the 2/1 staircase, this corresponds to a step-size on the order of 0.5 divided by the (ordinary) slope of the psychometric function at the threshold, which is $2\log(2)^{\frac{1-\beta}{\beta}}\alpha/\beta$ for a Weibull function.

Figure 11.3a–c (plate 8) assumes a reasonable initial starting value (about 0.65 dB above the true threshold) and step-size selection and shows (a) the bias of the staircase estimate as a function of the number of trials, (b) the precision of the staircase estimate as a function of the number of trials, and (c) the distribution of number of trials needed for a

10-reversal stop rule. Figure 11.3d–f (plate 8) shows how bias, precision, and numbers of trials vary as the starting value and step-size are changed. For this particular step-size rule (in which the step size is halved after the first, third and seventh reversal) and stop rule, it is very important that the initial step-size is large enough to achieve good accuracy. As mentioned previously, this should be about 0.5 divided by the slope of the psychometric function near threshold (which is graphed as 0 dB step-size). There is a trade-off for bias in decreased precision. The best starting values and starting step-sizes depend on the step-size reduction scheme and on the stop rule. A starting value and initial step-size for the staircase should be chosen to achieve desirable bias and precision by choosing appropriate values in the contour plots. For a cautious experimenter, a simulation study such as this one may prove very useful in planning a particular experiment or study.

11.2.2 The Accelerated Stochastic Approximation Method

Up–down staircases, as described in section 11.2.1, can only converge to a small number of particular values for selected performance levels determined by balancing the integer counts of correct and error trials. A different kind of staircase is needed to measure threshold at arbitrary performance levels. The accelerated stochastic approximation method,[4] or ASA staircase, is an adaptive procedure that converges to any specified accuracy level ϕ. It also does a good job, after the initial range finding, of controlling performance level throughout the course of training as long as the state of the observer remains stable.

The stimulus contrast used in each trial in the ASA is determined by the stochastic approximation procedure.[5] To converge to an arbitrary desired performance level ϕ, stimulus contrasts in the first two trials are given by

$$c_{n+1} = c_n - \frac{s}{n}(Z_n - \phi). \tag{11.3}$$

Here, n is the trial number, c_n is the stimulus contrast in trial n, $Z_n = 0$ (incorrect) or 1 (correct) is the response accuracy in trial n, c_{n+1} is the stimulus contrast for the next trial, and s is the prechosen step-size at the beginning of the trial sequence. From the third trial on, the sequence is "accelerated":

$$c_{n+1} = c_n - \frac{s}{2 + m_{\text{shift}}}(Z_n - \phi). \tag{11.4}$$

In these equations, m_{shift} is the number of shifts in the response category. These are switches from consecutive correct responses to incorrect responses and vice versa, or the number of reversals.

Working through the equations, the contrast (or value) on the next trial is equal to the contrast on the last trial adjusted up or down depending upon whether the response is below (incorrect) or above (correct) the target accuracy ϕ. The size of the adjustment decreases as testing goes on and differs for steps that are incorrect or correct. The stop rule in the ASA method is typically set by a lower limit on the final step-size. For example, if a target precision is 1.5 dB, one might choose a final step size of 0.50 dB.

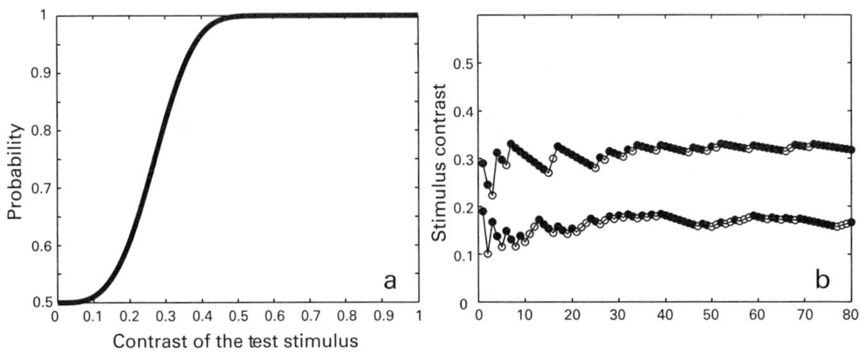

Figure 11.4
Psychometric function and sample accelerated stochastic approximation (ASA) staircases. (a) A psychometric function for discrimination. (b) Trial sequences of two ASA staircases, converging at 60% and 80% correct. The filled circles indicate "yes" responses; the unfilled circles indicate "no" responses.

Figure 11.4 shows (a) a psychometric function for two-alternative forced-choice discrimination and (b) sample trial histories corresponding to target probabilities of 60% and 80% for the ASA staircase.

In an influential review of adaptive psychophysical procedures, Treutwein[6] recommended the ASA as the best available adaptive procedure for measuring thresholds. Figure 11.5a–c (plate 9) shows the results of simulation studies of an ASA procedure for (a) bias, (b) precision, and (c) total trial number for a given starting value and step size. Graphs in Figure 11.5d–f (plate 9) show the sensitivity of (d) bias, (e) precision, and (f) trial number to the starting value and initial step-size. The simulation studies suggest that the optimal starting value c_1 is at the (unknown) threshold, and the optimal initial step-size s should traverse the psychometric function in one or two steps. Just as for the transformed staircases, these initial values may be informed by prior knowledge or pilot data.

The ASA staircases have many good properties, most notably the ability to converge on any arbitrary accuracy level relatively quickly. They also have a fairly large region of starting values in threshold and step-size that yield relatively unbiased estimates. The profile of bias and precision is somewhat better than that of the 2/1 or 3/1 transformed staircases tracking similar percent correct levels. Because the ASA staircase shifts to small step-sizes fairly quickly, it may be less robust to unfortunate selections of initial value, random errors early in the trial sequence, or drifts in the level of observer performance over a session. For this reason, the initial step-sizes should be chosen to be large enough.

11.2.3 Other Adaptive Staircase Methods

In sections 11.2.1 and 11.2.2, we considered two major adaptive staircase methods. A number of other adaptive staircase methods have also been developed and tested over

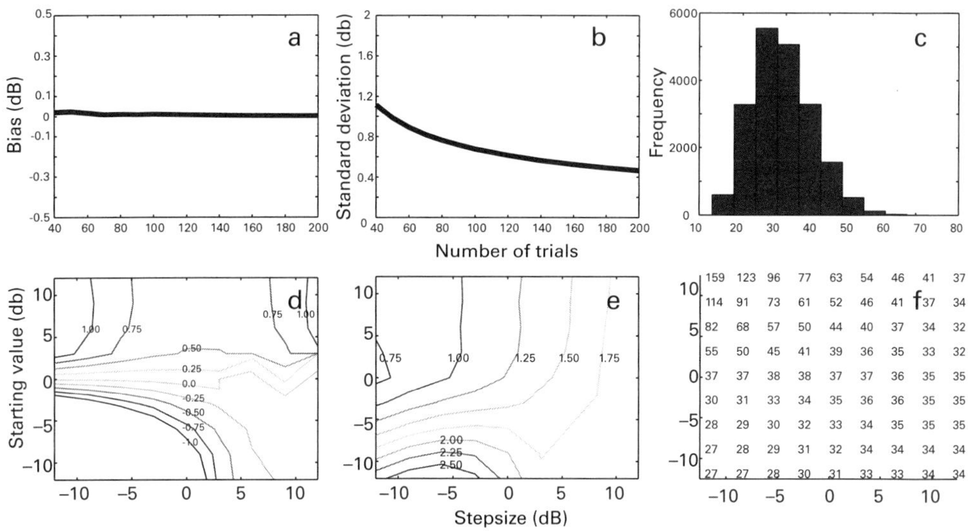

Figure 11.5 (plate 9)
Bias, precision, and trial requirements of the accelerated stochastic approximation (ASA) staircase tracking 75%, and its sensitivity to starting value and step-size. (a) Bias of the estimated threshold from an ASA staircase as a function of the number of trials. (b) Precision of the estimated threshold from the same staircase as a function of number of trials. (c) Histogram of the number of trials needed to reach 10 reversals in the staircase. (d) Contour plot of the bias of an ASA staircase as a function of the starting value and step-size. (e) Contour plot of the prevision of the staircase as a function of the starting value and step-size. (f) Number of trials needed to obtain 10 reversals in the staircase with 90% probability.

many years. These include a non-parametric up–down method,[7] weighted up–down,[8] weighted binary search,[9] and parameter estimation by sequential testing (PEST).[10] The most prominent of these was the PEST procedure. This method used several heuristics from sequential testing statistics to change the tested stimulus value. In this procedure, a given stimulus value is tested multiple times until the estimate of the probability exceeds a confidence interval based on the sample size so far and the target probability. After many years of extensive use, the PEST procedure was modified to incorporate maximum likelihood calculations to place the next stimulus level in a procedure called best PEST.[11] These and other methods led to the development of adaptive Bayesian methods for estimating thresholds and psychometric functions (see section 11.3).

11.2.4 Experimental Implementation of Staircases

There is an art to using adaptive methods in real experiments. Many laboratories incorporate unwritten best practices designed to improve the likelihood of successful estimation and application of the methods.

It is important in a real experiment to optimize the selection of the starting value and the step-size for a given stop rule and to choose a stop rule appropriate for the goal of the estimation. If the threshold and the slope of the psychometric function are known in advance, this is quite easy. Of course, in general, one does not know either, and the entire point of the adaptive procedure is to be able to do a quick estimate of threshold on the basis of a smaller number of trials. In such cases, it is useful to collect pilot data to increase the quality of the initial guesses from estimates of the first few observers.

Another important practical issue is the potential impact of trial-to-trial dependencies, often called sequential dependencies, on the staircase performance. Sequential dependencies—lack of independence in responses from trial to trial—arise due to fluctuations in alertness, or systematic variation in criteria for Yes/No procedures, or predictability of the stimulus, to name just a few possible causes. For this reason, experimenters often interleave several staircases in a single experimental session so that adjacent trials may be drawn from different independent staircases whose stimulus values will be decoupled from one another.[3] Sometimes these are staircases for different conditions. If an experiment is designed to measure only a single condition, it is often still a good idea to include several independent copies of staircases measuring the same target performance interleaved in a session.

In some cases, quick staircase estimates may be used to choose good values for a more trial-intensive measurement of the full psychometric function. For example, if 2/1 and 3/1 up–down staircases are used as a pretest to estimate the contrasts corresponding to the 70.7% and 79.4% performance levels, these in turn can be used to set contrasts to test for a 5-point psychometric function [c_1, c_2, c_3, c_4, c_5, where $c_2 = c_{70.7\%}$, $c_4 = c_{79.4\%}$, $c_1 = 0.5c_2$, $c_3 = 0.5(c_2 + c_4)$, and $c_5 = 2c_4$]. These five signal contrast levels represent a relatively efficient sampling of the psychometric functions when tested with a randomized method of constant stimuli.[12]

11.2.5 Bias, Precision, and Efficiency of Threshold Estimations

This section provides a brief quantitative introduction to definitions of bias, precision, and efficiency of threshold estimations. It should allow anyone wishing to simulate a particular procedure for the estimation of a given threshold to understand the formal definitions for evaluating these methods.

First, each threshold $v_{\text{estimated}}$ is estimated by a particular procedure. For example, the threshold estimated from a staircase that terminates with 10 reversals and based on the last 6 endpoints is

$$v_{\text{estimated}} = \left(\sum_{k=5}^{10} v_k \right) \Big/ 6 .$$

The bias index computes the mean of many such (simulated) threshold estimates for a given procedure and compares it to the true or actual threshold:

$$\text{bias} = \frac{\sum_{i=1}^{n_{\text{runs}}} \left(v_{\text{estimated}} - v_{\text{actual}} \right)}{n_{\text{runs}}}.$$

Ideally, a procedure will lead to an unbiased estimator, one where the bias $=0$ when n_{runs} approaches infinity. In practice, it may be sufficient to achieve a bias that approximately equals the precision of the measurement or slightly better.

The variability in the estimates yielded by the procedure is

$$s_{\text{estimated}}^2 = \frac{\sum_{i=1}^{n_{\text{runs}}} \left(v_{\text{estimated}} - \overline{v}_{\text{estimated}} \right)^2}{n_{\text{runs}} - 1}.$$

The higher the variability, the lower the precision of the estimate. Precision is defined as the inverse of variability of the estimates, or $1/s_{\text{estimated}}^2$, and this too is usually estimated by simulation.

As summarized in Treutwein,[6] the efficiency of the threshold procedure is related to the so-called sweat factor K, which is the product of the variance of the threshold estimate and the fixed number of test trials used to obtain the estimate,[10,13] or

$$K = n_{\text{trials}} s_{\text{estimated}}^2 = n_{\text{trials}} \frac{\sum_{i=1}^{n_{\text{runs}}} \left(v_{\text{estimated}} - \overline{v}_{\text{estimated}} \right)^2}{n_{\text{runs}} - 1}.$$

The sweat factor increases with the number of trials needed to provide the estimate and also increases with the variability of the estimate. Recovering a low variability estimate in few trials makes a procedure more efficient.

The efficiency of a psychophysical procedure is quantified by comparison to a standard or ideal procedure. The ideal procedure is usually benchmarked by the asymptotic variance of a Robbins–Monro process, which is

$$\sigma_{\text{RM}}^2 = \frac{\phi(1-\phi)}{n_{\text{trials}} \left(\frac{d\psi(x)}{dx} \Big|_{x=\text{threshold}} \right)^2},$$

where ϕ is the target probability to be estimated,

$$\frac{d\psi(x)}{dx} \Big|_{x=\text{threhsold}}$$

is the slope of the psychometric function at threshold, and n_{trials} is the number of trials in the estimation. So, $K_{\text{ideal}} = n_{\text{trials}} \sigma_{\text{RM}}^2$, and the efficiency $\eta_{\text{procedure}}$ of a particular procedure is the ratio of the ideal sweat factor to the sweat factor of a given procedure, or

$\eta_{\text{procedure}} = K_{\text{ideal}} / K_{\text{procedure}}$. By definition, because a given procedure can at best equal the ideal procedure (or else the ideal procedure isn't ideal), the efficiency is less than 1.

11.3 Adaptive Bayesian Procedures for Threshold and Psychometric Functions

11.3.1 General Framework

The staircase procedures and methods considered so far are non-parametric adaptive procedures for estimating thresholds. They do not assume a particular form for the psychometric function. In the rest of the chapter, we describe several adaptive Bayesian testing methods for estimating properties of psychological functions based on the selection of specific functional forms.

All of the adaptive Bayesian methods have several components. These include specification of the functional form, choosing the most informative stimulus to test on the next trial, updating the estimate of the parameters of the function on each trial, and defining a stop criterion. Each of these is considered in turn, followed by an explanation of the principles of Bayesian estimation and updating of probability distributions for estimated parameters.

The first stage of creating an adaptive Bayesian method defines the form or equation for the function being estimated. Here we illustrate the principles by considering the estimation of the psychometric function. Every function is defined in relation to a set of parameters. For example, if the Weibull is chosen as the functional form of the psychometric function, the two parameters are the threshold τ and the slope η. Other parameters, such as the minimum and maximum of the function, may be set a priori. Next, the adaptive procedures choose the most informative stimulus to test on the next trial. The adaptive Bayesian methods use a search algorithm to determine the value of the next test stimulus. After the test trial, the current estimate of the values of the function parameters is updated. Testing stops when a stop-rule criterion has been met. Then, the final estimates of the threshold and slope as well as other relevant statistics are calculated. Often, the stop rule is defined by having achieved a target level of precision of the estimates.

The core of these adaptive methods is the updating of the parameter estimates for the psychological function based on available evidence, using the Bayes rule. The Bayesian methods begin with a prior probability distribution for the parameters. They go on to use data that are collected in the experiment on each trial to update estimates of the probability of different parameter values given the data. Each adaptive Bayesian method first specifies the range and prior probability distribution for the parameter values of the function being estimated. Often, the prior probabilities are assumed to be uniform over a range of values. Alternatively, it is possible to use prior knowledge to focus more narrowly on likely values of the parameters. After each new piece of data is collected, the posterior probability distribution of the parameters is computed using the Bayes rule.

In the next section, we describe the Bayes rule and Bayesian updating before describing the remaining adaptive methods.

11.3.2 Bayes Rule, Bayesian Updating, and an Example

The Bayes rule specifies that if you know the prior probabilities of a set of n events $\{A_j, j=1:n\}$, that is, the $p(A_j)$, $j=1:n$, and you know the conditional probability of an event B given a particular A_i, labeled $p(B \mid A_i)$, then if event B occurs, we can update the posterior probability distribution of each A_i by the equation

$$p(A_i \mid B) = \frac{p(B \mid A_i)\,p(A_i)}{\sum_j p(B \mid A_j)\,p(A_j)}. \tag{11.5}$$

The Bayes rule has been an extremely important computational principle in computer science and in several areas of modern statistics. Starting with a set of prior probabilities, the Bayes rule provides a powerful method to improve or update the probability estimates using available observations or evidence. Posterior probability estimates will favor events that are more consistent with the observations.

Next, we show the Bayes rule in a simple computational example. We start with two urns. Each urn contains black and white marbles, but in different mixtures. At the beginning of a game, one urn is chosen, and all marbles are drawn from that urn throughout the game. The observer does not know which urn was chosen, but is told that the two urns will be chosen with equal probability. All of the marbles (samples) throughout the game are chosen from the urn that was chosen. The job of the observer is to guess which urn is being sampled.

The observers knows (was told) that the prior probability of the two urns is the same, or $p(\text{Urn}_1) = p(\text{Urn}_2) = 0.5$. They are told that Urn_1 has 25% white marbles, so that the conditional probability of a white marble from Urn_1 is 25%, or $p(\text{White} \mid \text{Urn}_1) = 0.25$ and $p(\text{Black} \mid \text{Urn}_1) = 0.75$. They are also told that Urn_2 has 60% white marbles, so that the conditional probability of a white marble from Urn_2 is 60%, or $p(\text{White} \mid \text{Urn}_2) = 0.60$ and $p(\text{Black} \mid \text{Urn}_2) = 0.40$ (figure 11.6). Now, the observer draws 10 marbles and observes that 7 are white and 3 are black marbles. Intuitively, Urn_2 is more likely to be the source than Urn_1 because it has more white marbles. The Bayes rule provides the equation to update the estimate of the probability of which urn was selected. It computes the "posterior probability" of Urn_1 and Urn_2 given the observations of 7 of 10 white marbles:

$$\begin{aligned} p(\text{Urn}_1 \mid W=7, B=3) &= \frac{p(\text{Urn}_1)[.7\times p(W \mid \text{Urn}_1) + .3\times p(B \mid \text{Urn}_1)]}{\sum_{\text{Urn}=1}^{2}[.7\times p(W \mid \text{Urn})p(\text{Urn}) + .3\times p(B \mid \text{Urn})p(\text{Urn})]} \\ &= \frac{0.50\times[.7\times 0.25 + .3\times 0.75]}{0.50\times[.7\times 0.25 + .3\times 0.75] + 0.50\times[.7\times 0.60 + .3\times 0.40]} \\ &= 0.425 \end{aligned} \tag{11.6a}$$

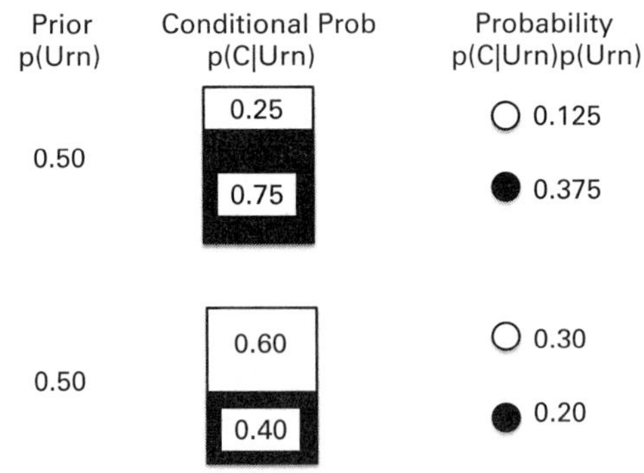

Figure 11.6
An illustration of the various probabilities entries used in the Bayes rule for a simple example.

$$
\begin{aligned}
p(\mathrm{Urn}_2 \mid W=7, B=3) &= \frac{p(\mathrm{Urn}_2)[.7 \times p(W \mid \mathrm{Urn}_2) + .3 \times p(B \mid \mathrm{Urn}_2)]}{\sum_{\mathrm{Urn}=1}^{2} [.7 \times p(W \mid \mathrm{Urn}) p(\mathrm{Urn}) + .3 \times p(B \mid \mathrm{Urn}) p(\mathrm{Urn})]} \\
&= \frac{0.50 \times [.7 \times 0.60 + .3 \times 0.40]}{0.50 \times [.7 \times 0.25 + .3 \times 0.75] + 0.50 \times [.7 \times 0.60 + .3 \times 0.40]} \\
&= 0.575.
\end{aligned} \tag{11.6b}
$$

Now, if the observer takes one more sample and gets a white marble, she can use the posterior probabilities resulting from the previous observations as the prior and update the probabilities to create a new posterior for the two urns using the Bayes rule again:

$$
\begin{aligned}
p(\mathrm{Urn}_1 \mid W=1, B=0) &= \frac{0.425 \times [1 \times 0.25 + 0 \times 0.75]}{0.425 \times [1 \times 0.25 + 0 \times 0.75] + 0.575 \times [1 \times 0.60 + 0 \times 0.40]} \\
&= 0.235
\end{aligned} \tag{11.7a}
$$

$$
\begin{aligned}
p(\mathrm{Urn}_2 \mid W=1, B=0) &= \frac{0.575 \times [1 \times 0.60 + 0 \times 0.75]}{0.425 \times [1 \times 0.25 + 0 \times 0.75] + 0.575 \times [1 \times 0.60 + 0 \times 0.40]} \\
&= 0.765.
\end{aligned} \tag{11.7b}
$$

This example shows how with new evidence, we can gradually improve the estimates of the probabilities of the true state of the world. The urns with different mixtures of white and black marbles correspond to a psychophysical function with different parameter values, or different states of the world. The samples of marbles are test samples in an experiment, and the Bayes rule specifies a computational method to update the probabilities of different parameter values.

11.3.3 QUEST

The first of the Bayesian methods developed for estimating the psychophysical threshold was the QUEST procedure.[14] QUEST assumes that the psychometric function is a Weibull expressed as a function of intensity in decibels (dB). This is a log form. This form of the Weibull differs slightly from the earlier forms provided in this book because the input, x, is expressed in decibels:

$$\Psi_T(x) = (1-\delta) - (1-\gamma-\delta)\exp\left[-10^{\left(\eta/20\right)(x-T-\tau)}\right]. \tag{11.8}$$

In this equation, γ specifies the chance level, which is $1/n$ for n-alternative forced-choice (i.e., 50% for two-alternative forced-choice) or the false alarm rate for a Yes/No paradigm. The parameter γ is determined by the testing paradigm. The value η specifies the slope of the psychometric function. The value T is the threshold or location parameter, and the value τ is associated with the proportion correct for the selected threshold performance level. The parameter δ is the lapse rate, which sets the highest performance level at less than 100%. Although in principle η might be estimated from the data, the QUEST procedure requires that it be specified in advance. It is usually assigned a value of 3.5 for two-alternative forced-choice paradigms. The only parameter to estimate is T.

The function $f_T(T)$ is the prior probability distribution for the threshold T. The experimenter sets this before the experiment. In QUEST, the priors are usually set to a Gaussian distribution with a large standard deviation and a mean that is chosen by the experimenter based on any prior experimental evidence that is available.

The stimulus value on the next trial, x_{n+1}, is set to be the mode of the QUEST function $Q_n(T)$. The QUEST function is the log posterior distribution of threshold T after trial n, based on the Bayesian update rule:

$$Q_n(T) = \ln f_T(T) + \sum_{i=1}^{n} \{r_i \ln \Psi_T(x_i) + (1-r_i)\ln[1-\Psi_T(x_i)]\}. \tag{11.9}$$

The value r_i is set to 1 for a success (correct trial) and 0 for a failure (error) on the trial. The term $\ln \Psi_T(x_i)$ or $\ln[1-\Psi_T(x_i)]$ is the log likelihood of the success or failure response on trial i. Before the first trial ($n = 0$), the first stimulus is set as the mode of the prior distribution.

In this scheme, the next value of the stimulus is the current maximum likelihood estimate of the threshold. As more samples are taken, the tested values converge on the estimated threshold value. The QUEST procedure sometimes uses a stop rule based on the confidence interval around the maximum likelihood estimate, which is set to a particular value (e.g., 1 dB). Alternatively, it uses a total number of trials as the stop criterion. At the end, the best estimate of the threshold is the final maximum likelihood estimate, or the maximum of the QUEST function on the last trial.

Display 11.1 shows a sample program, `QUEST.m`, that runs a simple experiment using the QUEST function from Psychtoolbox. The experiment uses a two-alternative forced-choice Gabor orientation discrimination task. For the `QUEST` function, the experimenter

Display 11.1

```
%%% Program Quest.m
function Quest

%% Experimental Module

% Specify the stimulus
p.stimSize = 4;          % image size in visual degrees
p.sf = 2; % cycles/degree
p.stimDuration = 0.2;    % stimulus duration in seconds
p.fixDuration = 0.2;     % seconds of fixation prior to stim
p.ITI = 0.5;
p.pThreshold = 0.75;     % performance level to measure the
threshold
nTrials = 80;
keys = {'left' 'right' 'esc'}; % keys to respond for left
                               % and right, and to break

% Compute stimulus parameters
ppd = pi/180 * p.ScreenDistance / p.ScreenHeight * …
      p.ScreenRect(4);  % pixels per degree
m = round(p.stimSize * ppd);    % image size in pixels
sf = p.sf / ppd;
sigma = m / 6;
fixLen = 1 / 4;          % cross length relative to half of
                         % image size
fixXY(1, :) = [-(1+fixLen) -1 1 1+fixLen 0 0 0 0]*m/2 + …
     p.ScreenRect(3) / 2;
fixXY(2, :) = [0 0 0 0 -(1+fixLen) -1 1 1+fixLen]*m/2 + …
     p.ScreenRect(4) / 2;
p.randSeed = ClockRandSeed;     % use clock to set seed for
                         % the random number generator

% procedural gabor allows us to change its parameters
% very quickly
tex = CreateProceduralGabor(windowPtr, m, m, 0, …
      [1 1 1 0] * 0.5, 1, 0.5);

% Initialize a table to set up experimental conditions
p.recLabel = {'trialIndex' 'tiltLeft' 'logContrast' …
      'respCorrect' 'respTime' 'Questmean' 'QuestSd'};
rec = nan(nTrials, length(p.recLabel));
       % matrix rec initialized with NaN
rec(:, 1) = 1 : nTrials;
       % count the trial numbers from 1 to nTrials
tiltLeft = ones(nTrials, 1);
tiltLeft(1 : round(nTrials / 2)) = 0;
rec(:, 2) = Shuffle(tiltLeft);
       % shuffle left and right orientations

% Initialize Quest. Parameters: thresholdGuess priorSd
% pCorrect eta delta gamma
```

Display 11.1 (continued)

```
q = QuestCreate(log10(0.5), 3.5, p.pThreshold, 3.5, 0.01, …
      0.5);

% Prioritize display to optimize display timing
Priority(MaxPriority(windowPtr));

% Start experiment with instructions
str = ['Press left and right arrow keys for top tilt '…
       'directions.\n\n' 'Press SPACE to start.'];
DrawFormattedText(windowPtr, str, 'center', 'center', 1);
% Draw Instruction text string centered in window
Screen('Flip', windowPtr);
       % flip the text image into active buffer
WaitTill('space');      % wait for SPACE to start
Secs = Screen('Flip', windowPtr);
p.start = datestr(now); % record start time

% Run nTrials trials
for i = 1 : nTrials
    % parameters for this trial
    tiltLeft = rec(i, 2);
    ang = (tiltLeft - 0.5) * 90; % tilt angle from vertical
    con = 10 ^ QuestQuantile(q); % get contrast
    con = min(0.5, con);
    WaitTill(Secs + p.ITI);

    Screen('DrawLines', windowPtr, fixXY, 3, 0);
    t0 = Screen('Flip', windowPtr, 0, 1); % show cross

    Screen('DrawTexture', windowPtr, tex, [], [], ang, 1,…
           1, [], [], 2, [180 sf sigma con 1 0 0 0]);
    t0 = Screen('Flip', windowPtr, t0 + p.fixDuration);
        % turn on stimulus
    Screen('Flip', windowPtr, t0 + p.stimDuration);
        % turn off stimulus

    [key Secs] = WaitTill(keys);  % wait till response
    if iscellstr(key), key = key{1}; end
        % take the first in case of multiple keys
    if strcmp(key, 'esc'), break; end % to stop
    correct = strcmp(key, keys{1}) == tiltLeft;

    % Update Quest and store data
    q = QuestUpdate(q, log10(con), correct);
    rec(i, 3:7) = [log10(con) correct Secs-t0 …
        QuestMean(q) QuestSd(q)];
end

p.finish = datestr(now);    % record finish time
save Quest_rst.mat rec p;   % save the results
```

provides an initial guess for the threshold T_{guess}, the standard deviation of the prior distribution, σ_T^{prior}, and the values for function parameters γ, η, and τ. For a two-alternative forced-choice paradigm shown in this example, $\gamma = 0.5$, $\eta = 3.5$, and $\tau = 1.5$ dB, which corresponds to 92% correct at threshold. Estimation of the threshold using QUEST usually requires 50–60 trials. Estimating a 75% threshold sets $\tau = -0.91$ dB, which is less efficient and requires more testing, perhaps as many as 100 trials.

Figure 11.7 shows (a) the assumed psychometric function, here graphed in decibels of contrast, and (b) and (c) two sample trial sequences corresponding with two different target threshold accuracies of 75% and 92% correct, respectively, for the QUEST procedure. Watson and Pelli[14] simulated the QUEST procedure for 4, 8, 16, 32, 64, and 128 trials. They estimated that convergence occurs by 64 trials in most circumstances.

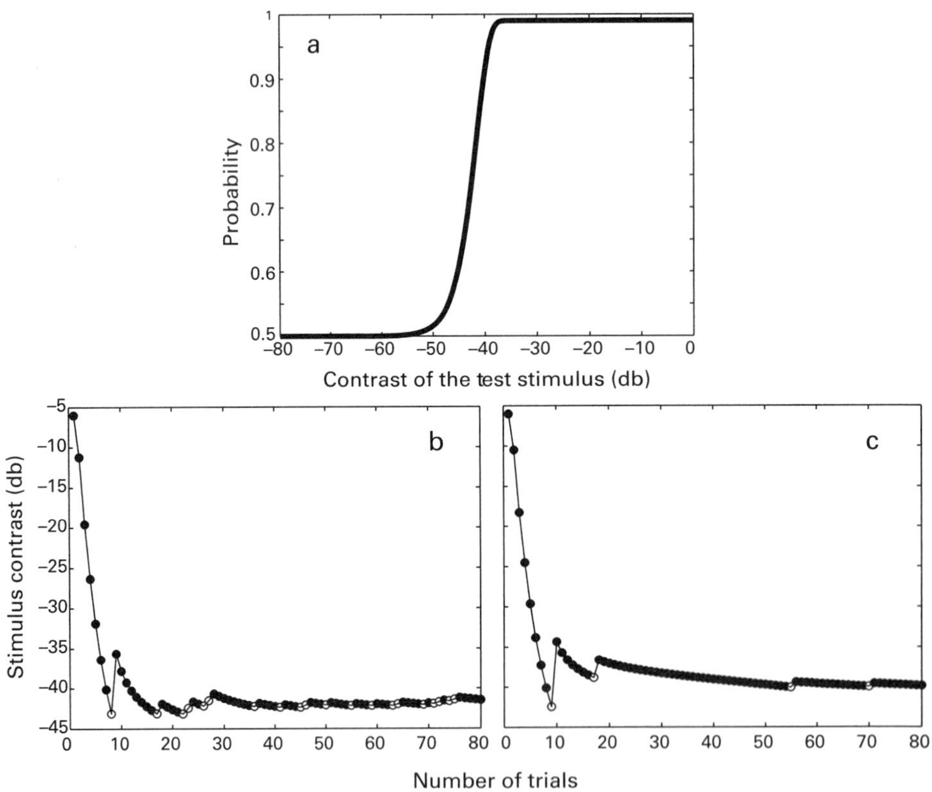

Figure 11.7
Assumed psychometric function and sample simulated trial sequences from the QUEST method for estimating two thresholds. (a) A psychometric function for a two-alternative forced choice task. (b, c) Two sample trial sequences from QUEST for estimating 75% and 92% correct thresholds.

QUEST is one of the most popular procedures for estimating the threshold if the approximate slope of the psychometric function is known in advance. An alternative to QUEST is ZEST,[15] which is identical except it uses the mean rather than the mode of the QUEST function $Q_n(T)$ to estimate the threshold.

11.3.4 Psi (Ψ) Method

The psi or Ψ method uses an adaptive Bayesian approach to estimate both the threshold and the slope of a psychometric function simultaneously.[16] The primary motivation is the observation that the slope of the psychometric function may depart significantly from the value of 3.5 usually assumed in the QUEST method. Using incorrect slope assumptions could yield incorrect estimates of the threshold. Therefore, it is useful to collect enough evidence to estimate both parameters of the psychometric function in situations where the slope is not known or where the slope can vary in different conditions. The assumed slope of 3.5 arose from empirical knowledge about contrast psychometric functions. However, psychometric functions may measure percent correct as a function of many other kinds of variables, such as orientation or color differences, and in these cases the slope in general can be expected to depart from 3.5.

The Ψ method[16] concurrently estimates threshold and steepness of the psychometric functions using an adaptive algorithm that minimizes the uncertainty about the estimated parameters of the psychometric function. It does this by minimizing the expected entropy on each trial—an approach that is described below.

The Ψ method was developed and tested for two-alternative forced-choice experiments. It assumes that the psychometric function is described by this equation:

$$\Psi_{a,b}(x) = \delta/2 + (1-\delta)G\left[\frac{1}{\sqrt{2}}\left(\frac{x}{a}\right)^b, 1\right]. \tag{11.10}$$

In this form of the psychometric function, x is the stimulus contrast, and G is the cumulative Gaussian with mean $(1/\sqrt{2})(x/a)^b$ and standard deviation of 1, evaluated at x. The parameter δ is the lapse rate—a probability with which the observer guesses, which limits the maximum percent correct at high values of x. The value of a is the threshold at the midpoint of the psychometric function, usually about 75% correct, and b is the slope. This corresponds to a psychometric function in which $d'(x) = (x/a)^b$. A slope of 2.8 in this equation is equivalent to the QUEST slope of 3.5 because of the different functional forms of the psychometric function.

Before the first trial, the experimenter defines the prior probability distribution for the two parameters, the threshold and slope of the psychometric function, as $p_0(a,b)$. The priors represent a best guess about these two parameters. A lapse rate such as 0.05 is typically assumed and not estimated.

Before every trial $t+1$, starting at $t=0$, the expected probability of a response r for any stimulus value x is

$$p_{t+1}(r \mid x) = \sum_a \sum_b p_t(a,b)\{r\Psi(x \mid a,b) + (1-r)[1-\Psi(x \mid a,b)]\}. \tag{11.11}$$

In this equation, $r=1$ is for a success (correct trial), and $r=0$ is for a failure (error). The probability of a correct trial is predicted by $\Psi(x \mid a,b)$, and the probability of an error is predicted by $1-\Psi(x \mid a,b)$. That is, the probabilities of a correct response and an error response are predicted by the location x along the psychometric function $\Psi(x \mid a,b)$. Every pair of threshold and slope values yields a different predicted probability of a correct or error response for a stimulus value x. The expected probability of response r given stimulus value x is computed by summing over all possible values of the two parameters weighted by their prior probability $p_t(a,b)$ as of trial t. This computation provides the expected probability of either response r for any stimulus x.

To select the stimulus for the next trial, the expected entropy is computed for any stimulus value x. To do this, the expected posterior probability distribution is computed for any stimulus value x for both potential responses r using the Bayes rule:

$$p_{t+1}(a,b \mid x,r) = \frac{p_t(a,b)\{r\Psi(x \mid a,b) + (1-r)[1-\Psi(x \mid a,b)]\}}{\sum_a \sum_b p_t(a,b)\{r\Psi(x \mid a,b) + (1-r)[1-\Psi(x \mid a,b)]\}}. \tag{11.12}$$

Then, the expected entropy for trial $t+1$ with any stimulus x, $E[H_{t+1}(x)]$ is

$$E[H_{t+1}(x)] = -\sum_{r=0}^{1} p_{t+1}(r \mid x) \sum_a \sum_b p_{t+1}(a,b \mid x,r) \log[p_{t+1}(a,b \mid x,r)]. \tag{11.13}$$

The entropy is a measure of the uncertainty associated with the values of random variables. Uncertainty is the opposite of predictability—the more we know about the random variables (here the parameters), the lower the entropy. To choose the next stimulus, the expected entropy (equation 11.13) is computed for all possible stimulus values, and the stimulus value with the smallest expected entropy is selected for presentation on the next trial.

This method uses a *greedy search algorithm* that looks one step ahead and asks the question—which stimulus value x is likely to reduce the variability in the posterior probability distributions for the two parameters a and b, which is defined as the stimulus value that minimizes the expected entropy. The "greedy" algorithms are heuristics that hope to find the true overall optimum by optimizing each step by itself.

After the stimulus value to test x_{t+1} has been selected, we run the trial and observe the response r_{t+1}. Then, the posterior distribution is updated using the Bayes rule for the observed data:

$$p_{t+1}(a,b \mid x_{t+1}, r_{t+1}) = \frac{p_t(a,b)\{r_{t+1}\Psi(x_{t+1} \mid a,b) + (1-r_{t+1})[1-\Psi(x_{t+1} \mid a,b)]\}}{\sum_a \sum_b p_t(a,b)\{r_{t+1}\Psi(x_{t+1} \mid a,b) + (1-r_{t+1})[1-\Psi(x_{t+1} \mid a,b)]\}}. \tag{11.14}$$

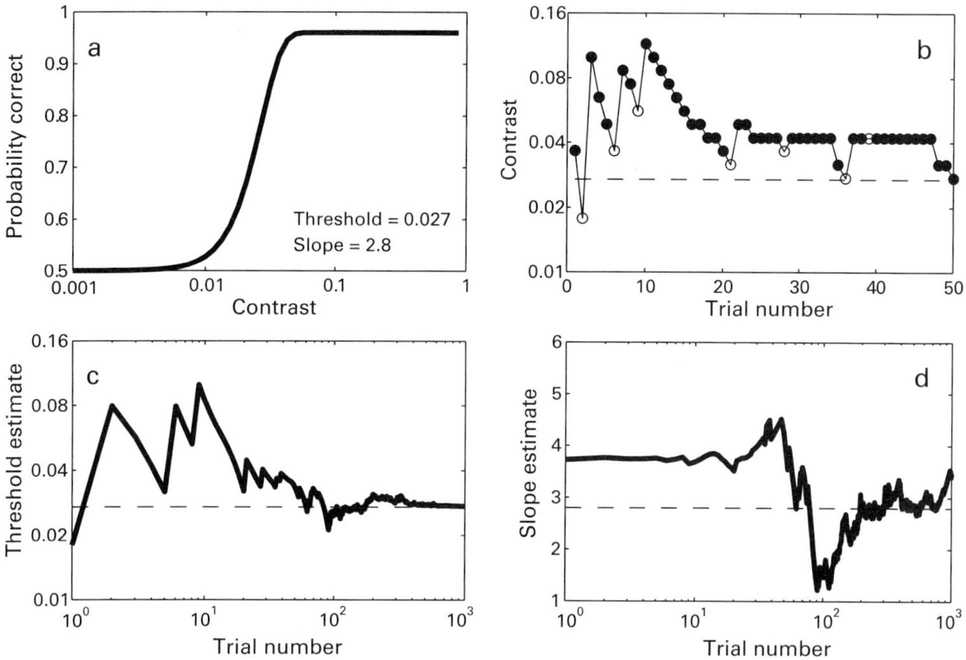

Figure 11.8
Assumed psychometric function and sample estimates of threshold and slope using the Psi (Ψ) procedure. (a) A psychometric function with threshold at 0.027, slope of 2.8, and lapse rate of 0.05. (b) A sample simulated trial sequence (c) The estimated threshold and (d) estimated slope of the psychometric function.

This provides the new posterior distribution. This procedure is reiterated until the criterion for the stop rule is met.

The standard stop rule for the Ψ method is based on the number of trials. Kontsevich and Tyler[16] simulated a two-alternative forced-choice experiment and determined that threshold estimation within 2 dB (or about 23%) precision usually required 30 trials if you can assume a given slope, whereas estimating the slope with a similar precision may take as many as 300 trials. Figure 11.8 shows a simulated trial sequence, estimated threshold, and slope of the psychometric function from the Ψ method.

11.3.5 Quick Yes/No Method

The previous adaptive procedures are traditionally applied to forced-choice tasks. Yet there may be circumstances when the simplicity and directness of the Yes/No tasks would be preferable. Forced-choice tasks require either memory for two-interval forced-choice or spatial attention for two-alternative forced-choice tasks with simultaneous stimuli. The Yes/No tasks test a single stimulus and interval and rely neither on memory nor on attention. Percent "yes" performance reflects both sensitivity and bias (see section

8.3 of chapter 8). One cannot interpret a threshold associated with a percent "yes" level such as 75% without considering decision bias.

An adaptive method, the quick Yes/No (q-YN) method, was developed by Lesmes et al.[17] to estimate threshold for Yes/No tasks. It takes bias into consideration by estimating the threshold for a fixed d' rather than a target percent "yes." The method combines signal-detection theory (SDT) with Bayesian adaptive inference to measure rapidly the sensitivity and decision parameters for the observer.

In the q-YN method, the d' psychometric function is assumed to be of this form:

$$d'(c;\tau,\gamma,\kappa)=\frac{\kappa(c/\tau)^{\gamma}}{\sqrt{(\kappa^{2}-1)+(c/\tau)^{2\gamma}}}. \tag{11.15}$$

The value τ is the sensitivity threshold at $d'=1$, γ controls the steepness of the d' contrast psychometric function, and κ is the maximum or saturating d' at high contrast. For applications with nearly perfect performance corresponding with $d'>4$, the asymptote parameter κ can be set to 5. With κ set, the q-YN method estimates the two free parameters that describe the d' psychometric function: the threshold, τ, and the steepness parameter, γ. To estimate the threshold that corresponds with d' of 1, the q-YN procedure must estimate a decision criterion, λ, or the false alarm rate when the signal is absent. It uses the estimated false alarm rate to construct the psychometric function for percent "yes":

$$\Psi_{yes}(c)=1-G(\lambda-d'(c)), \tag{11.16}$$

where $G(x)$ is the standard cumulative Gaussian function. If a lapse rate for inattention errors, δ (usually set at 4%), is included, the corresponding psychometric function is

$$\begin{aligned}\Psi'_{yes}(c)&=\delta+(1-\delta)\Psi_{yes}(c)\\&=\delta+(1-\delta)[1-G\left(\lambda-\frac{\beta(c/\tau)^{\gamma}}{\sqrt{(\beta^{2}-1)+(c/\tau)^{2\gamma}}}\right)].\end{aligned} \tag{11.17}$$

There are three different variants of the q-YN procedure: one for simple Yes/No detection, one to allow separate conservative and liberal criteria, and another to allow three response categories of Yes/No and unsure. The simple Yes/No procedure estimates three parameters: threshold τ, steepness of the psychometric function γ, and the false alarm rate λ. A procedure that cues either a strict or liberal criterion on different trials requires four parameters: τ, γ, and λ_{strict} and λ_{liberal} to define two different percent "yes" psychometric functions, one for the strict and one for the liberal criterion. These two functions both express the same d' psychometric function with different false alarm rates. A similar elaboration with multiple criteria describes performance in the rating procedure.

In the q-YN method, psychometric functions are defined with either three or four parameters. Before any testing, the experimenter defines a prior probability distribution for each of the parameters. The principle of estimation and selection of the next stimulus is the same as for the other Bayesian methods. The stimulus value for the next trial is selected to minimize the expected entropy and therefore to provide the most expected new information. A Bayesian update procedure is used to compute the posterior distributions for each parameter given the priors and the observed data. The posterior distribution of the threshold parameter is estimated during the q-YN procedure directly, so it is simple to use a stop rule defined as a target precision for the distribution of that parameter. Alternatively, a set sample size may also define the stop rule.

While the Bayesian principles are the same, the expansion from the two parameter distributions of Ψ, for example, to the three or four parameters of the q-YN applications pose significant additional demands on computer memory and computation that must be optimized to allow efficient on-line computation that occurs between each trial in choosing the next stimulus.

An example of the q-YN procedure applied to a simple Yes/No detection experiment is shown in figure 11.9. This shows (a) a d' psychometric function along with (b) the corresponding percent "Yes" psychometric functions for a strict criterion and a lax criterion. Figure 11.9c and d show sample testing sequences for the strict and lax criterion applications. The horizontal lines in figure 11.9c and d are the "true" threshold, which is known here because this is a simulation. The sample testing sequences include a number of very low contrast trials, which provide data to the Bayesian model related to false alarm rate.

The q-YN method has been tested by simulation and psychophysically compared with results from the method of constant stimuli.[17] The simple q-YN, the two-criterion, and the rating methods all showed excellent convergence. All estimates of threshold were closely equated to the threshold estimated by the method of constant stimuli, and to each other. The results suggest that about 25 trials may be sufficient to estimate a threshold corresponding to d' of 1 with a precision of about 2–3 dB.

The q-YN is a reliable new method for estimating the threshold and the bias in Yes/No paradigms. These Yes/No methods deliver criterion-free thresholds that have previously been exclusive to forced-choice paradigms and methods.

11.4 Adaptive Bayesian Procedures for Psychological Functions and Surfaces

Many of the fundamental properties of the visual system measured in visual psychophysics are expressed as functions or surfaces of stimulus properties. One example is the contrast sensitivity function, which captures the sensitivity to stimuli of different spatial frequencies and has been widely used as a "front end" in models of early visual processing. Another example might be sensitivity as a function of several characteristics such as spatial frequency and temporal frequency. The traditional ways of measuring these functions for individual observers are quite laborious. For example, for contrast sensitivity functions,

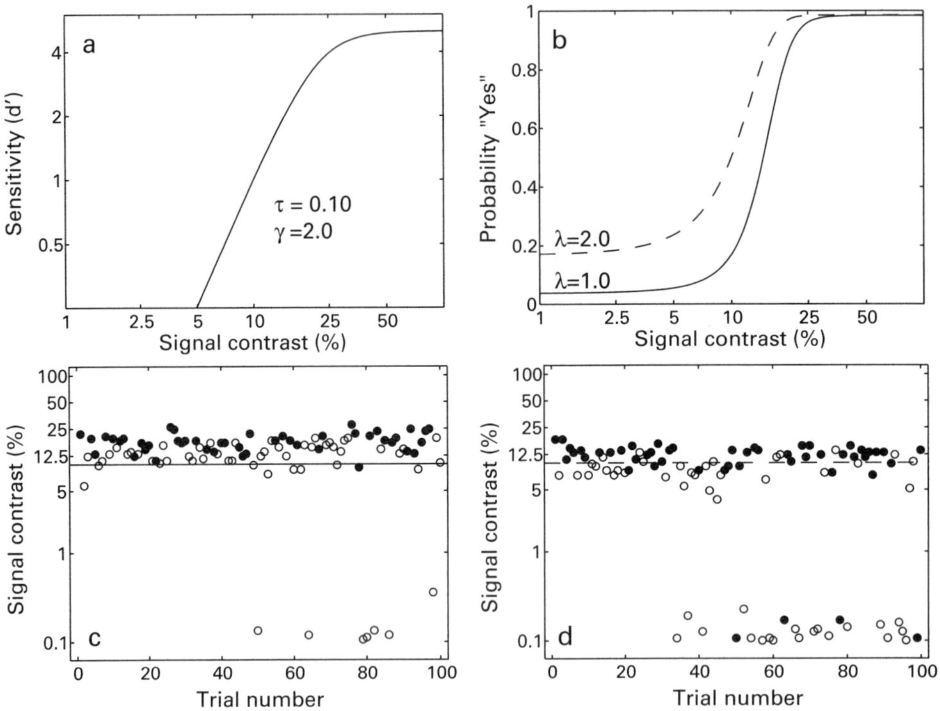

Figure 11.9
An illustration of a quick Yes/No (q-YN) method of threshold estimation. (a) A *d′* psychometric function and (b) corresponding percent "yes" psychometric functions for a strict criterion and a lax criterion. (c, d) Sample testing sequences for the strict and lax criterion. Filled circles are "yes" responses, and open circles are "no" responses.

we estimate contrast threshold at a sampling of different spatial frequencies and fit a functional form to the threshold estimations.

Traditional adaptive procedures reviewed in section 11.3 were developed to estimate threshold and in a few cases the slope of the psychometric function for a single stimulus condition. Using adaptive procedures to estimate the threshold at specified performance levels and then estimating the functional form from those thresholds is one way to measure perceptual functions. However, a new alternative idea is to use adaptive Bayesian procedures to estimate directly the parameters of the visual function.[18] The quick procedures use new computational methods in the Bayesian framework to estimate the parameters of some of the important functions of visual perception.[18–20]

There are several requirements for these new adaptive procedures. First, the form of the psychological function must be known or well approximated by a function with a small number of parameters. Next, this function must translate into predicted performance probability. The methods compute a trial-by-trial update of the posterior probability distributions of the parameters. The next stimulus that will be most informative for testing is selected by minimizing the one-step ahead expected entropy, and the procedure chooses an appropriate stop rule.

Quick adaptive methods for the estimation of parameters of known functions have the potential for significant reductions in the testing requirements in estimating these basic perceptual functions for individuals.

11.4.1 Contrast Sensitivity Function

As introduced in section 1.3.2 of chapter 1, sensitivity—or the minimum physical stimulus or difference that is just detectable[21,22]—is one of the most fundamental aspects of psychophysics that describes the limitations of the sensory system.[23] The contrast sensitivity function (CSF)[24,25] summarizes properties and limitations of the first stages of visual neurons[26,27] and is important in understanding clinical deficits in functional vision.[28]

Constructing the CSF usually involves estimating threshold at 5–10 different spatial frequencies of test grating, often requiring about 500–1000 trials, or 30–60 minutes of testing. The difficulty of measuring even a single CSF makes it less likely to be used in clinical settings. In the laboratory, often we wish to measure changes in the CSF as a function of other variables, such as temporal frequency, eccentricity, external noise, or practice. These require many times as many trials and many hours of testing.

The quick contrast sensitivity function, or q-CSF, was developed to measure CSF efficiently. The method assumes one of the good known functional forms, the *truncated log parabola*, as the shape to be estimated. As detailed in section 10.4.3 of chapter 10, sensitivity, $S_\theta(f)$, corresponding to 1/threshold, is defined with four parameters: (1) the peak gain (sensitivity) $\gamma_{\max}$; (2) the peak spatial frequency $f_{\max}$; (3) the bandwidth β, which describes the function's full-width at half-maximum (in octaves); and (4) ν, the truncation level at low spatial frequencies:

$$\log_{10}[S_\theta(f)] = \begin{cases} \log_{10}(\gamma_{\max}) - \kappa\left(\dfrac{\log_{10}(f) - \log_{10}(f_{max})}{\beta'/2}\right)^2, & if\ f \geq f_{\max} \\ log_{10}(\gamma_{\max}) - \nu, & if\ f < f_{\max}\ \&\ S_\theta(f) < \gamma_{\max} - \nu \end{cases}, \tag{11.18}$$

where $\kappa = \log_{10}(2)$ and $\beta' = \log_{10}(2\beta)$. Each set $\theta = \{\gamma_{\max}, f_{\max}, \beta, \upsilon\}$ of parameter values, together with this functional form, defines a different CSF.

The probability of a correct response for a grating of frequency f and contrast c is given by the log-Weibull psychometric function:

$$\Psi_\theta(f,c) = \min\{1-\delta,\quad 0.5[1-10^{\eta[-\log_{10}[S_\theta(f)]-\log_{10}(c)]}]\}. \tag{11.19}$$

The q-CSF method assumes that the steepness parameter of the psychometric functions does not change with spatial frequency. It is usually set to $\eta = 2$ based on the experimental evidence in this domain. The q-CSF allows for a small proportion of lapse errors, δ, usually set to 4%.[29,30]

After choosing this functional form, the experimenter sets a prior probability density for the four parameters of the CSF. To choose the stimulus to test on the next trial, a one-step ahead search finds the grating stimulus—a combination of spatial frequency and contrast—that minimized the expected entropy, or maximizes the expected information gain, about the CSF parameters. Data collected at one spatial frequency improve the estimates of the parameters of the whole CSF and so each trial of data improves the estimates across all spatial frequencies. The stop-rule for the q-CSF aims to achieve a certain precision for the parameter estimates. Alternatively, one could set the stop rule to a certain number of trials.

Although the Bayesian update procedure in the quick methods is analogous to that in the Ψ method,[16] it involves optimization of three or four parameters rather than two and may define several variations of the stimulus in multidimensional space (here spatial frequency and contrast).

The multidimensional parameter space is labeled as $\vec{\upsilon}$ where $\vec{\upsilon} = (f_{\max}, \gamma_{\max}, \beta, \nu)$, and the multidimensional stimulus space is labeled as $\vec{x}$ where $\vec{x} = (f, c)$ for spatial frequency and contrast of the test stimulus. Before the first trial, the experimenter defines the a priori probability distribution $p_0(\vec{\upsilon})$ that represents prior evidence about typical values of the parameters of the function describing the shape of the CSF.

Then, before every trial $t+1$, starting at $t=0$, the q-CSF computes the expected probability of each of the responses $r=0$ or $r=1$ in every possible stimulus condition $\vec{x} = (f, c)$:

$$p_{t+1}(r \mid \vec{x}) = \sum_{\upsilon} \{rp(r \mid \vec{\upsilon}, \vec{x}) + (1-r)[1 - p(r \mid \vec{\upsilon}, \vec{x})]\} p_t(\vec{\upsilon}) \,. \tag{11.20}$$

It applies the Bayes rule to compute the posterior probability distributions $p_{t+1}(\vec{\upsilon} \mid \vec{x}, r)$ following each possible response to each possible stimulus $\vec{x}$:

$$p_{t+1}(\vec{\upsilon} \mid \vec{x}, r) = \frac{p_t(\vec{\upsilon})\{rp(\text{correct} \mid \vec{x}, \vec{\upsilon}) + (1-r)[1 - p(\text{correct} \mid \vec{x}, \vec{\upsilon})]\}}{\sum_{\upsilon} p_t(\vec{\upsilon})\{rp(\text{correct} \mid \vec{x}, \vec{\upsilon}) + (1-r)[1 - p(\text{correct} \mid \vec{x}, \vec{\upsilon})]\}} . \tag{11.21}$$

It then computes the expected entropies of every possible stimulus following either a correct or incorrect response to$\vec{x}$:

$$E[H_{t+1}(\vec{x})] = \sum_{r=0}^{1} -p_{t+1}(r \mid \vec{x}) \sum_{\upsilon} \{rp_{t+1}(\vec{\upsilon} \mid \vec{x}, r) \log[p_{t+1}(\vec{\upsilon} \mid \vec{x}, r)] + (1-r)[1 - p_{t+1}(\vec{\upsilon} \mid \vec{x}, r)]$$
$$\log[1 - p_{t+1}(\vec{\upsilon} \mid \vec{x}, r)]\}. \tag{11.22}$$

The stimulus condition chosen for testing on the next trial is the stimulus that minimizes the expected entropy:

$$\vec{x}_{t+1} = \arg\min_{x} E[H_{t+1}(\vec{x})]\,. \tag{11.23}$$

The actual response on trial $t+1$ is used to compute the posterior probability function corresponding to $\vec{x}_{t+1}$ and used as the prior probability function for the subsequent trial:

$$p_{t+1}(\vec{\upsilon} \mid \vec{x}_{t+1}, r_{t+1}) = \frac{p_t(\vec{\upsilon})\{r_{t+1}p(\text{correct} \mid \vec{x}_{t+1}, \vec{\upsilon}) + (1-r_{t+1})[1-p(\text{correct} \mid \vec{x}_{t+1}, \vec{\upsilon})]\}}{\sum_{\upsilon} p_t(\vec{\upsilon})\{r_{t+1}p(\text{correct} \mid \vec{x}_{t+1}, \vec{\upsilon}) + (1-r_{t+1})[1-p(\text{correct} \mid \vec{x}_{t+1}, \vec{\upsilon})]\}}. \tag{11.24}$$

The estimation of one-step ahead expected entropy over four parameters in a four-dimensional parameter space for each stimulus specified in a two-dimensional space would require a prohibitive amount of computation if it were done exhaustively. The integration of modern computational algorithms, such as Markov chain Monte Carlo (MCMC),[31] into the q-CSF speeds this computation. MCMC methods are a class of algorithms that are based on sampling or estimating the posterior distributions as a function of the multidimensional parameter and multidimensional stimulus spaces. It is estimated that MCMC algorithms may reduce the computational load by a factor of 100 or more. Integration of the MCMC algorithms into the search for the next stimulus is what makes the q-CSF feasible in real-time testing.

Figure 11.10a shows a hypothetical underlying CSF along with marks showing the simulated sequence of test trials used by q-CSF. Figure 11.10b shows an actual empirical sequence of test trials from a q-CSF run along with the final best-fitting CSF for those data.

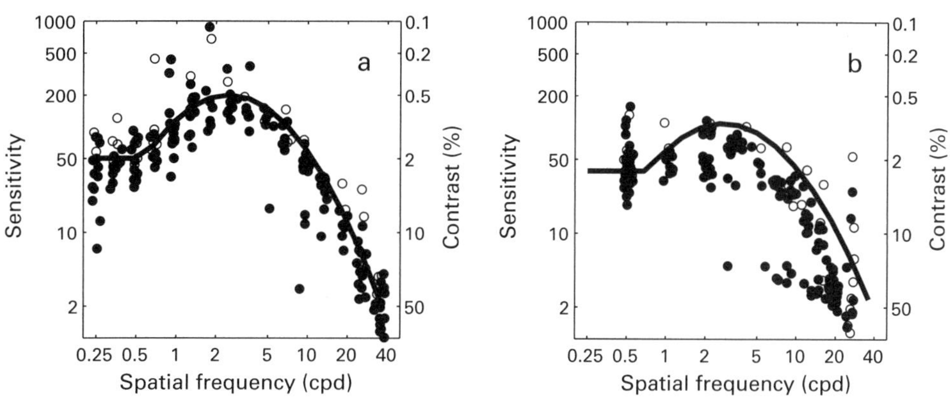

Figure 11.10
An illustration of a quick contrast sensitivity function (q-CSF) method. (a) A hypothetical underlying CSF along with marks showing the simulated sequence of test trials used by the q-CSF. (b) An actual empirical sequence of test trials from a q-CSF run along with the final best-fitting CSF for those data.

The q-CSF method was validated with a psychophysical study. The q-CSF with about 100 trials and 5 min of testing had excellent agreement with a CSF measured for the same individuals using classic methods, 1000 trials and 50 min of testing, to a precision of 2–3 dB across the different spatial frequencies.[19] These validation results were consistent with simulation analyses of the method.

The q-CSF method provides an effective and efficient method to measure the CSF in both normal and clinical applications.[20] The applications of this function are now far more accessible to integrate into larger studies of visual function.

11.4.2 Threshold versus External Noise Contrast Functions

The threshold versus external noise contrast (TvC) function measures the threshold contrast needed for either detection or discrimination as a function of the level of external noise added to the stimulus. As described in section 9.3.1 of chapter 9, measurements of TvCs at three or more performance levels provides the necessary information to specify the parameters of the underlying observer models and the corresponding functions. These are very useful in characterizing the visual systems of individual observers.

Traditional methods of measuring the TvC with the method of constant stimuli require about 3000–4000 trials, and staircase methods more often used for multiple estimates (over perceptual learning, for example) may require as many as 1500 trials. These high data requirements have limited the use of the TvC procedure. The *quick threshold versus external noise contrast*, or q-TvC, method was designed to minimize the number of trials needed for good estimation, and so extend the applicability of TvC measurements.[18] The q-TvC, or an equivalently rapid method of measurement, may be critical for practical applications to clinical populations.

The q-TvC method estimates threshold contrast as a function of external noise contrast in the stimulus for several levels of performance, 65%, 79%, and 92% correct. These three performance levels sample the psychometric functions and provide an estimate of their slope. It is known from previous research that these slopes are the same at different external noise levels.

Empirical regularities in the shape of the TvC functions lead to a three-parameter description in a simplified form. The three parameters are c_0, the contrast threshold at 79% correct in zero or low external noise; N_c, the critical noise at which the thresholds begin to depart from c_0, and η, the (common) slope of the psychometric functions. The simplified form of the TvC is bilinear and can be specified by c_0 and N_c because the high noise rising limb (on the log–log axes) is assumed to have a slope of 1.

At 79% correct, the threshold at different noise levels, $\tau(N_{\text{ext}})$, can be described by functions corresponding with $N_{\text{ext}} \leq N_c$, or the flat portion in low external noise, and $N_{\text{ext}} > N_c$, or the rising portion in higher external noise:

$$\tau(N_{\text{ext}}) = \begin{cases} \tau_0 \quad\;\; = \log(c_0), & \textit{if } N_{\text{ext}} \leq N_c \\ \tau_0 + \Delta\tau = \log(c_0) + \log(N_{\text{ext}}) - \log(N_c), & \textit{if } N_{\text{ext}} > N_c. \end{cases} \tag{11.25}$$

The corresponding expected percent correct as a function of external noise and signal contrast is defined using the log-Weibull psychometric function:

$$\psi(s,N_{\text{ext}})=\begin{cases}\gamma+(1-\gamma-\lambda/2)[1-\exp(-\exp\{\eta[\log(s)-\alpha_0]\})], & \text{if } N_{\text{ext}}\le \boldsymbol{N}_{\text{c}}\\ \gamma+(1-\gamma-\lambda/2)[1-\exp(-\exp\{\eta \\ [\log(s)-\alpha_0-\log(N_{\text{ext}})+\log(N_c)]\})], & \text{if } N_{\text{ext}}> \boldsymbol{N}_{\text{c}}.\end{cases} \quad (11.26)$$

The q-TvC uses the Bayesian adaptive procedure described in the previous section for q-CSF. In this case, the multidimensional parameter space is $\vec{v}=(c_0,N_c,\eta)$, the contrast in low noise, the critical value of external noise, and the common slope of the log psychometric function at all levels of external noise. And the multidimensional stimulus space is $\vec{x}=(N_{\text{ext}},c)$, the level of external noise and the signal contrast.

Figure 11.11 shows typical empirical results using the q-TvC method. The q-TvC method has been evaluated with both simulation and psychophysical validation experiments. Simulations showed that fewer than 300 trials are needed to estimate TvC functions at three widely separated criteria with a bias less than 5% and a precision of 1.5 dB or less.[18] The method showed excellent agreement between estimates of the TvCs with the q-TvC and a traditional method of constant stimuli in an orientation identification task, with correlations greater than 0.95. The q-TvC used 240 trials, and the method of constant stimuli used 1920 trials—a nearly 90% reduction in the amount of required testing.

The q-TvC provides a relatively efficient measurement of threshold versus external noise contrast functions characterization of observers. The rapidity of measurement makes

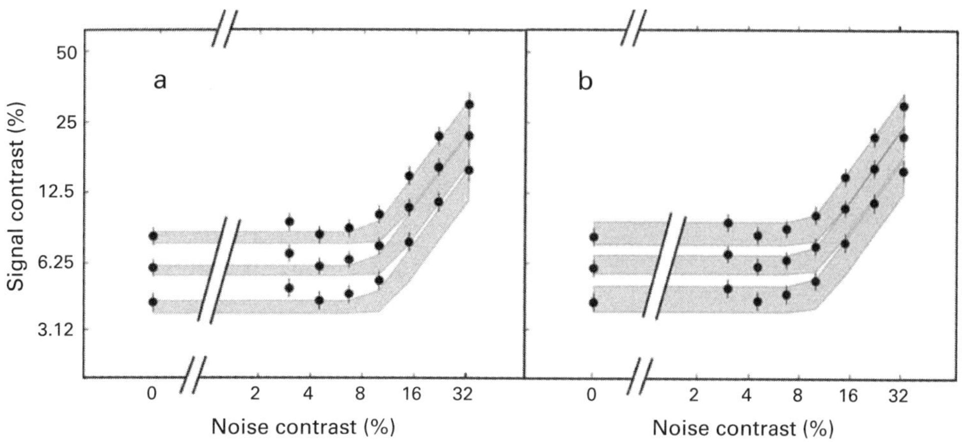

Figure 11.11
An illustration of the quick threshold versus external noise contrast (q-TvC) method. TvC functions (at 65%, 79%, and 92% correct) estimated by the q-TvC method from (a) 240 and (b) 480 trials. Shaded regions represent ±1 SD. The TvC functions collected with the method of constant stimuli are presented as circles, with error bars reflecting variability estimates (1 SD).

it a candidate for use in clinical populations, for children, and in situations in which this might be tested in multiple conditions.

11.5 Bayesian Approaches to Data Analysis

In this chapter, we covered adaptive methods for estimating parameters of psychological functions. Many of the methods used the Bayes rule for selecting stimulus conditions to test and to estimate parameters of key psychological functions or surfaces. The general Bayesian approach has been widely used not just in adaptive methods but in analysis of standard experiments as well. The purpose of Bayesian approaches for these standard experiments has been to derive summary statistics or estimate parameters of psychological models.[32–34]

The Bayesian approaches to model fitting and estimation provides an alternative method to the least-squares and maximum likelihood methods considered in chapter 10. The Bayesian methods provide the posterior distributions of key model parameters given the data and priors. This makes it relatively easy to estimate the confidence intervals or other measures of the variability of the parameters.

The Bayesian framework also provides an alternative method for model comparison.[35–37] The basic idea involves the computation of the *Bayes factor*:

$$K = \frac{p(\text{Data} \mid \text{Model}_1)}{p(\text{Data} \mid \text{Model}_2)}. \tag{11.27}$$

Bayes factors greater than 3 or 4 indicate that Model_1 is substantially better than Model_2. The Bayes factor is one component of the more general Bayesian formulation of the posterior beliefs about the two models:

$$\frac{p(\text{Model}_1 \mid \text{Data})}{p(\text{Model}_2 \mid \text{Data})} = \frac{p(\text{Data} \mid \text{Model}_1)}{p(\text{Data} \mid \text{Model}_2)} \frac{p(\text{Model}_1)}{p(\text{Model}_2)}. \tag{11.28}$$

The first factor following the equal sign is the Bayes factor, and the second factor takes into account the experimenter's priors about the two competing models. Posterior beliefs may disagree with the Bayes factor if the priors for one model over another are sufficiently strong. It has been claimed that Bayesian model selection automatically takes model complexity into consideration.[32] This is because models with more potential parameter values have more diffuse prior probability distributions. This contrasts to the approach of the Akaike Information Criterion (AIC) and Bayesian Information Criterion (BIC), which directly punish models for the number of free parameters.

Some researchers have used the Bayesian machinery to formulate an alternative to standard hypothesis testing. They have replaced traditional hypothesis tests such as *t*-tests, analysis of variance, and linear and nonlinear regression with a Bayesian approach focused on posterior distributions of the coefficients of a *generalized linear model* (GLM). A GLM claims that

$$y = f\left(\beta_0 + \sum_{i=1}^{M} \beta_i x_i\right), \tag{11.29}$$

where y is a dependent variable, the x_is are independent variables, the β_is are coefficients, and the function $f(.)$ is a linking function. The standard regression and analysis of variance problems can be recast in the GLM framework. Estimation of the posterior distributions of the coefficients (parameters) β_i can function somewhat like traditional hypothesis tests in assisting the researcher to make decisions about probable states of the world.[32]

Some argue that the Bayesian framework should replace most of the standard statistical developments. A number of textbooks introduce and discuss this new approach.[32,38–41]

11.6 Summary

The advantages of adaptive psychophysical procedures are only beginning to be mined in the area of visual perception. As these methods are improved, and as practitioners become more familiar with them, we anticipate that adaptive testing will be increasingly central to visual psychophysics and visual testing. The adaptive methods greatly reduce the burden of data collection. Efficient data collection can replace extensive testing of a few individuals in the dedicated laboratory with manageable testing protocols that bring sophisticated measurements of the visual system and visual function into reach for applications in the field.

This chapter begins with the classic adaptive methods, such as staircases, that estimated threshold. We illustrate how these approaches to the estimation of threshold were improved by more sophisticated sampling methods, and then expanded to consider threshold and slope. Finally, we showed how a parametric understanding of empirical functions—psychometric functions, contrast sensitivity functions, threshold versus external noise contrast functions—expand the ability to estimate psychological functions or surfaces. Several examples illustrated how this estimation can be approached with the machinery of Bayesian inference and effective sampling methods.

These new adaptive methods have been applied to a few key phenomena of visual perception, but by analogy could be developed for many other standard visual functions. The first requirement is a good understanding of a functional form that describes the behavior with a relatively small number of parameters. This functional form may first need to be established by more traditional methods and extensive validation testing. However, given that a parametric form provides a good characterization of the behavior, the remaining machinery of adaptive testing and estimation can easily be adapted. We expect that aspects of the sampling and computational methods will also continue to improve.

Adaptive methods have the potential to convert measurements that were the purview of a few laboratories and heroic observers into rapid testing protocols. This opens the field to consideration of individual differences in visual function and to measurement of individuals for whom long protocols are impossible, such as children, or special populations. In combination with mobile testing methods that use mobile devices (such as the iPad), we may take visual testing into the field.

An additional value for adaptive methods in the laboratory is the ability to measure changes in state, such as the effect of practice, in a smaller number of trials. It also allows us to consider measuring performance in a larger number of conditions.

The adaptive psychophysical methods offer important and wide-ranging advantages. There remain, however, issues with these methods that require care in application and that may be addressed in further development. One possible issue is that of sequential dependencies in testing, or predictability or incentive in the testing sequence. A related issue is the impact of lapse trials that may be overweighted in estimation, especially if they occur at particular points in testing. The introduction of a certain number of trials drawn more broadly from the stimulus range and so sampling more of the parameter space may guard against atypical early trials and generally improve the robustness of testing.

Another component that may be improved by future development involves the greedy one-step ahead methods that currently optimize sampling based on computations involving the next trial alone. As new algorithms are developed and computational speed is improved, we may expect that new methods may be able to consider more global optimization involving multiple steps ahead.

Finally, the parametric methods we described here assume a single functional form or estimate an index, for example the threshold, for one condition. Further work is under way to develop protocols using adaptive methods to help choose between several different *functional forms*, or optimize the testing for differences in *parameter values* (for the same form) between two or more conditions. The further development of adaptive testing is an important wave of the future that will contribute to new theory and to the application of measurement and estimation for individuals in biomedical and related research areas.

References

1. Dixon WJ, Mood A. 1948. A method for obtaining and analyzing sensitivity data. *J Am Stat Assoc* 43: 109–126.

2. Müller-Lyer FC. 1889. Optische urteilstäuschungen. *Archiv für Physiologie 2 (Suppl)*: 263–270.

3. Levitt H. 1971. Transformed up–down methods in psychoacoustics. *J Acoust Soc Am* 49: 467–477.

4. Kesten H. 1958. Accelerated stochastic approximation. *Ann Math Stat* 29: 41–59.

5. Robbins H, Monro S. 1951. A stochastic approximation method. *Ann Math Stat* 22(3): 400–407.

6. Treutwein B. 1995. Adaptive psychophysical procedures. *Vision Res* 35(17): 2503–2522.

7. Derman C. 1957. Non-parametric up-and-down experimentation. *Ann Math Stat* 28(3): 795–798.

8. Kaernbach C. 1991. Simple adaptive testing with the weighted up-down method. *Atten Percept Psychophys* 49(3): 227–229.

9. Tyrrell RA, Owens DA. 1988. A rapid technique to assess the resting states of the eyes and other threshold phenomena: The modified binary search (MOBS). *Behav Res Methods* 20(2): 137–141.

10. Taylor MM, Creelman CD. 1967. PEST: Efficient estimates on probability functions. *J Acoust Soc Am* 41: 782–787.

11. Pentland A. 1980. Maximum likelihood estimation: The best PEST. *Atten Percept Psychophys* 28(4): 377–379.

12. Green DM. 1990. Stimulus selection in adaptive psychophysical procedures. *J Acoust Soc Am* 87: 2662–2274.

13. Taylor M. 1971. On the efficiency of psychophysical measurement. *J Acoust Soc Am* 49: 505–508.

14. Watson AB, Pelli DG. 1983. QUEST: A Bayesian adaptive psychometric method. *Atten Percept Psychophys* 33(2): 113–120.

15. King-Smith PE, Grigsby SS, Vingrys AJ, Benes SC, Supowit A. 1994. Efficient and unbiased modifications of the QUEST threshold method: theory, simulations, experimental evaluation and practical implementation. *Vision Res* 34(7): 885–912.

16. Kontsevich LL, Tyler CW. 1999. Bayesian adaptive estimation of psychometric slope and threshold. *Vision Res* 39(16): 2729–2737.

17. Lesmes LA, Lu ZL, Tran NT, Dosher BA, Albright TD. 2006. An adaptive method for estimating criterion sensitivity (d') levels in yes/no tasks. *J Vis* 6(6): 1097.

18. Lesmes LA, Jeon ST, Lu ZL, Dosher BA. 2006. Bayesian adaptive estimation of threshold versus contrast external noise functions: The quick TvC method. *Vision Res* 46(19): 3160–3176.

19. Lesmes LA, Lu ZL, Baek J, Albright TD. 2010. Bayesian adaptive estimation of the contrast sensitivity function: The quick CSF method. *J Vis* 10(3): 17.1–21.

20. Hou F, Huang CB, Lesmes L, Feng LX, Tao L, Zhou YF, Lu ZL. 2010. qCSF in clinical application: Efficient characterization and classification of contrast sensitivity functions in amblyopia. *Invest Ophthalmol Vis Sci* 51(10): 5365–5377.

21. Weber EH. *De pulsu, resorptione, auditu et tactu*. Leipzig: Koehler; 1834.

22. Fechner G. *Elemente der psychophysik*. Leipzig: Breitkopf & Härtel; 1860.

23. Graham NVS. *Visual pattern analyzers*. Oxford: Oxford University Press; 2001.

24. Campbell FW, Robson JG. 1968. Application of Fourier analysis to the visibility of gratings. *J Physiol* 197(3): 551–566.

25. Enroth Cugell C, Robson JG. 1966. The contrast sensitivity of retinal ganglion cells of the cat. *J Physiol* 187(3): 517–552.

26. Movshon JA, Thompson ID, Tolhurst DJ. 1978. Spatial and temporal contrast sensitivity of neurones in areas 17 and 18 of the cat's visual cortex. *J Physiol* 283(1): 101–120.

27. Watson AB, Ahumada AJ. 2005. A standard model for foveal detection of spatial contrast. *J Vis* 5(9): 717–740.

28. Schwartz SH. *Visual erception: A clinical orientation*. New York: McGraw-Hill Medical; 2009.

29. Wichmann FA, Hill NJ. 2001. The psychometric function: I. Fitting, sampling, and goodness of fit. *Percept Psychophys* 63(8): 1293–1313.

30. Swanson W, Birch E. 1992. Extracting thresholds from noisy psychophysical data. *Percept Psychophys* 51(5): 409–422.

31. Gilks WR, Richardson S, Spiegelhalter DJ. *Markov chain Monte Carlo in practice*. Boca Raton: Chapman & Hall/CRC Press; 1996.

32. Kruschke JK. *Doing Bayesian data analysis: A tutorial with R and EUGS*. Burlington, MA: Academic Press; 2010.

33. Lee MD. 2008. Three case studies in the Bayesian analysis of cognitive models. *Psychon Bull Rev* 15(1): 1–15.

34. Lee MD. 2008. BayesSDT: Software for Bayesian inference with signal detection theory. *Behav Res Methods* 40(2): 450–456.

35. Raftery AE. 1995. Bayesian model selection in social research. *Sociol Methodol* 25: 111–164.

36. Pitt MA, Myung IJ, Zhang S. 2002. Toward a method of selecting among computational models of cognition. *Psychol Rev* 109(3): 472–491.

37. Shiffrin RM, Lee MD, Kim W, Wagenmakers EJ. 2008. A survey of model evaluation approaches with a tutorial on hierarchical Bayesian methods. *Cogn Sci* 32(8): 1248–1284.

38. Gelman A, Carlin JB, Stern HS, Rubin DB. *Bayesian data analysis*. Boca Raton, FL: CRC Press; 2004.

39. Sivia DS, Skilling J. *Data analysis: A Bayesian tutorial*. Oxford: Oxford University Press; 2006.

40. Carlin BP, Louis TA. *Bayesian methods for data analysis*.Boca Raton: Chapman & Hall/CRC Press; 2009.

41. Box GEP, Tiao GC. *Bayesian inference in statistical analysis*. New York: John Wiley; 1992.

12 From Theory to Experiments

A theory is a set of principles that explains the functioning of a system. From this set of theoretical principles, the experimenter must decide which aspects of the behavior of the system will be the focus of investigation and (ideally) develop a model of the system and generate a specific hypothesis and predictions. This is the process of starting with a theory and making that theory testable. Given a specific prediction or hypothesis, choices must be made about the general approach and the specific methods and design of testing. Chapter 10 discussed the methods of quantitative data analysis and model fitting. Chapter 11 discussed efficient testing procedures. This chapter considers the process of translating a theory to a hypothesis and then constructing a strong experiment. We focus on the development of experiments to provide strong tests of theories.

12.1 Start with a Theory: Generating Hypotheses

A theory of a domain or behavior—or at least a good hypothesis—is an ideal starting point for research. One dictionary definition of *theory* is "the general or abstract principle of a body of fact, a science, or an art" or "a plausible or generally acceptable general principle or body of principles offered to explain phenomena."[1] In the most developed science, we may have a very well-articulated theory. On the other extreme, we may simply have a good general idea.

In common language, a hypothesis may be a weaker form of theory—a possible explanation that is consistent with known facts so far, whereas a theory implies a more extensive set of evidence or greater range of conditions under which the principles may hold. A hypothesis often means a proposal or idea to be tested. The other side of a somewhat weaker form of theory is that more data and facts are required to generate a real test of the ideas.

In an ideal case, research starts with a theory or an idea and a goal to use this theory to understand behavior in a particular domain. It is often necessary to do some conceptual work to make a theory or idea testable in a specific domain—by further development, generalization, or specification of how the theory applies in that domain.

The researcher's next job is to make predictions and choose a paradigm to test those predictions. This is an iterative process. The selection of paradigm and predictions may be refined in successive attempts. The selection of a general approach to testing a theory may suggest a good paradigm for collecting data, but then the specific theoretical predictions need to be developed in more detail for the specific testing paradigm.

Next, researchers collect data or observations, summarize the results, and compare the observed results to the theory or model predictions. If the data are incompatible with the theory or model predictions, this requires a modification of the form of the model that applies in the particular domain or leads to a small or sometimes larger change in the theory itself. If the data are compatible with the predictions, this provides further support for the theory, and new predictions can be developed for further detailed testing. The next section provides several concrete examples.

12.2 From Theory to Experiment: Case Studies

12.2.1 Case 1: Mechanisms of Attention

One case study example of the transition from theory to experiment involves visual attention. From very early in the history of research on visual behavior—back to the late 18th century and the earliest writers in psychology—there have been several ideas about the role and function of visual attention. The earliest researchers recognized that information about the visual field was acquired best around the center of gaze or fixation and was influenced by the sequence of eye fixations. At the same time, early researchers realized that it was possible to attend away from the fixation of the eye to different locations in visual space or to different objects or features of objects.[2–5]

Early research on phenomena of visual attention, such as attention cued to a particular location in space, identified two potential mechanisms. One view saw attention operating like a filter to eliminate noise or distractors in the visual field in order to allow focused processing on one or several locations or objects.[6,7] This aligns with common usage in which attending to one thing implies trying to ignore all the other things that might compete for processing resources. Another view believed that attention improves the clarity or otherwise improves the representation of the attended object.[8,9] Early theories of both these mechanisms of attention were verbal theories or ideas—the ideas of filtering of noise or distractors and of clarification or increased strength of a target.

About a decade ago, we embarked on a research program to make these general verbal theories or ideas testable and quantifiable in a new way.[10] We began with a formal model of the observer, the perceptual template model (PTM) described in chapter 9. This model provides a good quantitative description of observer behavior for visual stimuli that vary in contrast or external noise. It provides a strong systematic set of predictions for how an observer in a single state operates in response to stimulus variation by estimating only a very small number of characteristic quantities, such as the gain of the perceptual template, the amounts of internal additive or multiplicative noise, or the nonlinearity in the visual system of an observer.

The insight was that this kind of model could be adapted to measure how the observer changed between *two or more states*. For example, if the state of attending versus not attending has no effect on visual perception, then the same system parameters would characterize performance in the two states. Or, if visual perception is different in the attended and unattended states, then the framework allows us to identify how attention changes the specific parameters of the system.

This strong theoretical framework of the PTM and external noise paradigm suggested testable predictions of the two verbal ideas about mechanisms by which attention might operate. The PTM framework created a natural way to develop performance signatures for the two mechanisms of attention for threshold versus external noise contrast (TvC) functions by measuring contrast thresholds in different amounts of external noise in the stimulus. Varying external noise provides a specific test of whether attention improves external noise filtering. External noise filtering by attention improves performance differentially in high-noise tests. In contrast, low-noise conditions allow us to ask whether attention still improves performance by making the stimulus somehow clearer even when there is no external noise to be filtered. We called this mechanism stimulus enhancement. External noise filtering was modeled as a reduction in the impact of external noise, whereas stimulus enhancement was modeled as a reduction in the internal additive noise.

Figure 12.1 shows the signature performance patterns for different mechanisms of attention as measured with external noise paradigms. It shows three distinct patterns or behavioral signatures, each corresponding to a particular mechanism of attention: (i) filtering out external noise by changing the template; (ii) enhancing the stimulus through reductions in internal additive noise; (iii) reductions in internal multiplicative noise or changes in nonlinearity parameter.

Effects of filtering in high external noise, or in the presence of visual distractors, correspond to the verbal idea of attention that operates like a filter by changing the template. Effects of attending in clear or noiseless displays at threshold correspond to the verbal idea of attention that operates by clarifying the visual representation. Both of these patterns have been observed in the data of attention experiments that measured TvC functions of spatially cued attention and of divided attention.[10–15] Of the two, filtering is the mechanism seen most often in attention experiments and may be the dominant mechanism across many different attention manipulations and paradigms. Sometimes, both mechanisms are found together in an experimental condition.[12,13] To date, there is no evidence for the third possible mechanism of attention through reduction of multiplicative noise or changes in nonlinearity that improve behavioral performance.

This analysis of visual attention started with verbal ideas or theories and created a framework for understanding attention effects by casting them within a quantitative model or theory of the observer. This in turn allowed the generation of testable predictions that

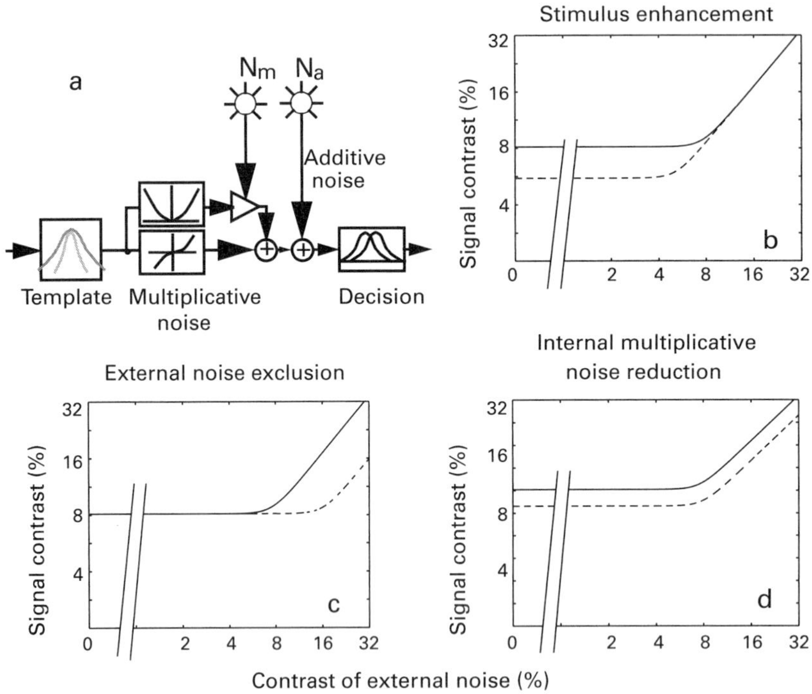

Figure 12.1
Mechanisms of attention within the perceptual template model (PTM) tested in TvC functions. (a) The perceptual template model. (b) Signature TvC changes by stimulus enhancement, where attention reduces contrast threshold in low external noise. (c) Signature TvC changes by external noise exclusion, where attention reduces contrast threshold in high external noise. (d) Signature TvC changes by internal multiplicative noise reduction, where attention alters the nonlinearity or multiplicative noise in the system affecting contrast threshold in all external noise conditions.

were signatures of each mechanism. This framework has provided one useful way to classify and think about effects of attention.[16,17]

12.2.2 Case 2: Mechanisms of Attention in Functional Magnetic Resonance Imaging

Case 1 investigated how to study the mechanisms by which covert attention might lead to performance improvements in perceptual tasks by studying attention within the PTM observer model. Another way to investigate the mechanisms of covert attention is to study the changes in neural response in early visual cortical areas using blood oxygenation level–dependent (BOLD) changes in functional magnetic resonance imaging (fMRI). One mechanism of attention, stimulus enhancement, predicts changes in response in the absence of external noise. These changes were modeled mathematically as relative reduction in additive noise but can be equivalently modeled as amplification of the signal. To relate the psychophysical mechanisms of attention to neural responses, fMRI was used to measure

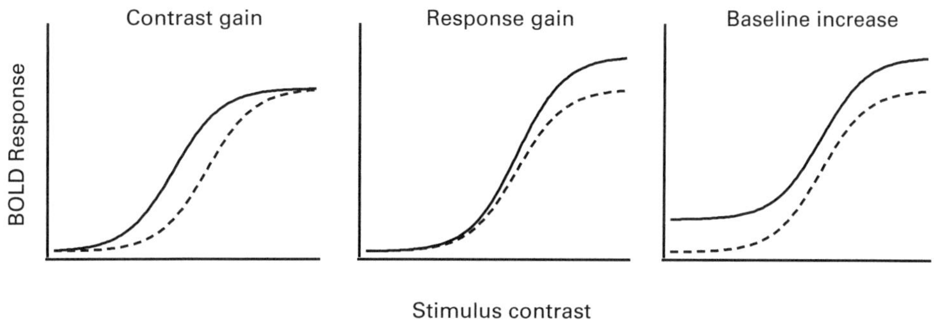

Figure 12.2
Potential effects of attention on the fMRI BOLD contrast response function: (a) contrast gain, (b) response gain, and (c) baseline increase. Solid lines show responses in attended conditions compared to unattended conditions shown with dashed lines.

the BOLD response to signal stimuli of different contrasts in several visual cortical areas: V1, V2, V3, V3A, and V4.[18] Cortical contrast response functions in the fMRI differed in attended and unattended conditions.

Using fMRI to measure neural responses to signals of increasing contrast in the absence of visual noise can discriminate several different effects of attention in the absence of external noise. Distinctive patterns, shown in figure 12.2, illustrate three functionally different ways in which attention can affect the contrast response of the visual system. Amplification of contrast increases responses only to stimuli with intermediate contrasts—a particular contrast when attended has the same effect as a larger contrast in the unattended condition. This pattern is labeled contrast gain in the physiology literature.[19,20] If neural responses are multiplied for attended stimuli, this leads to small differences at low contrasts and increasing differences between attended and unattended conditions at higher contrasts. This pattern is labeled response gain.[21] Finally, attention might simply raise the baseline fMRI response, shifting the response upward across the contrast response function.[19,21]

In one experiment we measured fMRI responses to sine waves of different contrasts. Observers either attended to a peripheral annulus containing a sine wave or did not attend to the sine wave and performed a task at fixation instead. Attention conditions were blocked together in this design. Each block of attended or unattended trials included a mixed sequence of trials of different contrast levels for the sine wave in an event-related design. The fMRI BOLD responses were measured for several seconds after each stimulus onset. This fMRI investigation discovered that attention does two things. It increases the responses at all contrasts through a baseline shift in activity. Attention also amplifies the response to contrast, showing a contrast gain pattern with increased responses for intermediate contrasts. The difference between attended and unattended responses were modeled as a combination of the effect patterns seen in figure 12.2a and c. The contrast gain was stronger in early cortical areas like V1 and somewhat smaller in higher cortical areas like V4. This study found an effect of attention on brain responses to visual stimulation that corresponds to stimulus enhancement within the PTM framework. It also shows evidence of baseline shifts that cannot be easily measured in the behavior alone.[18]

A companion fMRI experiment tested orientation discrimination for a peripheral sine-wave annulus in external noise. As in the previous case, the contrast of the signal was manipulated to measure contrast response functions, here in the presence of external noise. This experiment demonstrated evidence for external noise filtering in early visual areas, corresponding to the external noise filtering previously reported in psychophysical TvC experiments. If the sine-wave contrast is especially low, then the stimulus mostly consists of external noise, so a reduced response in this condition is a direct indication of external noise filtering. The responses in the earliest visual areas to external noise, with a very low contrast stimulus, was actually reduced in the attended condition.[22]

These measures of brain activity in different visual areas provide converging evidence that is consistent with the model-based behavioral signatures in the TvC functions and the PTM. This provides an especially strong interrelated body of evidence in support of these fundamental mechanisms of attention. The fMRI results specify how these mechanisms are embodied in the internal responses to the contrast stimuli in different visual cortical areas.

12.2.3 Case 3: Perceptual Learning and Visibility

In the attention example, we began with a long history of verbal theory and found a way to identify signature patterns for those ideas that made testable and unique predictions by using a quantitative model, the PTM, to investigate attention both behaviorally and using fMRI. Now, we consider a case where the initial level of theorizing begins with only a simple hypothesis and show how quantitative descriptive models have significant advantages that may allow you to generate new hypotheses to test and to develop theoretical understanding further.

Start with the general idea that practice will improve perceptual performance. Then, consider the case of how practice changes the contrast sensitivity function (CSF) of the observer. We begin with a simple verbal hypothesis that training or practice will improve the CSF of the observer in some way as yet unknown. Yet by choosing a quantitative form of an important test of visual processing, we stand a good chance of providing useful new information to improve the theory.

As described in section 10.4.3 of chapter 10 and section 11.4.1 of chapter 11, we have a descriptive model that approximates and describes the shape of the CSF. This is not a process model of the perceptual system, but merely a descriptive formula for the shape. The truncated log-parabola description of the CSF is repeated here:

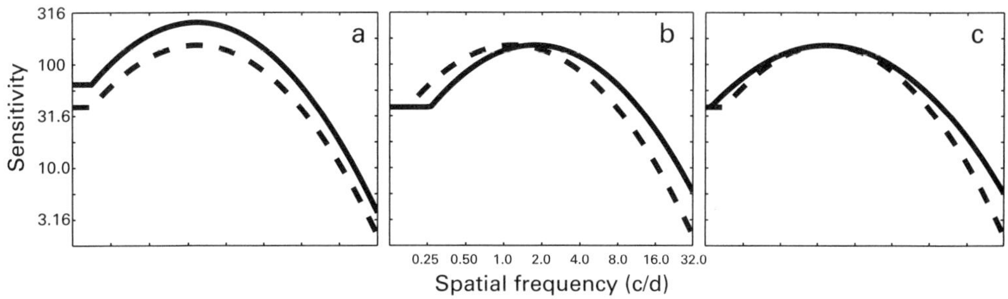

Figure 12.3
Potential effects of training on the contrast sensitivity function (CSF): (a) contrast gain, (b) increasing peak spatial frequency, and (c) increasing bandwidth.

$$\log_{10}[S(f)] = \begin{cases} \log_{10}(\gamma_{\max}) - \log_{10}(2)\left(\dfrac{\log_{10}(f) - \log_{10}(f_{max})}{\log_{10}(2\beta)/2}\right)^2, & \text{if } f \geq f_{\max} \\ log_{10}(\gamma_{\max}) - \delta, & \text{if } f < f_{\max} \,\&\, S(f) < \gamma_{\max} - \delta \end{cases} \qquad (12.1)$$

It has four parameters: the peak gain (sensitivity) $\gamma_{\max}$; the peak spatial frequency $f_{\max}$; the bandwidth β; and δ, the truncation level at low spatial frequencies.

By comparing the two CSFs before and after training, we can identify the ways in which a specific training protocol alters contrast sensitivity (figure 12.3). The experimenter devises a paradigm that measures the CSF before training, provides a training program, and then measures the CSF after training. Different forms of improvement are identified with changes in the parameters of the models. For example, if training alters the peak gain $\gamma_{\max}$, then the whole function should shift up. In this case, training makes everything better. In contrast, perhaps training that focuses on spatial frequencies just higher than the peak of the function may shift the peak spatial frequency $f_{\max}$ to be higher, or broaden the bandwidth β of frequencies to which the observer is sensitive. Several signature changes in the CSF are shown in figure 12.3a–c.

Just the availability of a strong paradigm with a good descriptive formulation can provide a scaffold from which it is possible to develop more serious tests of an initial vague hypothesis—and then generate more specific hypotheses for testing and evaluation. Together, these may stimulate the development of new process models of the human observer and of the nature of change of state due to training. At minimum, such a powerful quantitative test of a verbal hypothesis expands the data available for testing other generative models that will be more powerful and of higher quality.

12.3 Developing the Experiment

We have given several examples of starting with a theory or even a simple idea and developing testable predictions. We believe that it is especially useful to develop hypotheses using quantitative models or descriptive functions, as in the three case studies detailed earlier. In some situations, we already may have a good idea of what needs to be measured—the TvC curves or the CSF functions in the examples. In other cases, especially for broad or qualitative hypotheses, the choice of paradigm and design for testing allows a wide range of possible experiments. In either case, there are many detailed choices that must be made in determining an experimental design for testing hypotheses and theories.

This section describes a number of issues that arise while deciding on the design of an experiment. How do you go from specific predictions of a model to the details of an experimental design? How many conditions should be tested? What task should you use? Should multiple conditions be intermixed or tested in isolation? How many trials are needed for each condition of the experiment?

An experienced investigator in a particular area of research will have developed preferences for one kind of experiment over another and an intuitive sense of many of these aspects of design. Even so, whenever proposing an experiment or test that is new to you, it is important to understand the ways in which alternative experiments might reveal converging constraints on models or tests of hypotheses.

12.3.1 Choosing of the Experimental Setup

Starting from a theory or idea, the experimenter develops hypotheses to be tested. The approach to testing the hypothesis will determine the measurements required, and so the experimental setup. Experiments in psychophysics require the display of visual stimuli. Some experiments may include auditory cues or stimuli—or more rarely tactile or olfactory stimuli. In some cases, experimenters may choose to investigate theories using collateral measures to task performance, such as response times, electroencephalography (EEG) responses, eye movements, fMRI, and so on. These choices have obvious consequences for the experimental setups required.

Most cases considered in this book require high quality visual displays that can deliver accurately controlled grayscale or color contrasts. This requires checks for the display geometry, checks for pixel interactions, and calibration of contrasts using both psychophysical tests by the human eye and measurements by photometer or oscilloscope (see chapter 5).

If response times are collected and temporal precision is required, the experiment often involves the use of an external device such as an RTBox that collects manual responses and synchronizes the timing of those responses with the timing of a main computer controlling displays (see section 6.2 of chapter 6). Such devices allow the combined measurement of both speed and accuracy of behavioral responses.

An experimenter may decide to study a theory or hypothesis with augmented forms of data collection involving explicit and implicit behaviors or brain responses. For example, perhaps an experimenter wishes to measure the pattern of eye movements during a psychophysical task. Brain measures such as EEG, magneto-encephalography (MEG), and fMRI are of increasing interest for relating internal responses to the stimulus and to the decision and motor response. Each of these augmented data types—from eye movements to fMRI—are generally implemented with independent computers and measurement devices. In such cases, one important task of the experimenter involves the synchronization and time-stamping or time-marking of stimulus input displays, responses, and the data from other auxiliary measures (see section 6.4 of chapter 6).

Each of these experimental setups requires a choice of programming platform to carry out the experiment, coordinate multiple computers, and to collect and store all stimulus sequences and corresponding data. The early sections of this book were designed to support the selection and testing of display and measurement equipment and the programming of stimulus displays. The focus of this chapter is on how to define the experiment.

12.3.2 Defining the Experiment

What to measure in an experiment is dictated by the hypotheses to be tested. Here, we will see how to transform initial ideas into fully detailed experimental designs. In section 12.2, we started with classical verbal ideas of attention acting as a filter or as an amplifier—and cast this within the theoretical framework of the PTM (see section 9.2 of chapter 9). This translated verbal hypotheses into predictions about the impact of attention on the TvC functions. However, in order to test the theoretical predictions concerning the mechanisms of attention it is first necessary to choose a particular domain and task for the experiment. For example, testing the behavioral signatures of different attention mechanisms for simple orientation discrimination has been studied and documented in many experiments.[11,13] Other domains for testing the mechanisms of attention might include motion direction discrimination in the motion domain,[14] or letter or face identification, and many others.

Once we decide to test attention mechanisms in the domain of visual recognition using orientation discrimination, we must go on to choose a particular task. One possibility is an identification task in which a single stimulus is identified or discriminated as one of four spaced orientations, four-alternative forced-choice (4AFC). Alternatively, we might have chosen two-alternative forced-choice (2AFC) identification for opposite angles ±45° from vertical, or simple two-interval detection, and so on. Each of these tasks has its' own properties. For example, the 4AFC task maximizes the dynamic range of psychometric functions from chance at 25% correct to maximum accuracy near 100% correct and also limits the role of criterion for decision that is intrinsic to Yes/No tasks. The choice of task is informed by the prior research literature. Chapters 8 and 11 discuss the pros and cons of various paradigms, tasks, and methods.

Suppose that our theoretical goal requires measuring contrast thresholds at three widely separated accuracy levels and at different levels of external noise along the TvC function to constrain the estimates of the PTM fully (see section 9.3.1 of chapter 9). These measurements will allow us to decide whether attended and unattended conditions differ in external noise exclusion (filtering) or in stimulus enhancement. Suppose we have chosen to measure the thresholds with the method of constant stimuli. Measuring full psychometric functions using the method of constant stimuli leads to the reliable estimation of three widely separated threshold levels. Testing of psychometric functions often uses six to nine points across a range of stimulus contrasts selected appropriately for each level of external noise.

Next, we must decide how to measure the shape of the TvC function. The TvC approximates a bilinear function, so several points should test along the flat part of the function in low external noise to estimate its level, and several points should test along the rising part of the function in high external noise to estimate its position and slope. A point near the transition between the two bilinear parts might assist in constraining the point of transition. These intuitive considerations might lead us to measure the TvC function at seven or eight suitably positioned external noise levels. The number of points along a quantitative function needed for good estimation depends upon the functional form and how precisely we wish to estimate its shape and parameters.

Following on prior experiments, we choose eight levels of external noise sampled in approximately logarithmic spacing of σ_{external} equal to 0, 0.0156, 0.0313, 0.0625, 0.125, 0.16, 0.25, and 0.33. These values correspond with the standard deviation of random Gaussian pixel noise contrasts, so a value of 0 means there is no noise (clear tests), and a standard deviation of 0.33, which allows three standard deviations of the Gaussian distribution between contrasts of +1 and –1.

In case 1 (section 12.2.1), the researchers chose 4AFC orientation identification and measured attended and unattended conditions and 8-point psychometric functions at each of eight levels of external noise. The design has 2 (attention) × 8 (external noise) × 8 (contrast) conditions. This produces 128 conditions in which probability correct will be measured. Experience with measuring psychometric functions and with 4AFC suggests a target sample size of at least 50 trials. To balance the presentation of the four different orientation stimuli for identification, this number needs to be divisible by 4. So, including this as a hidden balancing factor, the number of conditions is 2 (attention) × 8 (external noise) × 8 (contrast) × 4, (orientations), or 512 conditions.

By estimating the duration of each test trial at 3 s each, we calculate that 512 trials can be completed in a relatively short testing session of 25–30 min, while a 1024-trial testing session would require 50–60 min. A session of 512 trials corresponds with a sample size of four trials per condition, pooling over the balancing factor of stimulus orientation. A session of 1024 trials generates eight trials per condition per session. If we need one session for practice, and eight sessions to generate sample sizes of 64 in each of the 128 conditions, the experiment will have 8192 test trials. This is a relatively expensive experiment of 9 full-hour testing sessions per observer.

For the PTM and the TvC functions, we want good estimates of attention effects within each observer. Yet evaluation of the generalizability of the effect requires that we measure attention in several observers. In visual psychophysics, an experimental design that is tested on each observer traditionally uses four to eight observers. This provides the final development of the full experimental design for the attention-TvC experiment.

Obviously, there is a trade-off between the length and expense of an experiment and the quality of the data and conclusions. To reduce the demands of the experiment, we might eliminate some conditions while hoping still to provide estimates that are sufficient for our research purpose. For example, reducing the number of external noise points along the TvC function from eight to six and reducing the number of contrast points along each psychometric function from eight to six would reduce the number of conditions to 2 (attention) × 6 (external noise) × 6 (contrasts), or 72 conditions, or 288 with the orientation balancing factor. Five copies of this design in each session yields a sample size of 20 per session from 1152 trials, so three experimental sessions, or 3456 trials generates a sample size of 60 per condition. This produces an experimental protocol that is about 40% as long as the first design.

Which version do we choose? Is the abbreviated design that takes just over 40% as long to test as the fuller design almost as good? If we have done many similar experiments in the past, we may know from experience whether the larger design is necessary to identify the mechanisms of attention of a typical size. Often, this sort of intuitive, experienced-based logic allows us to make critical design decisions. The best practice, however, is to use what you know about the domain, the variability of observers, and the typical size of effects to perform a "power analysis" of the experiment. The purpose of this analysis is to estimate whether the experiment as designed will collect enough data to yield significant effects if they are real. This important topic is discussed further in section 12.3.6.

Another example, case 3 (section 12.1.3) tested a qualitative prediction using a quantitatively specified functional form of behavior. The qualitative prediction is that practice improves visual perception, and the idea for testing was to measure the CSF before and after a period of training on a visual task. The general choice of stimuli and task is dictated by the definition of what a CSF measures—the ability to detect stimuli of different spatial frequencies. So, the task measures detection, and the stimuli vary in spatial frequency and contrast.

The standard way of measuring CSF is to measure detection in two-interval forced-choice, in which a signal stimulus appears randomly in one of two intervals, a paradigm that reduces biases in response. Detection threshold is measured by changing the contrast of the signal stimulus to achieve, say, a 70.7% choice accuracy through a 2-down, 1-up staircase. Each trial has two intervals of 100 ms, marked at the beginning by a brief tone, with a 500-ms interval between.

The stimulus is a sine-wave patch appearing within a spatial window. In this case, suppose that the experimenter chooses to manipulate the spatial frequency, or cycles per degree, of the stimulus by programming different images on a display screen. This has the advantage of allowing different spatial frequency stimuli to be displayed in any order just by changing the display image. It has the disadvantage that the display window shows different numbers of cycles of the sine-wave gratings for different spatial frequency conditions. Alternatively, the window size in the sine-wave image could change to equate the number of grating bars for different spatial frequencies. Another approach—which is more cumbersome—is to manipulate spatial frequency by changing the viewing distance of the observer to the same stimulus. The same stimulus at a greater viewing distance subtends fewer degrees of visual angle. Changing viewing distance requires changing the position of either the observer or the screen between trials or blocks of trials, but equates all the details of the stimuli in different spatial frequency tests. All of these methods have been used in the vision testing literature.[23]

From prior research, it is known that the visible range of spatial frequency for humans is from below 0.25 cycles per degree (c/d) to almost 60 c/d. The CSF is traditionally measured from about 0.5 c/d to 32 c/d. The bandwidth of the sensitivity of visual receptors to spatial frequency is about an octave,[24–27] and so it makes sense to test spatial frequencies with a logarithmic spacing within the tested range. An octave corresponds to a doubling of frequency. The experimenter measures sine-wave detection at spatial frequencies that are an octave apart at 0.25, 0.5, 1, 2, 4, 8, 16, and 32 c/d, corresponding to the eight values of 2^i for integers $i = -2$ to 5.

If sine-wave spatial frequency is manipulated by changing the image on the screen, a viewing distance in relation to the pixel resolution of the screen must be carefully selected to span this range of spatial frequencies with a reasonable presentation of sine variation for high spatial frequencies and a minimum number of cycles for low spatial frequencies. The display screen also needs sufficient contrast resolution to allow the measurement of the very low contrast thresholds at the peak of the sensitivity function, and the screen should be carefully calibrated (see section 5.2 of chapter 5).

The basic design includes two measurements of the CSF, one before and one after a practice manipulation. Each copy of the CSF is measured at eight spatial frequencies. The experimenter has decided—as is traditional in this area—to measure contrast threshold by staircase methods that are able to adjust to the range of visual sensitivity of observers. The 70.7% accuracy thresholds may be measured with the 2-down, 1-up staircase procedure (see section 11.2 of chapter 11). From testing and experience, these staircases are known to generally converge in 30–50 trials; the stop rule is set at 50 trials. Thresholds are often measured several times in this literature and averaged, so the measurements are repeated three times.

To measure the initial CSF will require 8 spatial frequencies × 50 trials × 3 repeated threshold measures, or 1200 trials. This can be tested in one long session with several breaks. The post-training CSF also requires 1200 trials.

The last and perhaps most important aspect of the experiment is the training protocol that may produce learned changes in the CSF. More elaborate experiments might compare and assess different training protocols in different groups of observers. One training method involves several, perhaps eight, sessions of 1000 training trials for detecting sine-wave patches at a high spatial frequency near the high-frequency cut-off of the contrast sensitivity function for the observer. This training protocol targets high spatial frequency limits on visibility and has in fact been used as an approach to rehabilitation in adult amblyopes, where it has been shown to improve the overall CSF, especially in the high-frequency limb.[28,29]

The value in assaying a quantitatively specified function of visibility such as the CSF as a measure of learning, as we have shown by example, is that it may lead to new signature changes attributed to learning that may then allow the refinement of hypotheses and theories (figure 12.3). A computational study of the power of the experimental design to discover different kinds of improvements could be useful. For example, we may discover that even more repetitions of the threshold measures are required to discriminate changes in the bandwidth of the measured CSF. This is considered again in section 12.3.6 on power analysis of experiments.

12.3.3 Other Issues in Design

Designing experiments reflects knowledge about the response of the human (or animal) in each domain. Experience and tradition can incorporate knowledge about many important factors and suggest whether a factor should be randomized or controlled. One set of hypotheses and tests will lead toward one set of choices, whereas another might lead in a different direction.

For example, one important aspect of vision, especially in detection tasks, is the state of adaptation of the system. Testing a mix of conditions with high average luminance provides information about a relatively light-adapted system, whereas mixtures of low-luminance conditions tested far apart in time in a dark room are required to test a dark-adapted system. The measurement goal will dictate the amount of ambient light in the testing setup, the frequency with which tests occur, and the mixture of types of test trials. Adaptation more generally tends to set the system for the average stimulus and can impact detection, discrimination, and sensory memory.

Another widespread and important phenomenon in psychophysical testing is sequential dependency between responses. Sequential dependencies are changes in the response to a stimulus that depend on the last several stimuli and responses. A number of causes may lead to sequential dependencies, including local adaptation, sensory priming, and changes in decision criteria or attention. Response times are often faster following several equivalent stimuli and responses, like a series of "A" responses in an "A/B" paradigm[30] in which the observer is asked to discriminate between two stimuli. Figure 12.4 shows results from an experiment that illustrates the impact of sequential dependencies in response time for a two-alternative choice task. It shows faster response times for sequences of the same response.[30]

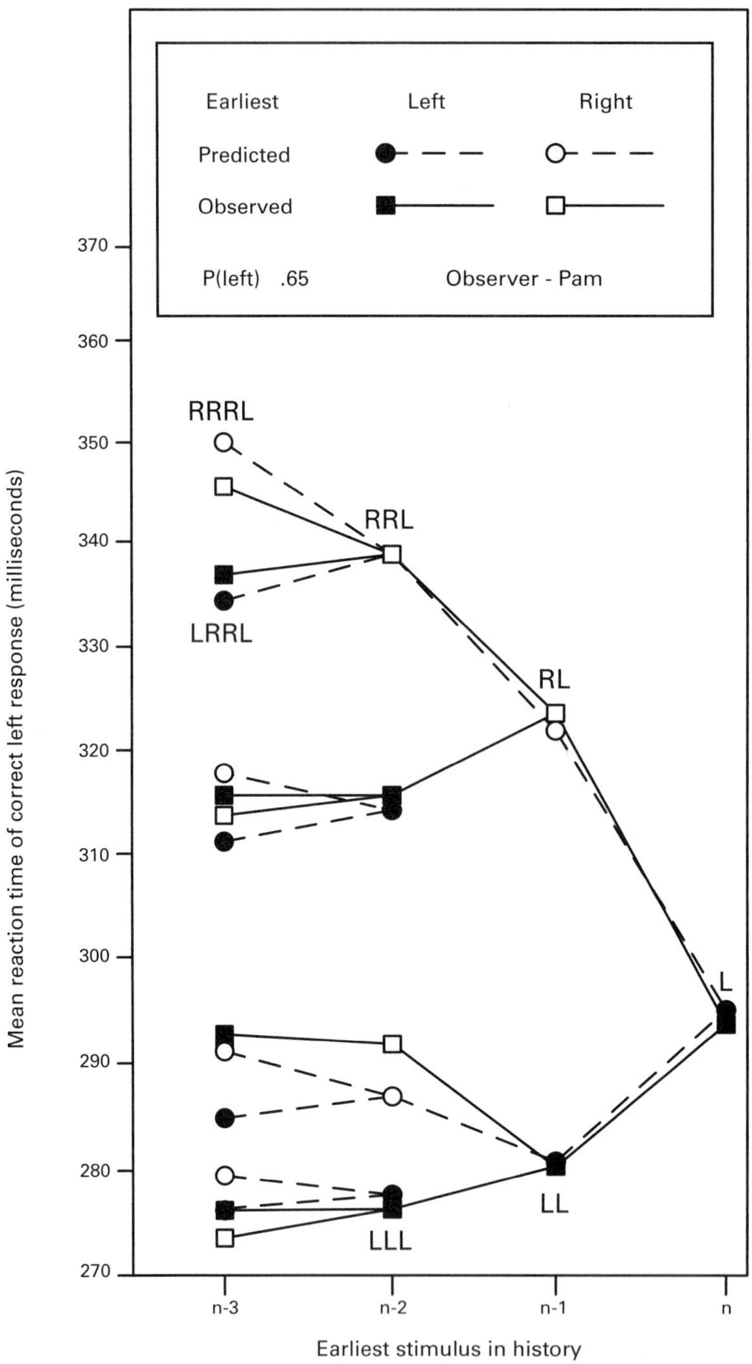

Figure 12.4
An illustration of sequential dependencies in choice reaction times (RTs) shown as a tree graph. The *x*-axis is the number of the earliest preceding trial in the conditional analysis (where the current trial is *n*), and the *y*-axis is the mean RT of a correct left response. Each node in the diagram is the mean correct RT conditionalized on the stimulus history of preceding trials, with three-trial histories on the left and the unconditional average on the right. The trial history (e.g., RRRL) is indicated for seven representative nodes. Data points are filled if the earliest trial in the history is left and unfilled if it is right (from Falmagne, Cohen, and Dwivedi[30]).

There are other examples of sequential effects. Observers may change criteria if they believe they have just made a mistake or to compensate for what they believe are too many same responses in a row. Observers running in a single simple staircase protocol may find the next stimulus predictable and adjust their criteria. Criteria for different scaling responses change as a function of the history of stimuli.[31,32]

Although both adaptation and sequential dependencies are phenomena that have themselves been studied, experimenters who are investigating other phenomena, such as attention or learning in our case examples, may seek to minimize such extraneous factors and choose an experimental design to achieve that goal. In particular, experimenters often choose *mixed designs* in which all conditions are mixed together for testing rather than *blocked designs* where many trials of one type are tested one after the other. This is certainly our general preference. The advantage of a mixed design is that—on average—the observer will be in the same state while testing each condition. A test of one condition is likely to be preceded by different conditions and only occasionally by the same condition. This reduces predictability and averages over sequential dependency patterns.

One example where a mixed design is especially critical is in testing observer models such as the PTM. The parameters of the model estimate properties of the visual system and observer, and these properties might be state-dependent. For situations such as this one, it is important for every test trial to occur while the observer is, on average, in the same state. If individual conditions, such as individual levels of external noise, are tested separately in blocks, then the system properties might change with the adaptive state. In that case, the parameters for each external noise level would define a different state of the system, and the blocked experiment would not measure a coherent TvC function.

Conversely, there may be some cases where it is important for certain trial variables to be blocked in a test design. For example, if a researcher is studying the effects of persistent attention set—as distinct from transitory shifts in attention—testing sets of 50–100 trials at a time with one attention set or another is a reasonable choice. In another example, blocks of trials with conservative or liberal response criteria may result in criterion setting with less variability than trial-by-trial cueing of decision state. If the goal is to measure the best performance of an observer in a given condition, then blocked testing may be appropriate. There are many other examples where blocked testing may be advantageous for a particular purpose.

Decisions about mixed versus blocked designs have direct implications for trial randomization. Mixed designs randomize the order of trials over all conditions, whereas blocked designs randomize trials over nonblocked stimulus manipulations.

12.3.4 Within-Observer and Between-Observer Designs

One key issue in experimental design is the choice of a within-observer design, a between-observer design, or a mixed design. In within-observer designs, the major theoretically motivated comparisons occur between performances of the same observer in distinct conditions. In between-observer designs, the major comparisons occur between performances of different observers in distinct conditions. Sometimes, the choice of design is tied directly to the theoretical hypothesis. For example, if one has a hypothesis about different groups of people, such as comparisons of older and younger observers, then of course a between-observer design will compare groups of older individuals and younger ones—possibly with a set of matching factors to limit extraneous variability. Another such example is the comparison of amblyopic individuals and individuals with normal vision.

Other hypotheses or tests lead naturally to within-observer designs. For example, if one is testing the effects of practice on performance, it is natural—though not absolutely necessary—to compare the performance of the same observers before and after practice. The alternative, which involves comparing one group of individuals before training to another group of individuals after training, is possible but fails to control for many sources of variability. Similarly, it is natural to compare performance with and without attention in the same observers.

Other cases may intrinsically involve mixed designs. For example, if the hypotheses involve differential learning abilities of older and younger observers, then the design might involve within-observer comparisons before and after practice, but also involve between-observer comparisons of older and younger groups. In some cases, the design is truly open for the experimenter to choose.

What are the benefits of within-observer and between-observer designs—assuming one could choose either? We almost always choose a within-observer design when possible. The within-observer design will include variability in performance within the observer, but will have controlled for the many ways in which observers might differ. The between-observer design includes within-observer variability between conditions but also variability between observers.

In psychophysics, we often design experiments that provide good estimates of the performance of each individual in the various conditions. Within-observer designs may also include mixed conditions so that every condition is tested when the observer is in the same average state (of adaptation, learning, etc.) for the testing of all conditions.

Within-observer designs may permit the use of smaller samples of well-characterized individuals. Between-observer designs almost always require larger samples of individuals. There is a close analogy to the simple statistical case of matched and independent *t*-tests. Figure 12.5 provides an example of a case that compares thresholds in two conditions, showing the values that would be obtained in a within-observer design and the two random samples of individuals in a between-observer design. In many cases, the ability to match performance in two conditions and partial out the variability between observers provides major increases in sensitivity to differences for the critical manipulation for the same amount of data collection.

Nonetheless, between-observer designs might be selected if the hypothesis tests individual differences and already requires large samples of people. In these circumstances,

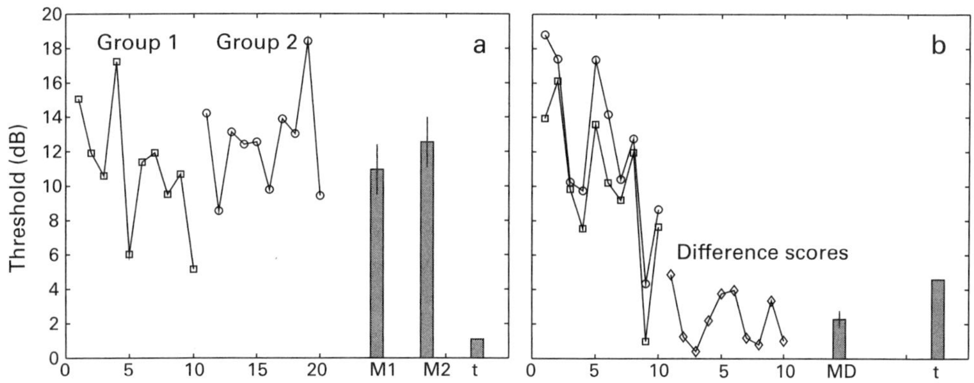

Figure 12.5
The advantage of within-observer designs. (a) Hypothetical data from a between-observer design. Each group consists of 10 observers. The mean and standard error of the two groups and the t statistic are graphed as gray bars (M1, M2, and t). (b) Hypothetical data from a within-observer design. Ten observers participated in both version of the hypothetical experiment. The difference scores of each observer in the two conditions are shown, along with the mean difference (MD) and t statistic.

between-observer designs may eliminate concerns about the order in which conditions are tested or about contaminating effects of learning or context.

12.3.5 Data Analysis and Modeling

Experiments should be designed with at least one plan for data analysis in mind. Often, the plan includes a quick analysis during the course of data collection so that the experimenter can monitor performance for inattention or major failures to perform the task. Once the data collection is complete, it is time to test the theory-driven hypotheses by planned statistical analyses. These planned tests could include simple statistical summaries such as the tabulation of proportions, means, and standard deviations. Other simple statistical tests may be used for comparison of conditions, such as χ^2 (chi-squared) tests, t-tests, analysis of variance, or regressions to compare groups or factors. We may be able to fit quantitative models to data and compare the fits of model variants, a topic treated extensively in chapter 10.

One very simple example is the measurement of psychometric functions in an attended and unattended condition and the effect of attention on performance accuracy in a task like 2AFC identification. Here, attention should increase accuracy in the threshold regions of the psychometric function. One plan is to test directly for significant differences between the proportions correct at relevant points of the two psychometric functions. Points near chance (near 50%) or at asymptote (which often converge toward 100% correct) are relatively nondiagnostic. Intermediate contrast conditions on the rising portions of the psychometric functions reveal meaningful differences. The attended condition should have higher percent correct at intermediate contrasts. A standard statistical test of the difference between proportions is

$$z = \frac{z_a - p_u}{\sqrt{\frac{p_a(1-p_a)}{n_a} + \frac{p_u(1-p_u)}{n_u}}}, \tag{12.2}$$

where p_a and p_u are the proportions correct for attended and unattended conditions.

Another direct way to estimate the effect of attention is to compare the full psychometric functions using a model. If attention has no effect, then the two psychometric functions should be the same and may be fit with the same Weibull function; if they are different, then we expect, for example, that the location (threshold) parameter of the attended condition should be smaller. This plan for statistical testing fits Weibull functions to the two psychometric functions (see section 10.4.2 of chapter 10) while comparing models in which the threshold, slope, and asymptotic levels are equal for the two conditions with models where, for example, the threshold levels differ. The data are the numbers of correct and incorrect responses, so maximum likelihood estimation is a natural choice. The χ^2 nested model contrasts evaluate whether there are significant differences in the threshold parameter. The size of the attention effect is summarized and understood by comparing the estimated thresholds for the attended and unattended conditions. Even if the significance of the difference is known from the nested model test, we may want to use bootstrap methods to estimate the variability in the estimated thresholds (see section 10.4.2). An example of this kind is considered in more detail in section 12.3.6 on power analysis.

Measuring the TvC functions and using the PTM to understand the mechanisms of attention in low and high noise requires a more complicated design. The aim is to test whether both external noise exclusion and stimulus enhancement, one or the other, or neither attention signature(s) occur in a particular experimental threshold data. Suppose we measured psychometric functions at each of eight external noise levels, with different contrast levels selected to be appropriate to span the entire function for each noise level. One could use the plan just detailed to compare the psychometric functions individually in each external noise level tested. The true aim, however, is to use the data to estimate and test the hypotheses within the context of the PTM.

There are (at least) two ways to carry out this analysis. One starts with estimating the thresholds at three different criterion accuracy levels at all external noise levels and then using a PTM to fit these data. The design has 16 psychometric functions, an attended and an unattended one at each of eight external noise levels. Each psychometric function is fit with a Weibull function by maximum likelihood methods, and the best-fitting Weibull is used to interpolate the estimates of three contrast thresholds corresponding to, say, 65%, 75%, and 85% correct. These 48 threshold estimates ($2 \times 8 \times 3$) are then fit with the PTM model using a least squares procedure. The basic PTM without attention has four parameters: template gain β, nonlinearity γ, internal multiplicative noise N_m, and internal additive noise N_a. There are two kinds of attention effects, implemented as multipliers (between 0 and 1): A_f that reduces or filters external noise ($A_f^2 N_{ext}^2$) and A_a that reduces internal additive noise ($A_a^2 N_a^2$) with attention. The first captures an attention effect in high-noise conditions, and the second captures an attention effect in low noise. The values are set to 1 in unattended conditions. Several model variants are compared: a basic model without either attention factor (all A values set to 1), a model with A_f but not A_a, a model with A_a but not A_f, and a model with both A_f and A_a. These different nested models are compared to determine whether adding each parameter yields a significant improvement in the fit of the model to the data. In the literature on attention in visual discrimination, we have found that central cueing often corresponds with an attention effect in filtering external noise but not in reducing internal additive noise.[11–13]

An alternative to first estimating thresholds at three accuracy levels and then fitting the PTM to the thresholds is to fit the PTM directly to the psychometric functions. This analysis is presented in more detail in section 12.3.6. One advantage of first estimating thresholds and then fitting the PTM is that graphs of the threshold and model fits show nice signature patterns of threshold changes that are easy to see and understand. In contrast, fitting the PTM directly to the family of 16 psychometric functions fits all the data directly using maximum likelihood methods, but the data pattern is more subtle, more dense to graph, and more difficult to visualize, but may in some circumstances provide a more powerful test of the model.

Experiments that collect physiological measures or other forms of data such as eye movements or reaction times must consider not just the data analysis and model testing for the task performance (i.e., accuracy). They must anticipate the special demands of analysis of the collateral measures. For example, if measuring response time in addition to response accuracy, the experimental design and the analyses depend upon whether the focus of testing a hypothesis is about mean response time, or the distributions of response times for correct and incorrect responses. EEG responses are averaged over trials to discover the waveform after preprocessing of the data to remove artifacts. Consideration and preplanning of the analysis of fMRI data is especially important and may drive many other considerations in the selection of the experimental design. Because these data are so expensive to collect and to analyze, careful advance consideration of the plan for analysis is especially important.

The analysis plan for an experiment should almost always be undertaken before choosing the final experimental design. Programs for preliminary data analyses that can be applied to pilot data may catch errors in programming or in design. Even more serious consideration of the experimental design may focus on its power to detect important effects, informing the selection of the sample sizes as well as the conceptual design of experiments. This topic is addressed next.

12.3.6 Power Analysis of Experiments

A topic that is often treated in very simple experimental designs, such as clinical trials that compare treatment A to treatment B, is the selection of sample size that can be expected to yield significance for an effect of a given size. This is a power analysis of the experiment. Power analysis is especially important in the context of medical testing, where a drug or an intervention may do some harm, and so we would like to test the smallest number of people for our purpose, at least initially. This often involves a simple calculation based on assumptions about error variance for a statistical test at different effect sizes and significance criterion.

In psychophysics, where active harm to the subject is essentially never the issue, a power analysis will suggest a more or less heroic amount of testing for an individual observer or overall data collection for the experimenter. The power analyses of psychometric tests that involve multiple conditions and measured functions are more complicated than the simple computations for comparing two conditions. These involve computational studies of an entire test design. Two examples are treated here.

The first example asks what sample size is necessary to detect a difference in threshold between two psychometric functions. The experiment measures two functions each at seven well-selected points of contrast along the psychometric function, for example between attended and unattended psychometric functions (section 12.3.5). Where to put seven contrast points along the psychometric function with a threshold and slope has been studied.[33] Our design begins with the standard placement.

Display 12.1a–c shows programs that carry out a power analysis for testing the difference between two psychometric functions. Our power computation simulates the experiment with different effect sizes and sample sizes. We assume a typical baseline threshold and slope for the experimental task (here, 0.012 and 2.84). A simulation generates many new samples of data and performs the nested Weibull model fits to test for significant differences between the psychometric functions. The simulation is carried out many times, here 2000 times.

The computations consider the effect size, or changes in threshold (0.5 to 2.5 dB), the sample size per point on each psychometric function (20–200), and a significance criterion, α. We chose $\alpha = 0.01$, setting a 1% chance of falsely rejecting the null hypothesis of no difference between the two psychometric curves.

The proportion of simulated experiments that finds a significant effect for each combination of effect size and sample size is shown in figure 12.6. The contour graph allows you to read off the sample size necessary to detect a particular effect size reliably. This can be used to guide the design of an experiment. Larger effect sizes require smaller sample sizes for an equal chance of correctly finding a significant effect. The graph suggests that small sample sizes in the 40s may be sufficient to detect a 3-dB effect size difference 90% of the time. Sample sizes approaching 110 may be sufficient to detect a 2-dB effect size 90% of the time. If you are hoping to detect a small effect size of 1 dB, then this is probably not the right experimental design as even 100-plus trials per point fails to find such a small effect almost half the time. Another computational study could check if adding points on the psychometric function would improve the situation of detecting small differences in the psychometric functions.

This analysis assumes the inductive logic of standard hypothesis testing.[34] An alternative Bayesian analysis might focus instead on the posterior distribution and the relative likelihood of different models.[35] If instead of significance, the focus is on the precision of the estimated thresholds, or the difference between thresholds, analogous computations could provide a corresponding analysis of precision.

A more complex example analyzes experiments that test a full quantitative model. Here, consider an experiment that measures full psychometric functions at different external noise levels and estimates attention effects using the PTM of the observer. The experiment tests 7-point psychometric functions at eight external noise levels in attended and unattended conditions. The purpose is to evaluate external noise exclusion.

Display 12.2a–c shows programs that carry out a power analysis of this attention study. It provides the programs for carrying out the power analysis for fitting the PTM model with template gain β, nonlinearity γ, internal multiplicative noise N_m, internal additive noise N_a, and attention effect A_f for filtering external noise ($A_f^2 N_{ext}^2$). The program simulates the response of a known PTM observer with given parameter values for several potential magnitudes of attention effects and different sample sizes. For each combination of these factors, it generates new data sets and fits them with the PTM to determine if the effect of external noise exclusion is significant. The PTMs are fit to the numbers of correct and incorrect trials. The output of the power analysis is the probability of rejecting the null hypothesis for combinations of set size and sample size.

The computational study considers effects sizes between 0.25 and 2.5 dB and sample sizes from 10 to 100 per point on the psychometric functions, and significance $\alpha = 0.01$. The results are shown in figure 12.7. Many conditions—16 psychometric functions rather than 2—constrain the PTM. Here, the power computations suggest that an effect as small as 1 dB can be detected with sample sizes of around 40 per condition.

In all of these computational studies of power, the more you know, the more accurate will be your power analyses of an experimental design. Knowing the typical threshold values, typical values of psychometric function slope or the normative PTM parameter values for the experiment being studied, and the typical effect sizes for attention mechanisms will all increase confidence in the selection of the design and the computation of power that determines the sample size. Knowing typical values of these factors will improve the validity of a power analysis. The more you know, the more you can learn through computational study.

Sometimes in highly studied areas, one omits a full power study because hundreds of very similar experiments have been carried out, and one can follow the (informal) power analysis encapsulated in those studies by simply following a typical experiment. But if the

Display 12.1a

```
%%% Program PowerAnalysis1.m
function beta = PowerAnalysis(effectSize, sampleSize, alpha)

xi = 0.50;
lamda = 0.02;
tau1 =  0.0112;                          % unattended
eta1 = 2.84;
tau2 =  0.0112 * 10 ^ (-effectSize/20); % attended
eta2 = 2.84;
totalN = 500;

% placement of stimulus contrast
p0 = [0.53 0.59 0.65 0.75 0.85 0.91 0.97];
c_pre = ((-1)*log((1 - lamda - p0)./(1 - lamda - ...
      xi))).^(1/eta1)*tau1;
c_post = ((-1)*log((1 - lamda - p0)./(1 - lamda - ...
      xi))).^(1/eta2)*tau2;

success = 0;
for i = 1 : totalN
    PFS(1, 1:7) = c_pre;
    PFS(2:3, 1:7) = 0;
    for j = 1 : 7
        prob = xi + (1 - xi - lamda) * (1 - exp(-...
                (c_pre(j)/tau1).^eta1));
        for k = 1 : sampleSize
            r = rand;
            if r < prob
                PFS(2, j) = PFS(2, j) + 1;
            else
                PFS(3, j) = PFS(3, j) + 1;
            end
        end
    end

    PFS(4, 1:7) = c_post;
    PFS(5:6, 1:7) = 0;
    for j = 1 : 7
        prob = xi + (1 - xi - lamda) * (1 - exp(- ...
              (c_post(j)/tau2).^eta2));
        for k = 1 : sampleSize
            r = rand;
            if r < prob
                PFS(5, j) = PFS(5, j) + 1;
            else
                PFS(6, j) = PFS(6, j) + 1;
```

Display 12.1a (continued)

```
            end
        end
    end

    NPF = 2;

    % full model fit
    data = PFS;
    guess = [0.01 0.01*10^(effectSize/20) 3];
    options = optimset('fminsearch');
    [psy, minus_maxloglikelihood] = …
       fminsearch('Weibullcostfunc1', guess, options, data);
    fullmodel_maxloglikelihood = - minus_maxloglikelihood;

    %reduced model
    data = PFS;
    guess = [(0.01+0.01*10^(effectSize/20))/2 3];
    options = optimset('fminsearch');
    [psy, minus_maxloglikelihood] = …
       fminsearch('Weibullcostfunc2', guess, options, data);
    reducedmodel_maxloglikelihood = …
       - minus_maxloglikelihood;

    chi2 = 2*(fullmodel_maxloglikelihood - …
       reducedmodel_maxloglikelihood);
    p = 1 - chi2cdf(chi2, 1);
    if p < alpha
        success = success + 1;
    end
end
beta = success/totalN;
```

experiment is expensive to run in any way, a preliminary computational study may save disappointment and aggravation.

Once an experiment is completed, information from that experiment may help to guide the design of follow-up experiments, and computational power analyses may sometimes be useful in determining the next successful experiment.

12.3.7 Experiments, Conclusions, and What Comes Next

Your theory led to a hypothesis. An experiment was developed to test the hypothesis. The experiment and analyses are done. Next, the results and analysis become the basis for a set of conclusions, suggesting new experiments and possibly revision of the theory.

Display 12.1b

```
%%% Program Weibullcostfunc1.m
function L = Weibullcostfunc1(guess, data)

NPF = size(data, 1)/3;
Nconditions = size(data, 2); % # of stimulus conditions
xi = 0.50;
lamda = 0.02;
L=0;
for n=1:NPF
    tau = guess(n);
    eta = guess(NPF+1);
    data1 =  data( (n-1)*3+1 : (n*3), :);
    for i=1:Nconditions
        p = xi + (1 - xi - lamda) * (1 - exp(- …
           (data1(1,i)/tau).^eta));   % Eq. 10.15
        if (p < 1/2/(data1(2, i)+data1(3, i)))
            % putting lower and upper boundaries on p
            p = 1/2/(data1(2, i)+data1(3, i));
        elseif (p> 1-1/2/(data1(2, i)+data1(3, i)))
            p = 1- 1/2/(data1(2, i)+data1(3, i));
        end
        L = L - (data1(2, i)*log(p) + …
              data1(3, i)*log(1-p));
    end
end
```

Display 12.1c

```
%%% Program Weibullcostfunc2.m
function L = Weibullcostfunc2(guess, data)

NPF = size(data, 1)/3;
Nconditions = size(data, 2); % # of stimulus conditions
xi = 0.50;
lamda = 0.02;
L=0;
for n=1:NPF
    tau = guess(1);
    eta = guess(2);
    data1 =  data( (n-1)*3+1 : (n*3), :);
    for i=1:Nconditions
        p = xi + (1 - xi - lamda) * (1 - exp(- …
             (data1(1,i)/tau).^eta));   % Eq. 10.15
        if (p < 1/2/(data1(2, i)+data1(3, i)))
            % putting lower and upper boundaries on p
            p = 1/2/(data1(2, i)+data1(3, i));
        elseif (p> 1-1/2/(data1(2, i)+data1(3, i)))
            p = 1- 1/2/(data1(2, i)+data1(3, i));
        end
        L = L - (data1(2, i)*log(p) + …
              data1(3, i)*log(1-p));
    end
end
```

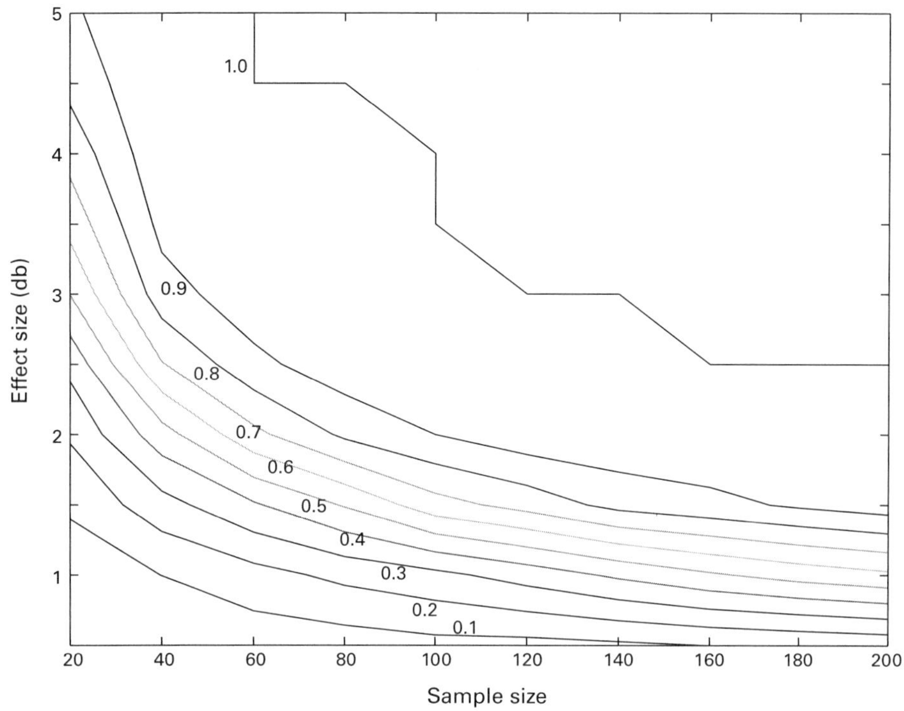

Figure 12.6
A power analysis for an experiment that compared two psychometric functions in attended and unattended conditions that may differ in threshold. The contour plot shows the probability of a significant effect when comparing two psychometric functions using different samples sizes and threshold effect sizes. See the text for the description of the experimental design.

From the perspective of hypothesis testing, there are two possible outcomes of the experiment: rejecting or failing to reject the null hypothesis. The null hypothesis usually assumes no difference between conditions whereas the theory or hypothesis predicts a difference or effect. For example, attention should improve performance, so we predict a difference (in a particular direction) between attended and unattended conditions, and so expect rejection of the null hypothesis. If the null hypothesis is rejected with a pattern consistent with the original prediction, then the data are consistent with the theory. The next step is to extend the theory to new predictions that might be falsified by new data or to extend the test of the theory to new experimental contexts.

If the data fail to reject the null hypothesis, then the experimenter must either decide that the experimental test had insufficient power for the size of the effect or that the theory or hypothesis should be modified, or that the particular application of the theory to the experiment was ill conceived. If there is a suspicion that the experiment had insufficient power or had inadequate control conditions, then the investigator will

Display 12.2a

```
%%% Program PowerAnalysis2.m
function beta = PowerAnalysis(effectSize, Next, sampleSize, ...
      alpha)

totalN = 500;    % total # of simulations
beta = 1.635;
gamma = 1.96;
Nm = 0.135;
Sa = 0.0095;
Af = 10^(-effectSize/20);
NnoiseLevels = length(Next);

% placement of stimulus contrast
p0 = [0.30 0.39 0.48 0.63 0.78 0.87 0.96];
dp0 = [0.18 0.49 0.78 1.25 1.81 2.26 3.05];
NperformanceLevels = length(dp0);
data1 = zeros(3*NnoiseLevels, NperformanceLevels + 1);
       % unattended
data2 = zeros(3*NnoiseLevels, NperformanceLevels + 1);
       % attended

for i = 1: NnoiseLevels
    data1(i, 1) = Next(i);
    data2(i, 1) = Next(i);
    for j = 1:NperformanceLevels
        logC1 = 1 / (2*gamma) * (2*log(dp0(j)) + ...
              log((1+Nm^2) * Next(i).^(2*gamma) + Sa^2) - ...
              log(1-Nm.^2*dp0(j).^2/2)) - log(beta);
        data1(i,  j + 1) = exp(logC1);
        % stimulus contrast level in the unattended condition
        logC2 = 1 / (2*gamma) * (2*log(dp0(j)) + ...
              log((1+Nm^2) * (Af*Next(i)).^(2*gamma) + Sa^2)...
              - log(1-Nm.^2*dp0(j).^2/2)) - log(beta);
        data2(i,  j + 1) = exp(logC2);
        % stimulus contrast level in the attended condition
    end
end

success = 0;
for m = 1 : totalN
    data1((NnoiseLevels+1):(3*NnoiseLevels), :) = 0;
    data2((NnoiseLevels+1):(3*NnoiseLevels), :) = 0;
    for i = 1 : NnoiseLevels
        for j = 1 : NperformanceLevels
            for k = 1 : sampleSize
```

Display 12.2a (continued)

```
                x = (-100:100)/10;
                dx = 0.10000000;
                prob = sum(normpdf(x - …
                       dp0(j)).*normcdf(x).^3)*dx;
                % compute percent correct from target dprime

                r = rand;    % simulate one trial in the
                             % unattended condition
                if r < prob
                    data1(i + NnoiseLevels, j+1) = …
                          data1(i + NnoiseLevels, j+1) + 1;
                else
                    data1(i + 2*NnoiseLevels, j+1) = …
                          data1(i + 2*NnoiseLevels, j+1) + 1;
                end

                r = rand;    % simulate one trial in the
                             % attended condition
                if r < prob
                    data2(i + NnoiseLevels, j+1) = …
                          data2(i + NnoiseLevels, j+1) + 1;
                  else
                      data2(i + 2*NnoiseLevels, j+1) = …
                            data2(i + 2*NnoiseLevels, j+1) + 1;
                 end
            end
        end
    end
    data = [data1; data2];

    % full model fit
    guess = [1.5 2 0.1 0.01 10^(-effectSize/20)];
    options = optimset('fminsearch');
    [ptm, minus_maxloglikelihood] = …
           fminsearch('PTMcostfunct1', guess, options, data);
    fullmodel_maxloglikelihood = - minus_maxloglikelihood;

    %reduced model
    guess = [1.5 2 0.1 0.01 10^(-effectSize/20)];
    options = optimset('fminsearch');
    [ptm, minus_maxloglikelihood] = …
           fminsearch('PTMcostfunct2', guess, options, data);
    reducedmodel_maxloglikelihood = - minus_maxloglikelihood;

    chi2 = 2*(fullmodel_maxloglikelihood - …
           reducedmodel_maxloglikelihood);
```

Display 12.2a (continued)

```
    p = 1 - chi2cdf(chi2, 1);
     if p < alpha
          success = success + 1;
     end
end
beta = success/totalN;
```

Display 12.2b

```
%%% Program PTMcostfunc1.m
function L = PTMcostfunc1(guess, data)

NnoiseLevels = size(data, 1)/6;
NperformanceLevels = size(data, 2) - 1;
        % # of stimulus conditions
beta = guess(1);
gamma = guess(2);
Nm = guess(3);
Sa = guess(4);
Af = guess(5);
data1 = data(1:3*NnoiseLevels, :);
data2 = data((3*NnoiseLevels+1):(6*NnoiseLevels), :);

L=0;
% unattended
for i = 1 : NnoiseLevels
    for j = 1 : NperformanceLevels
        Next = data1(i, 1);
        c = data1(i, j+1);
        m = data1(NnoiseLevels + i, j+1);
        k = data1(2*NnoiseLevels+i, j+1);
        dp = (beta*c)^gamma / sqrt(Next^(2*gamma)*(1+Nm^2) ...
            + Nm^2*(beta*c).^(2*gamma) + Sa^2);
         x = (-100:100)/10;
         dx = 0.10000000;
         p= sum(normpdf(x-dp).*normcdf(x).^3)*dx;
         % compute percent correct
        if (p < 1/2/(m+k))
            % putting lower and upper boundaries on p
            p = 1/2/(m+k);
        elseif (p> 1-1/2/(m+k))
            p = 1- 1/2/(m+k);
        end
        L = L - (m*log(p) + k*log(1-p));
```

Display 12.2b (continued)

```
    end
end

% attended
for i = 1 : NnoiseLevels
    for j = 1 : NperformanceLevels
        Next = data2(i, 1);
        c = data2(i, j+1);
        m = data2(NnoiseLevels + i, j+1);
        k = data2(2*NnoiseLevels+i, j+1);
        dp = (beta*c)^gamma / ...
             sqrt((Af*Next)^(2*gamma)*(1 + Nm^2) ...
            + Nm^2*(beta*c).^(2*gamma) + Sa^2);
        x = (-100:100)/10;
        dx = 0.10000000;
        p= sum(normpdf(x-dp).*normcdf(x).^3)*dx;
            % compute percent correct
        if (p < 1/2/(m+k))
            % putting lower and upper boundaries on p
            p = 1/2/(m+k);
        elseif (p> 1-1/2/(m+k))
            p = 1- 1/2/(m+k);
        end
        L = L - (m*log(p) + k*log(1-p));
    end
end
```

decide either to collect more data in the current experiment or to create a new and better one.

If instead the hypothesis and the corresponding prediction are not consistent with the data due to a convincing failure to reject the null hypothesis—or the null hypothesis was rejected but in an unexpected direction (i.e., attention unexpectedly damaged performance), then it is time to revise the theory.

12.4 From Experiment to Theory

No amount of experimentation can ever prove me right; a single experiment can prove me wrong.

A theory can be proved by experiment; but no path leads from experiment to the birth of a theory.
—attributed to Albert Einstein[36]; quoted in *Sunday Times*, July 18, 1976

The relationship between experiment and theory is a delicate and important one. Experiment and observations are the basis of developing the set of facts about the world—or

Display 12.2c

```
%%% Program PTMcostfunc2.m
function L = PTMcostfunc2(guess, data)

NnoiseLevels = size(data, 1)/6;
NperformanceLevels = size(data, 2) - 1;
        % # of stimulus conditions
beta = guess(1);
gamma = guess(2);
Nm = guess(3);
Sa = guess(4);
Af = 1.0;
data1 = data(1:3*NnoiseLevels, :);
data2 = data((3*NnoiseLevels+1):(6*NnoiseLevels), :);
L=0;
for i = 1 : NnoiseLevels  % unattended
    for j = 1 : NperformanceLevels
        Next = data1(i, 1);
        c = data1(i, j+1);
        m = data1(NnoiseLevels + i, j+1);
        k = data1(2*NnoiseLevels+i, j+1);
        dp = (beta*c)^gamma / sqrt(Next^(2*gamma)*(1+Nm^2) ...
            + Nm^2*(beta*c).^(2*gamma) + Sa^2);
        x = (-100:100)/10;
        dx = 0.10000000;
        p= sum(normpdf(x-dp).*normcdf(x).^3)*dx;
           % compute percent correct
        if (p < 1/2/(m+k))
            % putting lower and upper boundaries on p
            p = 1/2/(m+k);
        elseif (p> 1-1/2/(m+k))
            p = 1- 1/2/(m+k);
        end
        L = L - (m*log(p) + k*log(1-p));
    end
end

for i = 1 : NnoiseLevels   % attended
    for j = 1 : NperformanceLevels
        Next = data2(i, 1);
        c = data2(i, j+1);
        m = data2(NnoiseLevels + i, j+1);
        k = data2(2*NnoiseLevels+i, j+1);
        dp = (beta*c)^gamma ...
             / sqrt((Af*Next)^(2*gamma)*(1+Nm^2) ...
             + Nm^2*(beta*c).^(2*gamma) + Sa^2);
```

Display 12.2c (continued)

```
        x = (-100:100)/10;
        dx = 0.10000000;
        p= sum(normpdf(x-dp).*normcdf(x).^3)*dx;
                 % compute percent correct
        if (p < 1/2/(m+k))
            % putting lower and upper boundaries on p
            p = 1/2/(m+k);
        elseif (p> 1-1/2/(m+k))
            p = 1- 1/2/(m+k);
        end
        L = L - (m*log(p) + k*log(1-p));
    end
end
```

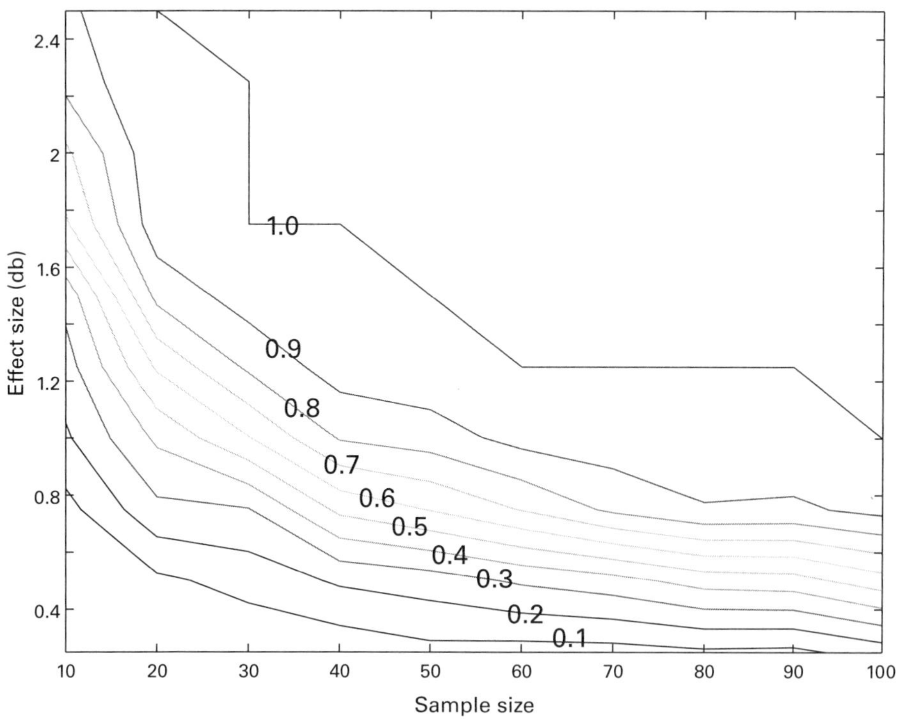

Figure 12.7
A power analysis for detecting attention effects on external noise exclusion in a set of psychometric functions at different noise levels defining a TvC. The contour plot shows the probability of rejecting the null hypothesis when comparing the psychometric functions in attended versus unattended conditions as a function of the size of the attention effect and the sample size. See the text for a description of the experiment.

about the human perceptual system—that theory must explain. Experiments generate data and tests of theory.

Scientific theories integrate a wide range of observations and phenomena. The better the theory, the more extensive the range over which predictions of the theory will hold. Stephen Hawking has said, "a theory is a good theory if it satisfies two requirements: It must accurately describe a large class of observations on the basis of a model which contains only a few arbitrary elements, and it must make definite predictions about the results of future observations."[37] Similarly, Einstein said, "A theory is the more impressive the greater the simplicity of its premises is, the more different kinds of things it relates, and the more extended is its area of applicability."[38]

The inductive logic behind the testing of theories holds that no amount of experimentation can prove a theory to be true, although our state of belief about its truth is stronger to the extent that it predicts or accommodates the widest range of observations. The more experiments of different kinds that have been carried out to test the theory *without falsification of it*, the more weight the theory holds. Again, in the inductive logic of theory testing, a single experimental inconsistency with a theory may be sufficient to reject it. In practice, this too is subject to interpretation. A single falsification of a powerful and successful theory may not lead to outright rejection of the theory, but instead to the addition of a footnote or special case process. Only the development of a better theory that accounts for all the known phenomena is likely to replace a strong original theory.

The development of theory in psychophysics has an added challenge. Like select areas in the physical sciences, theories of visual perception can be developed at several different levels—at the level of behavior, function, biological systems, neural systems, or cellular functions. Theories of visual perception at each of these levels have their own set of facts and phenomena to be explained, corresponding to different levels of theory. Each level has its own experimental tests. The challenge is in some ways greater as the sets of theories and hypotheses at different levels, like a set of nesting dolls, must fit together. The theory for visual perception should be integrated—or at least not inconsistent—from one level to the next. This integration between the neural, functional, and behavioral levels of analysis is the promise of the new area of neuropsychophysics.

Einstein said "no path leads from experiment to the birth of a theory." Yet the data provided by experimentation or observation are the observational ground from which the next theoretical invention will grow. Experiments should be driven by theory, while the next theory will be enriched by the right experiments.

References

1. Merriam-Webster online. Theory [definition 3]. *Merriam-Webster's collegiate dictionary*. Springfield, MA: Merriam-Webster; 2003. Available at: http://www.merriam-webster.com.

2. Wundt W. *Outlines of psychology*. Judd CH, trans. Leipzig: W. Engleman; 1902.

3. Mertens J. 1956. Influence of knowledge of target location upon the probability of observation of peripherally observable test flashes. *JOSA* 46(12): 1069–1070.

4. Posner MI, Nissen MJ, Ogden WC. Attended and unattended processing modes: The role of set for spatial location. In: Pick HL, Saltzman E, eds. *Modes of perceiving and processing information.* Hillsdale, NJ: Lawrence Erlbaum; 1978, pp. 137–157.

5. Dosher B, Lu Z-L. Mechanisms of visual attention. In: Chubb C, Dosher B, Lu Z-L, Shiffrin RM, eds. *Vision, memory, attention.* Irvine, CA: APA Press; 2013.

6. Davis ET, Graham N. 1981. Spatial frequency uncertainty effects in the detection of sinusoidal gratings. *Vision Res* 21(5): 705–712.

7. Shiu L, Pashler H. 1994. Negligible effect of spatial precuing on identification of single digits. *J Exp Psychol Hum Percept Perform* 20(5): 1037–1054.

8. Bashinski HS, Bacharach VR. 1980. Enhancement of perceptual sensitivity as the result of selectively attending to spatial locations. *Atten Percept Psychophys* 28(3): 241–248.

9. Downing CJ. 1988. Expectancy and visual-spatial attention: Effects on perceptual quality. *J Exp Psychol Hum Percept Perform* 14(2): 188–202.

10. Lu Z-L, Dosher BA. 1998. External noise distinguishes attention mechanisms. *Vision Res* 38(9): 1183–1198.

11. Dosher BA, Lu ZL. 2000. Noise exclusion in spatial attention. *Psychol Sci* 11(2): 139–146.

12. Dosher BA, Lu ZL. 2000. Mechanisms of perceptual attention in precuing of location. *Vision Res* 40(10): 1269–1292.

13. Lu ZL, Dosher BA. 2000. Spatial attention: Different mechanisms for central and peripheral temporal precues? *J Exp Psychol Hum Percept Perform* 26(5): 1534–1548.

14. Lu ZL, Liu CQ, Dosher BA. 2000. Attention mechanisms for multi-location first- and second-order motion perception. *Vision Res* 40(2): 173–186.

15. Lu ZL, Lesmes LA, Dosher BA. 2002. Spatial attention excludes external noise at the target location. *J Vis* 2(4): 312–323.

16. Carrasco M. 2011. Visual attention: The past 25 years. *Vision Res* 51: 1484–1525.

17. Logan GD. 2004. Cumulative progress in formal theories of attention. *Annu Rev Psychol* 55: 207–234.

18. Li X, Lu ZL, Tjan BS, Dosher BA, Chu W. 2008. Blood oxygenation level-dependent contrast response functions identify mechanisms of covert attention in early visual areas. *Proc Natl Acad Sci USA* 105(16): 6202–6207.

19. Reynolds JH, Pasternak T, Desimone R. 2000. Attention increases sensitivity of V4 neurons. *Neuron* 26(3): 703–714.

20. Treue S. 2002. Attentional modulation strength in cortical area MT depends on stimulus contrast. *Neuron* 35(2): 365–370.

21. Williford T, Maunsell JHR. 2006. Effects of spatial attention on contrast response functions in macaque area V4. *J Neurophysiol* 96(1): 40–54.

22. Lu ZL, Li X, Tjan BS, Dosher BA, Chu W. 2011. Attention extracts signal in external noise: A BOLD fMRI study. *J Cogn Neurosci* 23(5): 1148–1159.

23. Rovamo J, Franssila R, Näsänen R. 1992. Contrast sensitivity as a function of spatial frequency, viewing distance and eccentricity with and without spatial noise. *Vision Res* 32(4): 631–637.

24. Stromeyer III CF, Julesz B. 1972. Spatial-frequency masking in vision: Critical bands and spread of masking. *JOSA* 62(10): 1221–1232.

25. Henning GB, Hertz BG, Hinton J. 1981. Effects of different hypothetical detection mechanisms on the shape of spatial-frequency filters inferred from masking experiments: I. Noise masks. *JOSA* 71(5): 574–581.

26. Losada M, Mullen KT. 1995. Color and luminance spatial tuning estimated by noise masking in the absence of off-frequency looking. *JOSA A* 12(2): 250–260.

27. Lu ZL, Dosher BA. 2001. Characterizing the spatial-frequency sensitivity of perceptual templates. *JOSA A* 18(9): 2041–2053.

28. Zhou Y, Huang C, Xu P, Tao L, Qiu Z, Li X, Lu ZL. 2006. Perceptual learning improves contrast sensitivity and visual acuity in adults with anisometropic amblyopia. *Vision Res* 46(5): 739–750.

29. Huang CB, Zhou Y, Lu ZL. 2008. Broad bandwidth of perceptual learning in the visual system of adults with anisometropic amblyopia. *Proc Natl Acad Sci USA* 105(10): 4068–4073.

30. Falmagne J, Cohen SP, Dwivedi A. Two-choice reactions as an ordered memory scanning process. In Rabbit P, Dornic S, eds. *Attention and performance V.* New York: Academic Press; 1975, pp. 296–344.

31. Treisman M, Williams TC. 1984. A theory of criterion setting with an application to sequential dependencies. *Psychol Rev* 91(1): 68–111.

32. Petrov AA, Anderson JR. 2005. The dynamics of scaling: A memory-based anchor model of category rating and absolute identification. *Psychol Rev* 112(2): 383–416.

33. Green DM. 1990. Stimulus selection in adaptive psychophysical procedures. *J Acoust Soc Am* 87: 2662–2674.

34. Hays WL. *Statistics for the social sciences*, Vol. 410. New York: Holt, Rinehart and Winston; 1973.

35. Kruschke JK. *Doing Bayesian data analysis: A tutorial with R and BUGS.* Burlington, MA: Academic Press; 2010.

36. Calaprice A, ed. *The quotable Einstein.* Princeton, NJ: Princeton University Press; 1996.

37. Hawking SW. *The illustrated a brief history of time.* New York: Bantam; 1996.

38. Schilpp PA. *Albert Einstein: Philosopher-scientist*, La Salle, IL: Open Court; 1970.

V TAKING VISUAL PSYCHOPHYSICS OUT OF THE LABORATORY

13 Applications and Future Directions

The goal of this book is to enable the reader to become a practicing psychophysics researcher. This book takes an integrated theoretical approach to understanding perceptual systems and human information processing. Most of the examples in the book have focused on the measurement and testing of basic properties of visual functions. These examples show research principles that may be widely exploited by using the same or very similar test designs, analyses, and adaptive testing methods to research questions in many different applications. The current chapter outlines possible applications using a common philosophy, experimental approach, and computational method in a variety of domains and highlights future directions in neuropsychophysics that integrate psychophysics, neurophysiology, and computational modeling in vision research.

13.1 Clinical Vision and Biomedical Research

Standard clinical testing of vision includes tests of acuity, of perimetry (peripheral vision), depth perception, and color. All of these tests relate to scientific laboratory paradigms from visual psychophysics. The most obvious example is the widespread use of visual acuity testing. The letter eye charts or "tumbling E" eye charts used by optometrists and ophthalmologists use identification to measure acuity (figure 13.1). Although visual acuity testing is still the gold standard in clinical vision,[1] recent studies indicate that visual acuity is not the only—and may not even be the best—predictor of visual functions. Many patients who test at or near normal levels of 20/20 in visual acuity may nonetheless exhibit functional deficits of vision.[2–9]

Visual psychophysicists have developed many methods for measuring contrast sensitivity. Rather than measuring the limits of size resolution at very high contrast as in the standard Snellen letter chart, the Pelli–Robson charts measure the limitation in visibility due to low contrast of letters for a comfortable size.[10] This test detects the general contrast sensitivity deficits exhibited by cataract, macular degeneration, and diabetic retinopathy.[11] Yet the broadband letter stimuli do not provide information about frequency-specific defects.[4]

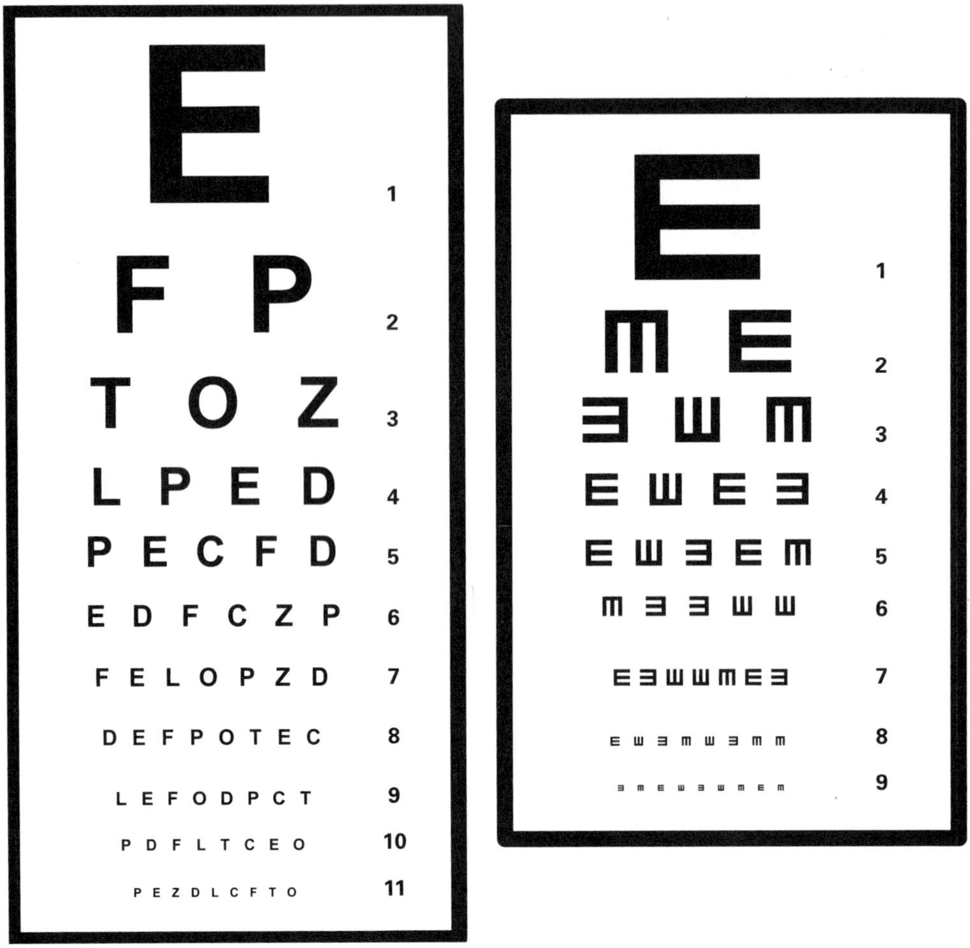

Figure 13.1
Applications of visual testing in the eye clinic: (a) Snellen and (b) tumbling E eye charts.

To identify frequency-specific deficits, we need to measure the contrast sensitivity function, or CSF. Many studies indicate that the CSF (see sections 2.3, 4.4.3, 10.4.3, 11.4.1, and 12.2.3 earlier in this book), provides a more comprehensive measure of performance and does a better job than simple acuity of predicting daily visual function in normal and clinical populations.[3,4,6,7] The CSF measures the minimum contrast required to perceive either gratings or letters of different spatial frequencies or sizes. Indeed, CSF measured in static and dynamic displays, in different lighting conditions, with glare, and at different locations in the visual field provides a comprehensive characterization of daily visual function. For example, the measurement of the CSF under glare can be an important predictive measure of driving performance.[12]

Many visual diseases exhibit specific alterations of the CSF even as tests of visual acuity are near normal or normal. For example, Huang et al.[5] found that clinically treated amblyopes, with 20/20 vision, remained deficient in spatial vision, especially at high spatial frequencies. Comprehensive visual testing also plays an increasing role in surgical planning and evaluation. Measuring the CSF presurgically in cataract patients provides critical information for the choice of lens correction and then serves as a baseline for comparison to postsurgical evaluation.[13–15]

A few contrast sensitivity tests (e.g., Arden cards, Vistech, FACT charts) that vary both the frequency and contrast of narrowband gratings have been developed.[16–18] For portability and ease of application, these tests use paper media and gratings with predetermined frequencies and contrast levels.[19] The development of adaptive paradigms to test contrast sensitivity such as the quick contrast sensitivity function (q-CSF),[20] described in section 11.4.1 of chapter 11, has the potential to make more comprehensive tests possible in clinical and other applications. In the future, we believe that incorporation of the more sophisticated tools of visual psychophysics can provide improved tests in many domains of clinical application.

13.2 Individual Differences in Vision

Even for individuals with apparently good visual acuity and normal color vision, significant individual differences exist in visual function. Certain individual differences in visual performance affect behavior in many domains, whereas others may be specific to a trained activity. For example, athletes in sports that rely on visual inputs, such as baseball players, seem to achieve better performance either in the speed or the ability to detect differences in motion patterns.[21] Video gamers may have special abilities in identifying visual patterns or in attending across the visual field.[22] In contrast, others such as those with dyslexia may have relatively poor abilities in some visual tasks although testing normal in standard vision tests.[23] Understanding individual differences in perception or perceptual task performance can be important in predicting many aspects of regular daily function.

We tend to think about visual perception in particular as being simply normative. Normal vision in individuals absent eye disease is characterized by 20/20 acuity, and the majority of individuals can be approximately corrected to normal with the appropriate lenses. On many other assessments, however, there is significant measurable variation in the level of performance both within and across visual tasks.[24–26]

Furthermore, training or experience may improve or modify the performance in visual behaviors. Nature leads to variation, and tasks select for natural variation in perceptual skills. And nurture—effects of practice or learning—may further influence individual differences in perceptual skill. Some athletes, for example, may have special visual abilities partly because being good at the sport early is easier for people at the top of the distribution of perceptual speed or accuracy. Good motion perception may improve your performance in baseball. The degree of selection in professional players may be substantial. Video gamers may also be self-selected for visual capacity or speed. The advantages of self-selection are almost sure to be advanced by disproportionate practice and training. This in turn can amplify the initial level of individual differences.

Visual psychophysical methods, tasks, and analyses presented in this book provide a basic approach to the careful assessment of visual functions and individual differences. The models from visual psychophysics, such as the perceptual template model (PTM),[27] aim to go beyond qualitative predictions to a full characterization of the basic limitations in vision for each individual and then use these estimates to make predictions over a wide range of situations. The strategy advocated in this book involves the measurement of a few fundamental properties of the individual's system, such as the CSF, or the threshold versus external noise contrast (TvC) function, to predict performance in many other tasks.

Another approach to understanding individual differences is to measure performance in many tasks and domains and observe the correlation between different task performances. This can improve our understanding of normative performance, the variance and range in the expression, and the nature of individual differences. For example, a multifunction study was carried out on a relatively large population of amblyopes (individuals with "lazy eye"), each assessed on eight different visual tasks.[28] Analyses of the performance on each task and the correlations between tasks showed two major factors that determined patient performance: visual acuity and contrast sensitivity. These methods of analysis could be extended to other visual deficits. Information about the population characteristics of individuals with normal vision and about the correlation structure between tasks is becoming increasingly important in understanding and predicting functioning in daily tasks such as reading or driving.[29–31]

Many apparently visual tasks assess more than just pure vision. For example, consider the useful field of view test.[32] Detection or identification is measured along different rays emerging from central fixation at different distances into the periphery; that is, different eccentricities. The test assesses the visual processing of a briefly displayed stimulus in different spatial locations without eye or head movements (figure 13.2). Performance on

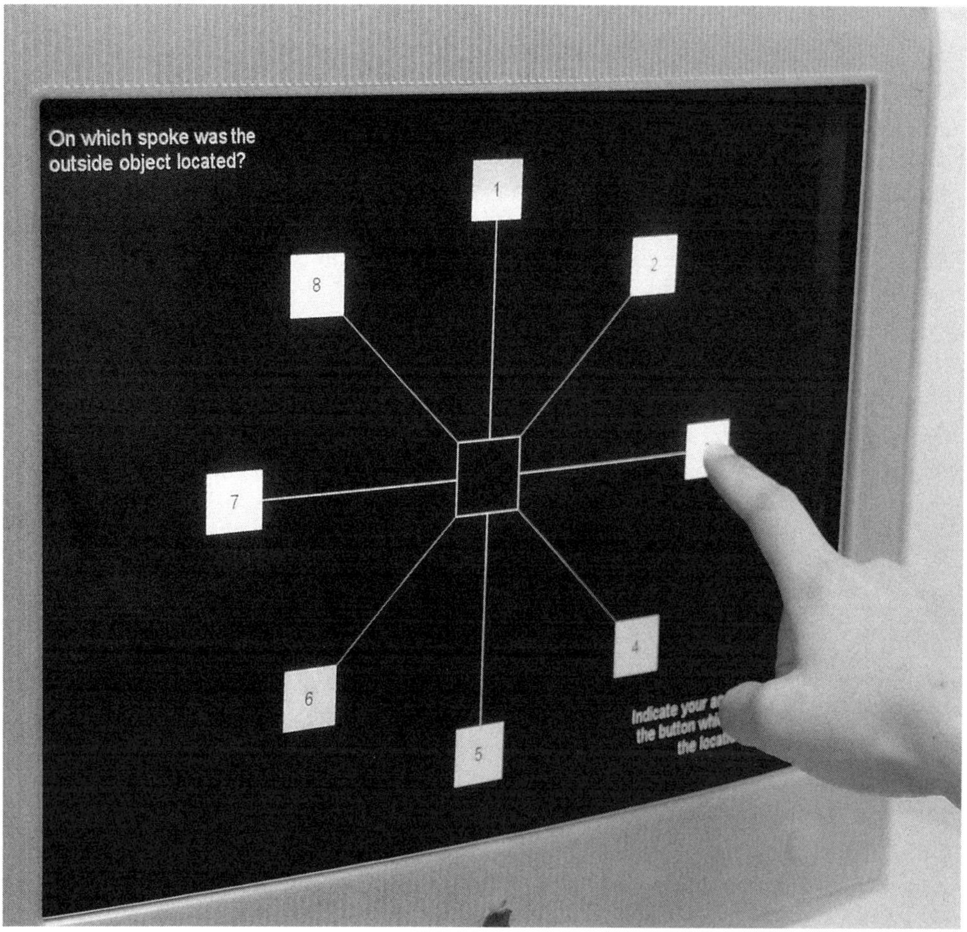

Figure 13.2
An illustration of a useful field of view test.

such tasks would of course depend on eccentricity, contrast, optical limitations, and any other noisiness or difficulty in early visual processing stages. However, useful field of view is also affected by attention limitations. Peripheral deficits in useful field of view with aging, for example, may also reflect fundamental limitations in the deployment of attention in addition to limitations in optics or visual transmission per se.[33–35] It is important to assess pure vision in order to interpret the results of the test correctly. Performance at a location in the periphery from a useful field of view test can be compared to single-location detection or discrimination in the same location to determine the relative contributions of vision and attention.

Good visual psychophysics is quite important in the assessment and interpretation of individual differences in visual performance. Conversely, individual differences can make important contributions to the study of visual perception. The existence of key individual differences may provide important clues about the existence of multiple pathways of visual perception.[36–38] For example, different individuals sometimes perceive ambiguous or multistable stimuli in different ways. Stimuli in which global and local structures coexist can lead to different percepts, with some observers favoring global analysis and others local analysis. The existence of multistable state perception is a clue to distinct levels of visual analysis that could be selected and weighted differently in the performance of different individuals.[39] Although currently in the beginning stages, the study of individual differences in vision and perception is an open area for exploration and subsequent application.[40]

13.3 Applications of Psychophysics in Special Populations

Reports of visual or other sensory deficits in clinical populations are now quite common. For example, reduced responses to contrast of visual stimuli has been cited in depression,[41] and some researchers have reported that schizophrenic patients are less susceptible to certain classes of visual illusions.[42] These reported relationships between sensory processing and clinical conditions can arise in a number of ways. A sensory deficit or atypical response pattern may be a causal factor in the clinical condition, altered sensory processing may be a consequence of the clinical condition, or there may be a common causal mechanism that is implicated in both.

In many examples, sensory deficits are *not* among the primary criteria for classification of the clinical population. Whether altered sensory processing is causal or merely an epiphenomenal effect associated with a particular clinical condition, measurements of sensory function could in the future provide converging evidence in initial assessment or an added method of evaluating the efficacy of a treatment. And, even if the sensory deficit is neither causal nor core to a clinical condition, sensory deficits bring their own consequences that must be managed, and it may be possible to either treat or overcome the sensory deficit with visual prosthetics or through training.

One very interesting example is the role of sensory deficits in developmental dyslexia, or what is sometimes more broadly termed reading deficiency (RD).[23,43–49] Patients with developmental dyslexia, or RD, have a range of reading deficits, including reduced scores in general reading ability, phonological processing and awareness, and orthographic skills. One classic claim is that developmental dyslexia reflects abnormal function of the magnocellular (M) visual pathway.[23,43–46] The M-pathway in the visual cortex is dominant in the processing of luminance stimuli of low spatial frequency and high temporal frequency or motion—at least in comparison to the parvocellular (P) visual pathway, which is more dominant in the processing of color, form, and high spatial frequency in low temporal frequency.[50,51] Some studies have reported reduced performance in tasks associated with M-pathway processing in those with dyslexia or RD. Other reports focus on temporal processing of sensory stimuli, whether visual or auditory.

An alternative view argues that—although phonological deficits may be the proximal cause of poor reading—the core issue is reduced identification or recognition from noisy displays.[52] Much of the evidence favoring reduced M-pathway performance in those with dyslexia utilized noisy displays, or very low luminance displays in which the signal-to-noise ratio is dominated by internal noise. Unusual noise susceptibility may be directly related to deficits in development of sophisticated phonological recognition templates and word templates.[53–56] Sensory deficits may be only one part of a more complicated picture. Individuals with dyslexia often show collateral impairments not just in visual processing but also in motor processing, selective attention, and social and affective issues.[57]

Another example of a special population with modified perceptual phenomena is schizophrenia.[42] The vast majority of the research in schizophrenia has focused on cognitive aspects of the condition.[58] However, the possible sensory dysfunction in the disease is now widely discussed. Atypical responses of schizophrenic patients to visual testing are of two broad types: reduction in gain control[59–61] and limited long-range interactions.[62–64] In gain control, surrounding patterns with contrast reduce the perceived contrast of a target region. A central texture is perceived to be of lower contrast when surrounded by textures of high contrast of similar spatial frequencies and orientations.[60] Contrast-gain reduces or silences the response to a pattern that is like other patterns and enhances the perception of unique patterns in the visual field. Schizophrenic patients show much reduced effects of contrast-gain reduction. Their perception of the central pattern in such tests may be more accurate, but they are more susceptible to interference from other patterns in noisy fields (figure 13.3a).

Schizophrenic patients also show deficits in long-range pattern integration. One test asks observers to detect the presence of orientation "snakes"—lines or curves of patches of Gabors with locally similar orientation and scale among a cluttered field of Gabor patches (figure 13.3b).[62] Longer spatial gaps between Gabor patterns and jitter in local orientation make the integration snakes harder to perceive. An ability that continues to develop into adolescence, contour integration is significantly reduced in the schizophrenic population.[64] There have been parallel, though less well documented effects of schizophrenia in auditory perception.

The ties between sensory and perceptual function and clinical classification have been most strongly documented in dyslexia and schizophrenia. More diffuse or less well understood perceptual anomalies have been cited in other clinical populations. Issues with contrast perception have been reported for individuals with clinical depression.[41] Deterioration of visual function over and above normal aging is reported in Alzheimer's disease and some dementias.[65,66] Diffuse issues of visual function, many involving dysfunctions of eye movements or pupil response, are cited in Parkinson's disease patients.[67,68] These

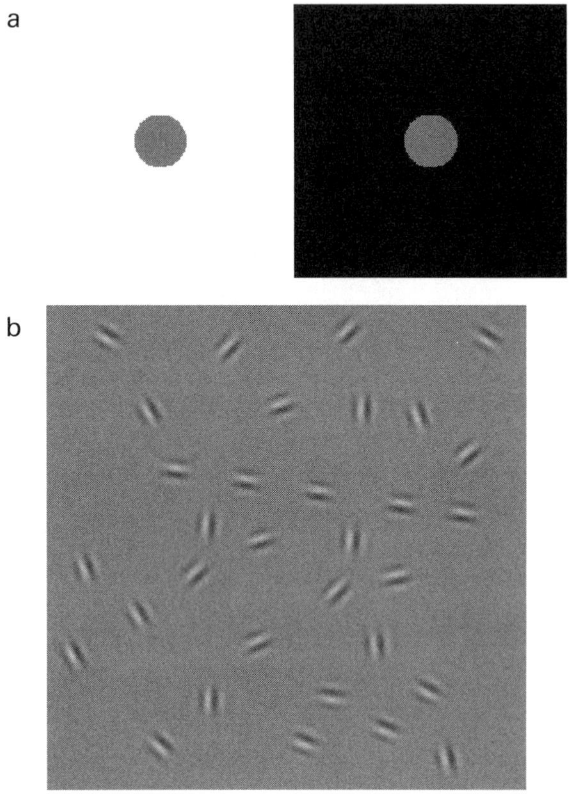

Figure 13.3
Examples of contrast gain control and long-range interaction phenomena in vision. (a) The contrast–contrast illusion. (b) Visual contour integration.

issues with visual function may reflect decoupling of vision with motor control. Some visual deficits have been cited in attention deficit hyperactivity disorder (ADHD).[69]

In each of these examples, rapid and efficient diagnosis of perceptual deficits with good psychophysical measures may be one important tool in the arsenal of assessment. Further research in assessment is needed to determine whether the visual or perceptual dysfunctions are causal or epiphenomenal. However, regardless of the causal relationship, it is important to understand perceptual aspects of these conditions in order to address a full range of coping behaviors.

Understanding perceptual deficits in clinical conditions might suggest very practical interventions that improve visual or auditory perceptual behaviors. One example is the role of contrast enhancement in compensating for visual losses in Alzheimer's disease. One study showed that contrast enhancement can equate letter detection in early Alzheimer's disease with control groups.[70] Another study showed that contrast enhancement on the plate improved food and liquid intake in late-stage Alzheimer's disease.[71]

13.4 Human Factors

Psychophysics, in particular visual psychophysics, has played a historic role in determining the specification for instrument and interface design. Obvious examples are the role of color vision science in determining international color standards[72] and the creation of television and movie standards that respect the parameters of human visual persistence and support the perception of continuous motion from discrete image frames.[73,74] The science of incorporating knowledge of the limits of human perception, cognition, and skeletal-motor capabilities into industrial and artifact design is called *human factors*.[75] This multidisciplinary field integrates testable properties or limits of human interaction from psychology and psychophysics with industrial and graphic design in practical engineering.

The purpose of human factors is to design systems that support good-quality human performance and human comfort in a target environment. This environment may involve digital or other visual displays and information display protocols such as in control stations in power plants or air traffic control centers, or sustained visuomotor interaction with the environment in activities such as driving a car, navigating a boat, or flying a plane. Human factors approaches generally involve at least three components: the analysis of human performance, the principles of technical design, and expertise in how humans interact with computers.

Human performance analysis incorporates sensory systems such as vision or audition, knowledge of motor systems and interactions, and general cognitive properties of the human operator such as limits in attention, memory, and inference. Human performance testing resolves how different designs either enhance or damage human performance. All this information is then used to improve technical design of systems. Human–computer interaction focuses on designing both the physical attributes of the operating environment and the interfaces for programs in computer-based systems. In most if not all of these applications, visual and auditory displays play a major role—and the psychophysicist and sensory psychologist have a fund of knowledge of sensory systems and a toolkit of tests for detection, discrimination, and appearance that can be used in the development of new systems and devices.

One example of the role of psychophysics in human factors design concerns the development of the JPEG image standards.[76,77] *JPEG* stands for Joint Photographic Experts Group, a group that developed compression standards for digital photographic color images. The JPEG standards achieved a typical compression factor of 10 to 1, or image size reduction to about one-tenth—a very substantial reduction in storage and transmission

Figure 13.4
JPEG compression losses. (a) The original image in high resolution. (b, c, d) The same image saved in JPEG with high, medium, and low image qualities.

demands. The JPEG image formats are commonly used in digital camera applications and as the format for color photographs on the Web.

Any compression system must lose information relative to a high-quality digital color image (figure 13.4). JPEG compression was designed to minimize loss for human perception. The JPEG algorithms were optimized for photographs or realistic paintings, rather than line drawings or computer renderings, where the compression may introduce artifacts or aliasing. The compression algorithms take advantage of the typically smooth variations in color and intensity in scenes of the natural world. The JPEG algorithms exploit the relative insensitivity of human vision to variation in color compared to luminance and relative insensitivity to high spatial frequencies. They reduce the number of bits for color and high spatial frequencies. Psychophysical tests for perceptible loss of information quality in human users guided the selection of the compression algorithm and specified the circumstances where JPEG compression is appropriate.[78,79]

The compressed JPEG images are not in general appropriate input for other image-processing functions such as threshold manipulations, cropping, shifting, or resizing. Successive applications of image-processing manipulations can reveal or even highlight the losses introduced by JPEG compression. For this reason, uncompressed or raw image formats such as tagged image file format (TIFF) are used in programs such as Adobe Illustrator or Photoshop or special-purpose image-manipulation programs. Original source images of high quality may also be important in the processing and generation of derived images in support of virtual reality environments or forms of augmented vision. Many clinical or industrial applications require display systems that match or exceed human visual capacity.[80,81]

High-fidelity images render real-world stimulus situations given direct viewing and retain many details of the outside environment. Display systems of sufficient fidelity should have adequate spatial, temporal, and gray-level resolution. For some applications, replication of a wide field of view or small pixel resolution may be important. Understanding the resolution limits of the human visual system assists in design of display systems that just exceed those limits, without wasting excess fidelity that cannot be perceived by the human operator.[82]

Luminance resolution and linearity have been seriously considered in the area of medical X-ray and diagnostic imaging, where the *perceived* luminance information codes physical tissue density.[83] Assessment of the adequacy of image and display fidelity for particular purposes relies on assessment of human performance in specially constructed tests, informed by visual psychophysics.[84]

Human factors testing is especially important in designing the layout and content in dense information displays such as those in large airplane cockpits.[85] Optimizing operator or pilot performance in such dense displays incorporates knowledge about limitations of the visual system and the human ability to attend to relevant information (figure 13.5, plate 10). It includes how best to visualize and code information about the flight systems, routing information, and multiple augmented displays such as multiview environmental cameras and route maps. Such applications in human factors go beyond simple visual information limits. Many of the designs for testing human performance developed for sensory psychophysics can be easily extended to testing these more complex situations.[86]

One especially important class of applications of human factors design and testing is aimed at optimizing performance and training in synthesized flight simulators or driving simulators.[87] There is also an increasing interest in remote-operator systems for clean-up of hazardous materials, undersea exploration, or in military applications where some scenarios revolve around remote pilots. Research will determine to what degree environmental fidelity is necessary to optimize performance or training effectiveness. Systems development teams in these areas may include perceptual, psychophysical, and testing expertise as a part of engineering and design teams.[88]

At the other end of the spectrum, knowledge of basic sensory psychophysics under certain extreme environmental circumstances is lacking, and simple tests of discrimination and detection can quantify human performance in unusual circumstances. For example, researchers at NASA are now testing basic visual abilities in high-G environments. The experimental challenges have to do with carrying out vision testing while subjects are

Figure 13.5 (plate 10)
A picture of the Airbus A380 cockpit (from http://www.flickr.com/photos/83823904@N00/64156219).

exposed to high-G simulations in large centrifuges that mimic perception during the critical high-G moments of takeoff.[89,90]

13.5 Midlevel and High-Level Vision, Cognition, and Action

The methods developed so far were applied to basic visual tasks, such as simple detection, CSFs, or orientation identification—usually classified as *low-level* vision. However, similar or slightly more complicated displays, together with related paradigms of detection or discrimination and other methods, can be used to study *midlevel* and *high-level* vision.[91,92] These methods are also applicable to somewhat more complex stimulus domains including the processing of faces, objects, and scenes or of other objects critical to daily function, such as letters or words. Questions in these domains often parallel questions asked in low-level vision. How easy are these objects to detect or discriminate? And what is the scaled similarity of their internal representations among different examples?

As in basic visual domains, careful description and measurement of midlevel and high-level stimuli is very important—although often more complex. Progress in these areas depends critically upon a sophisticated understanding of the stimuli. For example, low-level descriptions of faces at the level of pixel or spatial frequency content are not adequate to account for the complex configural properties of perception of these stimuli. Indeed, the heart of the problem of understanding face, object, or scene perception is developing an understanding of how to specify the stimulus space.

To make this point more explicit, consider the recognition of faces or facial expression. Images of faces may of course be described and analyzed in terms of standard image attributes, such as contrast, spatial frequency, or orientation content. However, at another level, face recognition systems extract facial features and configurations, such as the distance between the eyes or the distance from the eyes to the mouth, and many others.[93] Alternatively, some systems use computer algorithms to analyze large sets of face images mathematically. One example uses principal components analysis. Each face is described as a mixture of an average face plus some weighting on other face components or dimensions.[94,95]

Figure 13.6 illustrates one such computational system for face recognition. Figure 13.6a shows component faces with decreasing importance for describing a set of real face images, the so-called eigenfaces. An algorithm applied to a large set of training images extracts these descriptors. Each individual face is represented as a weighted sum of the eigenfaces. Figure 13.6b shows that by successively adding back each weighted eigenface, a more and more realistic image of the original face is created.[96]

In each of these cases, each face stimulus is quantified with a set if weights on features within a feature space. This serves as the basis for manipulation in psychophysical investigations. Manipulations of feature dimensions can be tested using standard psychophysical paradigms described in earlier chapters. Related methods have been used to identify the critical feature dimensions in the recognition of gender, age, and aspects of emotion from images of faces.[93–95,97]

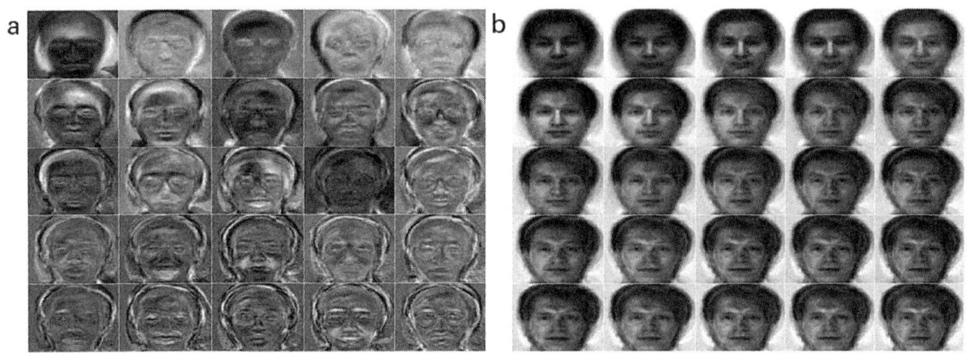

Figure 13.6
Describing natural faces as a weighted combination of "eigenfaces." (a) A set of eigenfaces derived from a face database by performing singular value decomposition on a set of training faces. Each eigenface accentuates certain characteristics. (b) Each new image from left to right adds one more eigenface in the face reconstruction. The face of a particular person becomes recognizable around the seventh or eighth image (from http://www.cs.princeton.edu/~cdecoro/eigenfaces/).

Figure 13.7
Piazza San Marco, Venice, Italy. Two pictorial cues are available to define the slant and tilt of the plaza with respect to the line of sight of the camera, linear perspective, and the texture gradient formed by humans, pigeons, other miscellaneous objects and their shadows (from Oruc, Maloney, and Landy[128]).

Another challenge in midlevel and high-level vision is to model how cues or aspects of the stimulus, once separated in early visual analyses, are combined to lead to an overall perception. One approach to this issue measures how multiple cues to perception are combined by the perceptual system. Figure 13.7 is a natural image that shows several different cues to depth in the real world, most obviously linear perspective and texture gradients. The issue of cue combination has been brought into the laboratory. Straightforward extensions of simple psychophysical experiments manipulate multiple cues in a stimulus and observe the resulting behavior. For example, there has been an extensive study of what kinds of cues are used to infer three-dimensional (3D) depth from the two-dimensional (2D) images that are projected on the eye. These cues include manipulations of stereo, but also of motion parallax, or texture, or luminance. Careful independent or partially independent manipulations of the different cues together can be used to test the importance of each cue in determining the perceived depth of different objects or of different parts of the same object. These investigations can also reveal the principles of cue interaction or cue integration using quantitative models of how the information from each cue is integrated—cue integration.[98–100] The same approach has been used in multimodal cue integration, studying the impact of simultaneous auditory cues on visual perception, or vice versa.[101–103] The issue has been important in speech perception, where extensive work has studied the biasing of speech perception by the visual cues from movements of the mouth.[104,105]

Many cognitive processes start with perception. The theories we develop in perception become the front end for cognitive processes. In addition, the effects of cognitive manipulations and functions can be measured through their impact on perception. Finally, the paradigms of visual psychophysics have been imported into other areas where they provide a rigorous framework for experiments in the cognitive domain.

Psychophysical methods have been extremely important in studying attention and perceptual learning. Cognitive manipulations of attention are sometimes measured through tests of the accuracy of low-level visual perceptual tasks. A substantial literature has combined manipulations of visual attention such as spatial cueing with classic measures of contrast threshold, orientation discrimination, or the CSF.[106–109] A similar analysis has incorporated visual psychophysical testing into the analysis of practice effects and of practice in video game playing.[22,110,111]

The theoretical innovations in psychophysics, such as the observer models, provide a strong framework for understanding these cognitive processes that begins with the sensory inputs and ends with the decision mechanisms. Perceptual processes are an integral part of models of many cognitive processes. We must consider the functionally important elements in sensory and perceptual processes in the development of complete cognitive models.[112,113] For example, comprehensive models of reading must incorporate the perceptual systems that take in the written information. These integrated models may improve our understanding of reading deficits.[114]

Finally, the decision architectures, such as signal detection and other choice theories that guide decision in cognitive tasks,[115] were in many cases first developed in the domain of either auditory or visual psychophysics and remain highly influential in the study of human cognitive abilities.

13.6 Future Directions: Neuropsychophysics

The goal of classic psychophysics is to quantify the relationship between stimuli and responses and to understand the nature of the internal representations of the external world. Neuropsychophysics aims to develop integrated computational models of psychophysics and brain responses that support behavior. Advancements in neurophysiology and brain imaging are yielding new insights about internal representations and the brain networks involved in representing stimuli, making decisions, and producing responses. The stimulus and task specifications of psychophysical tasks and new knowledge about neural responses will be central in developing new theories in both psychophysics and sensory and cognitive neuroscience. Developments and advancements in neuropsychophysics may also play an increasing role in practical applications.

Studies of physiology using cellular recording, studies of brain response using electrical or magnetic activity signatures in electroencephalography (EEG) and magnetoencephalography (MEG), and brain-imaging responses in functional magnetic resonance imaging (fMRI) will help us to understand the neural basis of human behavior. Behavioral psychophysics has measured the relationship between the stimulus and the human response and specified the properties of the internal representations and decision. These inferred properties of the internal representations suggest the nature of the responses we expect in a candidate neural internal representation. For example, responses in early visual cortex to variation in stimulus contrast provide candidates for internal representations that may underlie psychometric contrast discrimination functions.[116,117]

Very recent research in brain imaging is seeking homologs, or direct structural parallels, between multidimensional representations of different objects based on behavioral measures of similarity or confusion in identification and the similarity of the patterns of brain activity in response to these same stimuli.[116,118–120] Other research has examined the relationship between successful recognition and the similarity of the neural response to an initial stimulus and the neural response to the same stimulus at a delay.[121–123] Multivariate pattern analysis associates certain spatial patterns of responses in fMRI voxels to different classes of input stimuli. Indeed, the broad use of multivariate pattern analysis to understand the patterns of activity in different voxels in fMRI responses to different stimuli provides further applications and tests of some of the methods of scaling developed in psychophysics.

Another important aspect of neuropsychophysics tests our understanding of internal representations, decisions, and precursors to behavioral responses. In open head methods such as cellular recording, electrical stimulation to key neurons in a neural circuit or localized injection of drugs may be used to alter neural response or block the activity in a possible node within a hypothesized neural circuit.[124] Alternative forms of intervention that briefly disable or disrupt a local brain region use transcranial magnetic stimulation (TMS)[125] and transcranial direct current stimulation (tDCS)[126]. Active stimulation is used to test the role of a certain brain region either singly or in a hypothetical circuit of regions thought to be central in generating specific task behaviors.[127] With major advancements in neuroscience and brain imaging, it is increasingly possible to identify candidate neural systems that mediate perception of the stimulus, selection and generation of response, and implementation of a goal structure for behavior.

All of these new developments in neuropsychophysics open up the possibility of an enriched understanding of the visual system at multiple levels, from the sensory input to brain responses and behavior. In this new frontier of vision research, psychophysics will provide the stimuli, the tasks, the experimental paradigms, and a systematization of the relationship between the stimulus and the behavioral response. At the same time, the insights about the neural coding of internal sensory and cognitive representations derived from physiology and brain imaging will improve our understanding of the brain substrates for behavior. Together, both psychophysics and neuroscience will contribute to the development and refinement of increasingly realistic and detailed models of representation and processes.

The future of neuropsychophysics will lead to the development of processing models of the visual system and human behavior, provide the basis for a better understanding of cognitive functions, and generate new tests in clinical vision and biomedical research. The new paradigm will also improve our understanding of the mechanisms leading to individual differences and various clinical conditions and improve design for human factors applications.

With parallel technical and theoretical developments in visual psychophysics, neurophysiology and brain imaging, and computational modeling, a solution to Fechner's original challenge to quantify mental processes is finally within reach. We hope that this book will contribute to the development of a new generation of researchers working on the frontiers of vision science.

References

1. Cline D, Hofstetter HW, Griffin JR. *Dictionary of visual science*. Radnor, PA: Chilton Book Company; 1980.

2. Wood JM, Owens DA. 2005. Standard measures of visual acuity do not predict drivers' recognition performance under day or night conditions. *Optom Vis Sci* 82(8): 698–705.

3. Comerford J. 1983. Vision evaluation using contrast sensitivity functions. *Am J Optom Physiol Opt* 60(5): 394–398.

4. Ginsburg AP. 2003. Contrast sensitivity and functional vision. *Int Ophthalmol Clin* 43(2): 5–15.

5. Huang C, Tao L, Zhou Y, Lu ZL. 2007. Treated amblyopes remain deficient in spatial vision: A contrast sensitivity and external noise study. *Vision Res* 47(1): 22–34.

6. Faye EE. 2005. Contrast sensitivity tests in predicting visual function. *Int Congr Ser* 1282: 521–524.

7. Jindra LF, Zemon V. 1989. Contrast sensitivity testing: A more complete assessment of vision. *J Cataract Refract Surg* 15(2): 141–148.

8. Ginsburg A, Tedesco J. 1986. Evaluation of functional vision of cataract and Y AG posterior capsulotomy patients using the Vistech contrast sensitivity chart. *Invest Ophthal Vis Sci* 27(3) (Suppl): 107.

9. Kurzer AR. 1986. Contrast sensitivity signals pituitary adenoma. *Rev Opt* 123(4): 119.

10. Pelli D, Robson J, Wilkins A. 1988. The design of a new letter chart for measuring contrast sensitivity. *Clin Vis Sci* 2(3): 187–199.

11. Ismail GM, Whitaker D. 1998. Early detection of changes in visual function in diabetes mellitus. *Ophthalmic Physiol Opt* 18(1): 3–12.

12. Holladay JT, Dudeja DR, Chang J. 1999. Functional vision and corneal changes after laser in situ keratomileusis determined by contrast sensitivity, glare testing, and corneal topography. *J Cataract Refract Surg* 25(5): 663–669.

13. Rubin GS, Adamsons IA, Stark WJ. 1993. Comparison of acuity, contrast sensitivity, and disability glare before and after cataract surgery. *Arch Ophthalmol* 111(1): 56–61.

14. Muñoz G, Albarrán-Diego C, Montés-Micó R, Rodríguez-Galietero A, Alió JL. 2006. Spherical aberration and contrast sensitivity after cataract surgery with the Tecnis Z9000 intraocular lens. *J Cataract Refract Surg* 32(8): 1320–1327.

15. Kurz S, Krummenauer F, Thieme H, Dick HB. 2007. Contrast sensitivity after implantation of a spherical versus an aspherical intraocular lens in biaxial microincision cataract surgery. *J Cataract Refract Surg* 33(3): 393–400.

16. Arden G, Jacobson J. 1978. A simple grating test for contrast sensitivity: Preliminary results indicate value in screening for glaucoma. *Invest Ophthalmol Vis Sci* 17(1): 23–32.

17. Ginsburg A. 1984. A new contrast sensitivity vision test chart. *Am J Optom Physiol Opt* 61(6): 403–407.

18. Ginsburg AP. Next generation contrast sensitivity testing. In: Rosenthal B, Cole R, eds. *Functional assessment of low vision*. St. Louis: Mosby Year Book; 1996: pp. 77–88.

19. Owsley C. 2003. Contrast sensitivity. *Ophthalmol Clin North Am* 16(2): 171–178.

20. Lesmes LA, Lu ZL, Baek J, Albright TD. 2010. Bayesian adaptive estimation of the contrast sensitivity function: The quick CSF method. *J Vis* 10(3): 17.1–21.

21. Williams AM, Davids K, Williams JGP. *Visual perception and action in sport*. Philadelphia: Taylor & Francis; 1999.

22. Green CS, Bavelier D. 2003. Action video game modifies visual selective attention. *Nature* 423(6939): 534–537.

23. Stein J, Walsh V. 1997. To see but not to read; the magnocellular theory of dyslexia. *Trends Neurosci* 20(4): 147–152.

24. Webster MA, MacLeod DIA. 1988. Factors underlying individual differences in the color matches of normal observers. *JOSA A* 5(10): 1722–1735.

25. Neitz J, Jacobs GH. 1986. Polymorphism of the long-wavelength cone in normal human colour vision. *Nature* 323: 623–625.

26. Wilmer JB, Nakayama K. 2007. Two distinct visual motion mechanisms for smooth pursuit: Evidence from individual differences. *Neuron* 54(6): 987–1000.

27. Lu ZL, Dosher BA. 2008. Characterizing observers using external noise and observer models: Assessing internal representations with external noise. *Psychol Rev* 115(1): 44–82.

28. McKee SP, Levi DM, Movshon JA. 2003. The pattern of visual deficits in amblyopia. *J Vis* 3(5): 380–405.

29. Ginsburg A. 1987. Contrast sensitivity, drivers' visibility, and visions standards. *Transportation Research Record* 1149: 32–39.

30. Lovegrove WJ, Bowling A, Badcock D, Blackwood M. 1980. Specific reading disability: Differences in contrast sensitivity as a function of spatial frequency. *Science* 210(4468): 439–440.

31. Brown B. 1981. Reading performance in low vision patients: Relation to contrast and contrast sensitivity. *Am J Optom Physiol Opt* 58(3): 218–226.

32. Ball KK, Beard BL, Roenker DL, Miller RL, Griggs DS. 1988. Age and visual search: Expanding the useful field of view. *JOSA A* 5(12): 2210–2219.

33. Scialfa CT, Kline DW, Lyman BJ. 1987. Age differences in target identification as a function of retinal location and noise level: Examination of the useful field of view. *Psychol Aging* 2(1): 14–19.

34. Myers RS, Ball KK, Kalina TD, Roth DL, Goode KT. 2000. Relation of useful field of view and other screening tests to on-road driving performance. *Percept Mot Skills* 91(1): 279–290.

35. Sekuler AB, Bennett PJ, Mamelak M. 2000. Effects of aging on the useful field of view. *Exp Aging Res* 26(2): 103–120.

36. Peterzell DH, Teller DY. 2000. Spatial frequency tuned covariance channels for red-green and luminance-modulated gratings: Psychophysical data from human adults. *Vision Res* 40(4): 417–430.

37. Billock VA, Harding TH. 1996. Evidence of spatial and temporal channels in the correlational structure of human spatiotemporal contrast sensitivity. *J Physiol* 490(Pt 2): 509–517.

38. Morrone MC, Burr DC, Pietro SD, Stefanelli MA. 1999. Cardinal directions for visual optic flow. *Curr Biol* 9(14): 763–766.

39. Kanai R, Bahrami B, Rees G. 2010. Human parietal cortex structure predicts individual differences in perceptual rivalry. *Curr Biol* 20(18): 1626–1630.

40. Wilmer JB. 2008. How to use individual differences to isolate functional organization, biology, and utility of visual functions; with illustrative proposals for stereopsis. *Spat Vis* 21(6): 561.–579

41. Bubl E, Kern E, Ebert D, Bach M, Tebartz van Elst L. 2010. Seeing gray when feeling blue? Depression can be measured in the eye of the diseased. *Biol Psychiatry* 68(2): 205–208.

42. Butler PD, Silverstein SM, Dakin SC. 2008. Visual perception and its impairment in schizophrenia. *Biol Psychiatry* 64(1): 40–47.

43. Breitmeyer BG, Ganz L. 1976. Implications of sustained and transient channels for theories of visual pattern masking, saccadic suppression, and information processing. *Psychol Rev* 83(1): 1–36.

44. Cornelissen P, Hansen P, Hutton J, Evangelinou V, Stein J. 1998. Magnocellular visual function and children's single word reading. *Vision Res* 38(3): 471–482.

45. Lovegrove W, Martin F, Slaghuis W. 1986. A theoretical and experimental case for a visual deficit in specific reading disability. *Cogn Neuropsychol* 3(2): 225–267.

46. Ramus F, Rosen S, Dakin SC, Day BL, Castellote JM, White S, Frith U. 2003. Theories of developmental dyslexia: Insights from a multiple case study of dyslexic adults. *Brain* 126(4): 841–865.

47. Cavanagh SHMBP. 1996. Low level visual processing skills of adults and children with dyslexia. *Cogn Neuropsychol* 13(7): 975–1016.

48. Hulme C. 1988. The implausibility of low-level visual deficits as a cause of children's reading difficulties. *Cogn Neuropsychol* 5(3): 369–374.

49. Skottun BC. 2000. The magnocellular deficit theory of dyslexia: The evidence from contrast sensitivity. *Vision Res* 40(1): 111–127.

50. Zeki S. *A vision of the brain*. New York: Wiley; 1993.

51. Zihl J, Von Cramon D, Mai N. 1983. Selective disturbance of movement vision after bilateral brain damage. *Brain* 106(2): 313–340.

52. Sperling AJ, Lu ZL, Manis FR, Seidenberg MS. 2005. Deficits in perceptual noise exclusion in developmental dyslexia. *Nat Neurosci* 8(7): 862–863.

53. Boets B, Wouters J, Van Wieringen A, De Smedt B, Ghesquiere P. 2008. Modelling relations between sensory processing, speech perception, orthographic and phonological ability, and literacy achievement. *Brain Lang* 106(1): 29–40.

54. Ziegler JC, Pech-Georgel C, George F, Lorenzi C. 2009. Speech perception in noise deficits in dyslexia. *Dev Sci* 12(5): 732–745.

55. Boets B, Vandermosten M, Poelmans H, Luts H, Wouters J, Ghesquière P. 2011. Preschool impairments in auditory processing and speech perception uniquely predict future reading problems. *Res Dev Disabil* 32(2): 560–570.

56. Sperling AJ, Lu ZL, Manis FR, Seidenberg MS. 2006. Motion-perception deficits and reading impairment. *Psychol Sci* 17(12): 1047–1053.

57. Ramus F. 2003. Developmental dyslexia: Specific phonological deficit or general sensorimotor dysfunction? *Curr Opin Neurobiol* 13(2): 212–218.

58. Nuechterlein KH, Barch DM, Gold JM, Goldberg TE, Green MF, Heaton RK. 2004. Identification of separable cognitive factors in schizophrenia. *Schizophr Res* 72(1): 29–39.

59. Butler PD, Zemon V, Schechter I, et al. 2005. Early-stage visual processing and cortical amplification deficits in schizophrenia. *Arch Gen Psychiatry* 62(5): 495–504.

60. Dakin S, Carlin P, Hemsley D. 2005. Weak suppression of visual context in chronic schizophrenia. *Curr Biol* 15(20): R822–824.

61. Tadin D, Kim J, Doop ML, Gibson C, Blake R, Lappin JS, Park S. 2006. Weakened center-surround interactions in visual motion processing in schizophrenia. *J Neurosci* 26(44): 11403–11412.

62. Keri S, Kelemen O, Benedek G, Janka ZN. 2005. Lateral interactions in the visual cortex of patients with schizophrenia and bipolar disorder. *Psychol Med* 35(7): 1043–1051.

63. Uhlhaas PJ, Phillips WA, Mitchell G, Silverstein SM. 2006. Perceptual grouping in disorganized schizophrenia. *Psychiatry Res* 145(2): 105–117.

64. Silverstein SM, Kovács I, Corry R, Valone C. 2000. Perceptual organization, the disorganization syndrome, and context processing in chronic schizophrenia. *Schizophr Res* 43(1): 11–20.

65. Calderon J, Perry R, Erzinclioglu S, Berrios G, Dening TR, Hodges J. 2001. Perception, attention, and working memory are disproportionately impaired in dementia with Lewy bodies compared with Alzheimer's disease. *J Neurol Neurosurg Psychiatry* 70(2): 157–164.

66. Lu ZL, Neuse J, Madigan S, Dosher BA. 2005. Fast decay of iconic memory in observers with mild cognitive impairments. *Proc Natl Acad Sci USA* 102(5): 1797–1802.

67. Mosimann UP, Mather G, Wesnes K, O'brien J, Burn D, McKeith I. 2004. Visual perception in Parkinson disease dementia and dementia with Lewy bodies. *Neurology* 63(11): 2091–2096.

68. Demirci M, Grill S, McShane L, Hallett M. 1997. A mismatch between kinesthetic and visual perception in Parkinson's disease. *Ann Neurol* 41(6): 781–788.

69. Barkley RA. *Attention-deficit hyperactivity disorder: A handbook for diagnosis and treatment*, New York: The Guilford Press; 2006.

70. Gilmore GC, Cronin-Golomb A, Neargarder SA, Morrison SR. 2005. Enhanced stimulus contrast normalizes visual processing of rapidly presented letters in Alzheimer's disease. *Vision Res* 45(8): 1013–1020.

71. Dunne TE, Neargarder SA, Cipolloni P, Cronin-Golomb A. 2004. Visual contrast enhances food and liquid intake in advanced Alzheimer's disease. *Clin Nutr* 23(4): 533–538.

72. CIE. *Commission Internationale de l'Eclairage proceedings, 1931*. Cambridge, UK: Cambridge University Press; 1932.

73. Elsaesser T, Barker A. *Early cinema: Space, frame, narrative*. London: BFI Publishing: 1990.

74. Edgerton GR. *The Columbia history of American television*. New York: Columbia University Press; 2007.

75. Sanders MS, McCormick EJ. *Human factors in engineering and design*. New York: McGraw-Hill; 1987.

76. Wallace GK. 1991. The JPEG still picture compression standard. *Commun ACM* 34(4): 30–44.

77. Pennebaker WB, Mitchell JL. *JPEG still image data compression standard*. Berlin: Springer; 1993.

78. Watson AB. 1993. DCT quantization matrices visually optimized for individual images. *Proc SPIE* 1913: 202–216.

79. Rosenholtz R, Watson AB. Perceptual adaptive JPEG coding. The IEEE International Conference on Image Processing, 1996, pp. 901–904. 1996.

80. Steuer J. 1992. Defining virtual reality: Dimensions determining telepresence. *J Commun* 42(4): 73–93.

81. Cruz-Neira C, Sandin DJ, DeFanti TA. 1993. Surround-screen projection-based virtual reality: The design and implementation of the CAVE. *ACM SIGGRAPH* 135–142.

82. Bowman DA, McMahan RP. 2007. Virtual reality: How much immersion is enough? *Computer* 40(7): 36–43.

83. Samei E, Badano A, Chakraborty D, Compton K, Cornelius C, Corrigan K, et al. 2005. Assessment of display performance for medical imaging systems: Executive summary of AAPM TG18 report. *Med Phys* 32: 1205–1225.

84. Burgess AE, Jacobson FL, Judy PF. 2001. Human observer detection experiments with mammograms and power-law noise. *Med Phys* 28: 419–437.

85. Coombs L. *Control in the sky: The evolution and history of the aircraft cockpit*. South Yorkshire, UK: Pen & Sword Aviation; 2005.

86. Hooey BL, Foyle DC, Andre AD. A human-centered methodology for the design, evaluation, and integration of cockpit displays. In: *Proceedings of the NATO RTO SCI and SET Symposium on Enhanced and Synthetic Vision Systems*. September 10–12, 2002. Ottawa, Canada.

87. Foyle DC, Ahumada AJ, Larimer J, Townsend Sweet B. 1993. Enhanced/synthetic vision systems: Human factors research and implications for future systems. *SAE Transactions* 101: 1734–1741.

88. NASA. Human factors. Available at: http://human-factors.arc.nasa.gov/.

89. Lackner JR, DiZio P. 2000. Human orientation and movement control in weightless and artificial gravity environments. *Exp Brain Res* 130(1): 2–26.

90. Clément G, Clément G, Bukley AP. *Artificial gravity*, Vol. 20. Berlin: Springer; 2007.

91. Ullman S. *High-level vision: Object recognition and visual cognition*. Cambridge, MA: The MIT Press; 2000.

92. Wang JYA, Adelson EH. 1994. Representing moving images with layers. *IEEE Transactions on Image Processing* 3(5): 625–638.

93. Wiskott L, Fellous JM, Kuiger N, von der Malsburg C. 1997. Face recognition by elastic bunch graph matching. *IEEE Transactions on Pattern Analysis and Machine Intelligence* 19(7): 775–779.

94. Turk M, Pentland A. 1991. Eigenfaces for recognition. *J Cogn Neurosci* 3(1): 71–86.

95. Belhumeur PN, Hespanha JP, Kriegman DJ. 1997. Eigenfaces vs. fisherfaces: Recognition using class specific linear projection. *IEEE Transactions on Pattern Analysis and Machine Intelligence* 19(7): 711–720.

96. de Coro C. Face recognition using Eigenfaces. Available at: http://www.cs.princeton.edu/~cdecoro/eigenfaces/.

97. Valentine T. 1991. A unified account of the effects of distinctiveness, inversion, and race in face recognition. *Q J Exp Psychol* 43(2): 161–204.

98. Dosher BA, Sperling G, Wurst SA. 1986. Tradeoffs between stereopsis and proximity luminance covariance as determinants of perceived 3D structure. *Vision Res* 26(6): 973–990.

99. Landy MS, Maloney LT, Johnston EB, Young M. 1995. Measurement and modeling of depth cue combination: In defense of weak fusion. *Vision Res* 35(3): 389–412.

100. Hillis JM, Watt SJ, Landy MS, Banks MS. 2004. Slant from texture and disparity cues: Optimal cue combination. *J Vis* 4(12): 967–992.

101. Welch R, Warren D. Intersensory interactions. In: Boff K, Kaufman L, Thomas JP, eds. *Handbook of perception and human performance*. New York: Wiley; 1986, pp. 25.1–25.36.

102. Calvert GA. 2001. Crossmodal processing in the human brain: Insights from functional neuroimaging studies. *Cereb Cortex* 11(12): 1110–1123.

103. Stein BE, Meredith MA. *The merging of the senses*. Cambridge, MA: The MIT Press; 1993.

104. McGurk H, MacDonald J. 1976. Hearing lips and seeing voices. *Nature* 264: 746–748.

105. Massaro DW. *Speech perception by ear and eye: A paradigm for psychological inquiry*. Hillsdale, NJ: Lawrence Erlbaum; 1987.

106. Dosher BA, Lu ZL. 2000. Noise exclusion in spatial attention. *Psychol Sci* 11(2): 139–146.

107. Lu Z-L, Dosher BA. 1998. External noise distinguishes attention mechanisms. *Vision Res* 38(9): 1183–1198.

108. Yeshurun Y, Carrasco M. 1998. Attention improves or impairs visual performance by enhancing spatial resolution. *Nature* 396(6706): 72–75.

109. Smith PL, Ratcliff R, Wolfgang BJ. 2004. Attention orienting and the time course of perceptual decisions: Response time distributions with masked and unmasked displays. *Vision Res* 44(12): 1297–1320.

110. Dosher BA, Lu Z-L. 1998. Perceptual learning reflects external noise filtering and internal noise reduction through channel reweighting. *Proc Natl Acad Sci USA* 95: 13988–13993.

111. Dosher BA, Lu Z-L. 1999. Mechanisms of perceptual learning. *Vision Res* 39(19): 3197–3221.

112. Meyer DE, Kieras DE. 1997. A computational theory of executive cognitive processes and multiple-task performance: Part I. Basic mechanisms. *Psychol Rev* 104(1): 3–65.

113. Anderson JR. *The architecture of cognition*, Vol. 5. Hillsdale, NJ: Lawrence Erlbaum; 1995.

114. Harm MW, Seidenberg MS. 1999. Phonology, reading acquisition, and dyslexia: Insights from connectionist models. *Psychol Rev* 106(3): 491–528.

115. Wixted JT. 2007. Dual-process theory and signal-detection theory of recognition memory. *Psychol Rev* 114(1): 152–176.

116. Boynton GM, Demb JB, Glover GH, Heeger DJ. 1999. Neuronal basis of contrast discrimination. *Vision Res* 39(2): 257–269.

117. Li X, Lu ZL, Tjan BS, Dosher BA, Chu W. 2008. Blood oxygenation level-dependent contrast response functions identify mechanisms of covert attention in early visual areas. *Proc Natl Acad Sci USA* 105(16): 6202–6207.

118. Ress D, Backus BT, Heeger DJ. 2000. Activity in primary visual cortex predicts performance in a visual detection task. *Nat Neurosci* 3(9): 940–945.

119. Ress D, Heeger DJ. 2003. Neuronal correlates of perception in early visual cortex. *Nat Neurosci* 6(4): 414–420.

120. Haushofer J, Livingstone MS, Kanwisher N. 2008. Multivariate patterns in object-selective cortex dissociate perceptual and physical shape similarity. *PLoS Biol* 6(7): e187.

121. Cohen MR, Kohn A. 2011. Measuring and interpreting neuronal correlations. *Nat Neurosci* 14(7): 811–819.

122. Xue G, Dong Q, Chen C, Lu L, Mumford JA, Poldrack RA. 2010. Greater neural pattern similarity across repetitions is associated with better memory. *Science* 330(6000): 97–101.

123. Kriegeskorte N, Mur M, Bandettini P. 2008. Representational similarity analysis–connecting the branches of systems neuroscience. Front Systems Neurosci 2: 4

124. Newsome WT, Britten KH, Movshon JA. 1989. Neuronal correlates of a perceptual decision. *Nature* 341(6237): 52–54.

125. Wassermann E, Epstein CM, Ziemann U. *The Oxford handbook of transcranial stimulation*. Oxford: Oxford University Press; 2008.

126. Nitsche MA, Cohen LG, Wassermann EM, et al. 2008. Transcranial direct current stimulation: State of the art 2008. *Brain Stimulat* 1(3): 206–223.

127. Xue G, Juan CH, Chang CF, Lu ZL, Dong Q. 2012. Lateral prefrontal cortex contributes to maladaptive decisions. *Proc Natl Acad Sci USA* 109(12): 4401–4406.

128. Oruç I, Maloney LT, Landy MS. 2003. Weighted linear cue combination with possibly correlated error. Vis Res 43: 2451–2468.

Index